Astronomical Python

An introduction to modern scientific programming

Online at: https://doi.org/10.1088/2514-3433/acfa9a

AAS | AMERICAN ASTRONOMICAL SOCIETY   IOP ebooks™

AAS Editor in Chief

Ethan Vishniac, Johns Hopkins University, Maryland, USA

About the program:

AAS-IOP Astronomy ebooks is the official book program of the American Astronomical Society (AAS) and aims to share in depth the most fascinating areas of astronomy, astrophysics, solar physics, and planetary science. The program includes publications in the following topics:

GALAXIES AND COSMOLOGY

INTERSTELLAR MATTER AND THE LOCAL UNIVERSE

STARS AND STELLAR PHYSICS

EDUCATION, OUTREACH, AND HERITAGE

HIGH-ENERGY PHENOMENA AND FUNDAMENTAL PHYSICS

THE SUN AND THE HELIOSPHERE

THE SOLAR SYSTEM, EXOPLANETS, AND ASTROBIOLOGY

LABORATORY ASTROPHYSICS, INSTRUMENTATION, SOFTWARE, AND DATA

Books in the program range in level from short introductory texts on fast-moving areas, graduate and upper-level undergraduate textbooks, research monographs, and practical handbooks.

For a complete list of published and forthcoming titles, please visit iopscience.org/books/aas.

About the American Astronomical Society

The American Astronomical Society (aas.org), established 1899, is the major organization of professional astronomers in North America. The membership (~7,000) also includes physicists, mathematicians, geologists, engineers, and others whose research interests lie within the broad spectrum of subjects now comprising the contemporary astronomical sciences. The mission of the Society is to enhance and share humanity's scientific understanding of the universe.

Astronomical Python

An introduction to modern scientific programming

Imad Pasha

Department of Astronomy, Yale University, New Haven, CT, USA

IOP Publishing, Bristol, UK

ISBN   978-0-7503-5147-8 (ebook)
ISBN   978-0-7503-5145-4 (print)
ISBN   978-0-7503-5148-5 (myPrint)
ISBN   978-0-7503-5146-1 (mobi)

DOI   10.1088/2514-3433/acfa9a

Supplementary material is available for this book from https://doi.org/10.1088/2514-3433/acfa9a.

Version: 20240501

AAS–IOP Astronomy
ISSN 2514-3433 (online)
ISSN 2515-141X (print)

British Library Cataloguing-in-Publication Data: A catalogue record for this book is available from the British Library.

Published by IOP Publishing, wholly owned by The Institute of Physics, London

IOP Publishing, No.2 The Distillery, Glassfields, Avon Street, Bristol, BS2 0GR, UK

US Office: IOP Publishing, Inc., 190 North Independence Mall West, Suite 601, Philadelphia, PA 19106, USA

Contents

Preface

Across a vast and breathtaking universe, gas first formed in the cooling afterglow of the Big Bang finds itself eventually funneled into a vast cosmic web of galaxies, each themselves rich in substructure—they are factories, which process this pristine gas within stellar furnaces powered by nuclear fusion. The processes by which this gas collapses into stars, shines as starlight, or is expelled via winds, supernovae, and AGN outflows, all produce photons that race across the universe and, if we are lucky, land upon the mirrors of our telescopes, whether they be the *Hubble Space Telescope* and its ilk above the atmosphere or Keck Observatory and its companions below it.

Bouncing off of primary, secondary, and even tertiary mirrors, passing through fiber optic cables, and focusing, finally, at the planes of detectors, this light is finally captured by our best technology—technology that has advanced from glass plates to charge-coupled devices, and may soon advance again to CMOS and other chip technologies. This capture—photons striking a detector, releasing a cascade of electrons that can be measured—passes through wires and cables and eventually finds itself being written onto the hard drive of a computer disk, a collection of pixels each with a simple measure—counts—all, in the end, represented by simple ones and zeros. Only then does a `SAOImage DS9 window`[1] update on an observer's screen as they peer at the images and try to interpret the results.

It is a pretty picture, public notions of eyepiece-driven astronomy aside, but also an increasingly incomplete one. We are in an age of multimessenger astronomy, and our view of the universe has expanded to include waves of gravity we can "hear," or neutrinos requiring vast underground detectors to detect. In a completely different sphere, our theoretical knowledge is pushed forward by vast supercomputers running simulations of the universe, advancing time moment by moment as our best-guess at the physical laws of the universe guide them forward.

Nevertheless, one fundamental truth connects *every single* astronomical observation, simulation, and even instrumentation build in the modern era: *code*.

Code to run the telescopes and instruments, code to run the simulations, code to analyze, interpret, and plot the results. With every passing year, the importance of coding to astronomy—the degree to which at least *some* programming is needed for even basic tasks, and *advanced* programming is needed for pushing the cutting edge of the field forward—grows.

This text is intended to help anyone interested in astrophysics, and particularly data analysis, to push away from shore, and take the leap into programming. It has also been a long time coming. The origins of this book extend far back, into the early 2010s, to a casual, pass no-pass course taught by students, for students, at UC Berkeley. It is a testament to the rapid shifts in our field that even recently,

---

[1] If you have taken data at an observatory of late and this did *not* happen, then this edition of this book is likely horribly out of date.

undergraduates in astrophysics received no formal coding training, and at most, this type of "underground" education.

I am nevertheless indebted to that particular class as my first introduction to programming. In 2013, the so-called Python Decal was taught by Pauline Arriaga and Baylee Bordwell, and it was they who at the end of the semester approached me about teaching it after their graduation that year. That simple request kick-started my long and winding path through the niche of "Pythonic Pedagogy for Astronomy." Despite being woefully under-qualified at first, I taught the Python Decal (with Christopher Agostino as a co-instructor) for the next three years of my undergraduate career, learning more about Python, and pedagogy, with each iteration, and supplanting the course with skills I learned at REUs, in individual research projects, and more.

By 2015, I had realized that no satisfactory textbook existed which met my needs as an instructor for this course. I needed, in short, a text which introduced Python at the most basic level, but which focused on the rather specific subset of tools, libraries, and algorithms useful to an astronomer. Since few to none existed... I wrote one. That book, which is still floating around online, was the proper forbear to the text you are reading now.

Now, nine years after its original "publication," I've returned with what I hope is an improved, expanded, and comprehensive introduction to the modern world of computing for data analysis. I will be the first to admit there are areas of this wide, and rapidly growing field, that this book does not cover. For example, our focus is not on implementing ordinary differential equations in Python to solve physics problems. My focus, as it were, is on *data*, and the analysis of that data. The data may come from observations or simulations (or models), but the techniques I'll discuss fall firmly into the category of "data analysis" or, now, "data science." Still, this is *most* people's introduction to astronomy, and is a valuable, and tractable, way to start learning to program for this field. I endeavour to include several research subfields in this text, though some (like simulations) pose logistical (file size) challenges, while others I simply have too little experience with to convey with proper nuance. That said, compared to the previous text, you will find multiple new sections, revised coverage of others, and many new examples using real data in a realistic context.

Besides Pauline and Baylee, I'm indebted to all of my colleagues and instructors who have improved my own coding and knowledge therein. I'd like to particularly thank Juan Guerra, my roommate, who has engaged in (or suffered) countless hours of whiteboard discussions about code, and who always pushes for better-written code. I'd also like to thank Christopher Agostino, Tim Miller, Joel Leja, Johnny Greco, Dan Foreman-Mackey, Ryan Trainor, Michael Tremmel, David Hogg, and Adrian Price-Whelan for either their input on this book, or for conversations or materials that aided and advanced my own knowledge.

Finally, I would express un-quantifiable gratitude to Marla Geha, who co-taught a Python course with me, and has has been both a colleague and mentor in more ways than I can describe. I would not have even attempted writing this text without her support and encouragement.

To anyone reading this book and setting out on this journey, I have a single piece of advice: do not let anyone tell you that programming is a "talent" skill. Progress as a programmer is made not by intelligence or genius, but simply by persistence. Stick with it, keep practicing, find the projects that bring you joy, and you'll make it through.

Good luck.

Acknowledgements

This book would not be possible if I were not here, and I would not be here without my mother, who supported my lofty space-bound aspirations from the earliest ages and always pushed me to accomplish those goals. I want to thank Malena Rice and Carla Nicholson, my oldest friends and closest confidantes who have been involved in every step of not only this book, but every teaching and scientific endeavour I have undertaken. I want to give a long-standing thanks to Chris Agostino, who taught the very first version of this material with me in 2014; as well as to Mariska Kriek, Joel Leja, Nir Mandelker, Reinout van Weeren, Felipe Andrade-Santos, and John Blondin, who as advisors and mentors in various contexts made me the scientist I am today. In particular, I want to acknowledge my PhD advisor, Pieter van Dokkum, and Marla Geha, both of whom have encouraged my pursuit of both science and pedagogy throughout my time in graduate school.

I also want to *deeply* thank a group of friends and colleagues who graciously read and commented on drafts of this book, without whom it would certainly not exist: Chloe Neufeld, Yasmeen Asali, Sebastian Monzon, Zili Shen, H Perry Hatchfield, and Michael Tremmel. Their comments and suggestions dramatically improved this final product. I want to give special recognition to Chloe and Yasmeen, who in addition to beta reading and commenting were critical to the development and implement of some of the examples shown in this text and its organization.

Finally, a thanks to Leigh Jenkins at IOP Publishing, without whom this book wouldn't be a reality and would only be a spark in the back of my mind, Robbie Trevelyan, who oversaw the submission and publishing of this text, and the rest of the IOP production team.

About the Author

Imad Pasha

Imad Pasha is an NSF Graduate Research Fellow and PhD candidate at Yale University. Before Yale, he earned Bachelors degrees in Physics and Astrophysics from the University of California, Berkeley, as well as a minor in Creative Writing. He worked as a reporter, senior editor, and photographer at *The Daily Californian*, the newspaper of record in Berkeley, CA. At Yale, his research has focused broadly on the processes driving galaxy evolution. He is interested in particular in how gas is accreted onto galaxies from the cosmic web, processed into stars, and (partially) expelled back out into the intergalactic medium, to be potentially later re-accreted.

Astronomical Python
An introduction to modern scientific programming
Imad Pasha

Chapter 1

Introduction

It is of great surprise to many non-academics, and indeed to many undergraduate students deciding to pursue astronomy and astrophysics majors, that the day-to-day life of an astronomer is dominated not by writing out long physics derivations on chalkboards (as seen in classes or in popular media), but rather by *programming*.

Modern astrophysics, whether it be focused on theory, simulation, or observations, relies almost exclusively on the use of programming to carry out novel research. In observational astronomy, ever since the deployment of the charged coupled device (CCD) in place of photographic emulsion plates, even basic visualizations and calculations of astronomical imaging and spectroscopy have relied explicitly on the use of computers. Similarly, our ability to *simulate* the universe—to establish some initial conditions and advance a universe forward in time to see what emerges— has *always* been a field reliant on computers. As a large degree of flexibility is needed in almost every calculation, our field has shied away from proprietary software in the form of GUIs (graphical user interfaces), e.g., Microsoft Excel, and instead entrusts astronomers to write their own programs from scratch— a choice driven primarily by the need of customization needed in a code that can do science.

While the need for computing and programming has been a feature of astronomical fields for roughly 40 years now, the nature of that programming—both the hardware systems and programming languages used—has changed dramatically. What began with punch cards fed into machines to carry out calculations has since evolved into advanced data science techniques currently at the forefront of both science and industry (in particular, the "tech" industry). Several languages have served over the years as the *workhorse* language of astronomy. Fortran and C++ were (and still are) used for simulations and other calculations that require the highest possible speeds. On the other hand, for what I'll call "data science" roles (the cleaning, reduction, and analysis of data), popular languages included IRAF (Image Reduction and Analysis Facility)—less a full language and more a collection of well-written scripts and algorithms—and IDL (Interactive Data Language), a proprietary language that led to much headache over departmental licenses.

Between its release in 1995 and today, Python has become the *de facto* language for data science in astronomy. This rise has been mirrored outside of astronomy, as Python has become the world's most popular language overall, and has become particularly dominant in the regime of data science. Python is a mature (now nearing version 3.12), dynamically-typed language that is both easy to read and easy to learn (compared to many other languages). It is not particularly *fast* computationally, a point we will return to throughout this text, but this drawback does not often impede the types of calculations we carry out as astronomers.[1] Python's broad popularity means that numerous packages (libraries of useful tools and functions) have been written to carry out a variety of common tasks. Very rarely must we program a significant mathematical "algorithm" from scratch. Its near universal adoption within astronomy *specifically* has also been fortuitous: astrophysics-specific pack-ages such as `Astropy` have made many common tasks we must carry out as astronomers significantly more straightforward. Best of all, these libraries are "open source," meaning a large community of astronomers are using, contributing to, double checking, and fixing bugs across these softwares. That said, there are several other languages worth learning (or considering) when beginning to pursue astron-omy. Julia, for example, is an increasingly popular language for carrying out statistical modeling at high computational speed. Rust is also growing in popularity as a successor to C++ (and C++ and Fortran are still in use, particularly for the construction of large simulations).

But let's take a step back. What *is* programming? If you are just starting out in astrophysics and have no prior programming experience, even the concept of programming can seem mysterious (and intimidating). Popular media rarely portrays coding accurately (see any hacker in any film). Our understanding of what program-ming is, and what it can do, will grow throughout this text. But to begin, I suggest the following analogy: when we program, we are using our computer as a glorified calculator. Adding, subtracting, multiplying, dividing, exponentiating, and logging numbers (or vectors of numbers) is a fundamental task of any astronomical program, and a built-in feature of most languages. Where using Python on a computer is better than, say, a trusty TI-84 hand calculator, is in its ability to perform (1) many calculations in a row in an automated fashion (2) perform different calculations based on an analysis of the inputs and (3) the ability to save and store the outputs of its calculations into something usable (such as a file on your computer). To give a concrete example: if I asked you to use the formula

$$m_1 - m_2 = -2.5 \log(F_1/F_2)$$

to calculate the magnitude difference between star A and star B, this would be easy to do with a hand calculator. But if I asked you to use this formula to get the

---

[1] For example, given a typical script, the difference might be several milliseconds. There are classes of problem, which we will discuss, that require highly-optimized execution, and we will mention several ways to achieve those speeds within and outside of Python.

magnitude differences of a list of 150,000 stars or 1,500,000 stars, and to write those outputs down in a table, a hand calculator would immediately become impossible to use. It turns out what I've asked for is nearly trivial to carry out in Python, and requires only one line of code. If I gave you your list of star fluxes as long lists of numbers, `stars_a` and `stars_b`, we could calculate everything in one shot like this:

```
import numpy as np
mag_differences = -2.5 * np.log10(stars_a/stars_b)
```

The above is an example of how programming can be used for "automation," making a repetitive task much easier to complete (or at least, saving the user time spent actually performing the repetitive task). To build on this example, let's add a *control flow*: say I asked again for this magnitude calculation for many thousands of stars, but I wanted to only calculate the value if the flux of star A is *larger* than some value, N. This makes the task even more untenable on a calculator (or at least adds a manual step of sorting all the values), but adds only a single line of code to our program:

```
ind, = np.where(stars_a > N)
mag_differences = -2.5 * np.log10(stars_a[ind]/stars_b[ind])
```

We'll learn how these above lines of code work (and much more!) in this book. But they highlight how computers can help us make data analysis feasible in the regime of large data sets. And of course, one cannot open, look at, or measure anything about an astronomical *image* on a calculator—but to Python, an image is simply an array of numbers that has two axes instead of one.

Throughout the course of this text, we will learn the basic principles of programming in Python, with a particular focus on how these principles are used in an astrophysics setting.[2] This text assumes *no* prior programming experience; we will be starting from the very beginning. By the end of this text, and after working through its examples and problems, students should feel comfortable beginning a research project in astronomy—for example, a summer project with a faculty member or at a summer internship. Some astrophysics knowledge will be assumed (or covered in summary) throughout, but extensive coursework in astronomy is not necessary to follow the examples in this text; particularly those in the earlier chapters. For those who are interested in seeking out this astronomy background before (or during) the completion of this text, the quintessential offering is

---

[2] I am an observational astronomer who studies galaxy evolution, so there will be some bias toward imaging and spectroscopy examples from this field, but I have endeavoured to include several other subfields of astronomy throughout.

An Introduction to Modern Astrophysics by Carroll and Ostlie (2006). Another often-used text is *Foundations of Astrophysics* by Ryden and Peterson (2020).

This book will cover the native data types and operations, and how the language, interpreter, and operating system work together to carry out commands. The book will lean heavily on standard packages (libraries of functions and classes) used in our field, including `Numpy`, `SciPy`, `Matplotlib`, and `Astropy`. After discussing the installation and basic structure of the language and libraries, the text will move into a discussion of script writing, conditional statements, loops, and other code structures that allow for complex outcome management. The text will then discuss the creation and use of functions and classes within Python, which enables unit-testing and more robust and flexible code creation, and use these tools in a data science context on an astronomical survey. The text will then cover the creation of packages, and the methods for re-using, importing, and otherwise standardizing code. Our examples will focus on data from astronomical observations, but the core skillset in Python is equally applicable to working with simulation data; this choice was primarily made because simulation data files tend to be large in size and require a bit of extra handling to get into Python.

Finally, this book will contain several higher-level chapters that carry students from the beginner stage of programming into the intermediate, providing research-level instruction on the use of algorithms such as Markov Chain Monte Carlo (MCMC). Advanced resources on these topics exist, and the final chapters of this book would be oriented around bridging the gap between the programming level achieved in the textbook to that point and these advanced resources, all of which are becoming necessary for modern research in astrophysics. There are numerous "next step" texts to explore once you have mastered the material included here, including *Statistics, Data Mining, and Machine Learning in Astronomy* (Ivezić et al. 2020), along with freely accessible *Python Data Science Handbook* by Jacob T. VanderPlas[3] (2016).

1.1 How to Use This Book?

This text was designed to serve either as a reference, or as an introductory guide. It is not designed to be a comprehensive listing of every function, class, and tool available to astronomers, but rather to provide a selection of a few highly useful tools from each library and to walk through real-world examples with real astronomical data that use those tools. The goal is to provide a scaffold illustrating the way code is built and to direct you toward where the functions you need for *your* application can be found.

As a strong note of caution: to keep the examples throughout this book short enough for a textbook, straightforward enough that extensive explanation is unnecessary, and focused on the learning goal at hand (usually some library or tool), I have almost always simplified the scientific problem or solution in a way that would not pass muster in a formal scientific setting. You should not consider these as

---

[3] Available on Github at: https://github.com/jakevdp/PythonDataScienceHandbook.

polished, publishable examples, but rather as sketches that might represent your first pass through some data. Such a pass would be followed up iteratively with improved and refined analysis. I have tried to denote locations where I make a particularly egregious simplification or downright incorrect assumptions, but in general you should assume that a more accurate and nuanced method exists to carry out each of the tasks we will cover.

Because this text is a practical, rather than theoretical introduction, I highly recommend having Python open as you read. I provide numerous code examples that can be copied and pasted directly into your own code, and I recommend following along in each chapter and running the code yourself, then stopping to explore the data at each turn before continuing. There are also exercises scattered throughout the text which are generally short, and ask you to in some way interact with the data we have been analyzing in that chapter.

1.2 Data Availability

There are numerous data sets used throughout this textbook. All are publicly available (that is, you can download them yourself from where they are hosted). In the interest of longevity and portability, however, I have "frozen" one set of all the data used in this textbook into a single repository, which is now hosted at **zenodo**. If you download that, you should be able to follow any example in the text. That archive is hosted at https://zenodo.org/records/10732223.

References

Carroll, B. W., & Ostlie, D. A. 2006, An Introduction to Modern Astrophysics and Cosmology (San Francisco, CA: Pearson)

Ryden, B., & Peterson, B. M. 2020, Foundations of Astrophysics (Cambridge: Cambridge Univ. Press)

Ivezić, Ž., Connolly, A. J., VanderPlas, J. T., & Gray, A. 2020, Statistics, Data Mining, and Machine Learning in Astronomy. A Practical Python Guide for the Analysis of Survey Data, Updated Edition (Princeton, NJ: Princeton University Press)

VanderPlas, J. T. 2016, Python Data Science Handbook (Sebastopol, CA: O'Reilly Media)

Part I

Unix and Basic Python

AAS | IOP Astronomy

Astronomical Python
An introduction to modern scientific programming
Imad Pasha

Chapter 2

Essential Unix Skills

Before we can dive into Python, it is worth spending time learning about the ecosystem in which Python (and any programming language) lives. That ecosystem is the *operating system* (OS) of whatever computer is being used. Understanding how the code we write interacts with the operating system that actually carries out those commands is incredibly important, and OS skills are also needed to handle the installation and maintenance of the code we use day to day.[1]

Most of us, before we learned how to program, interacted with our computers' operating systems primarily via a GUI—a graphical user interface. This means we see a visual representation on the screen (like images of folders), and then use a mouse to point and click on folders of which we want to open and view the contents. We also used software that had a GUI written for it, e.g., Microsoft Excel for data entry and calculations.

Before the age of these GUIs, operating systems were interacted with via the *command line* (also known as a *shell* or *terminal*) and this line by line, text-only-based interaction was the only way of navigating a computer and carrying out tasks (like copying files from location A to location B). We would issue commands to the terminal in a *syntax* (specific wording) the computer understood, and it would carry out the command or show some results (all in the form of text). That syntax is not universal; it exists as part of a particular language—the same way Python is a language. The program which "translates" the input syntax into computer language is known as the shell/terminal, and the type of shell defines which language we use when interacting with it.

As it turns out, the command-line/terminal interface for interacting with Linux, Windows, or MacOS systems is an integral part of programming in *any* language, as our code uses text-based (rather than point-and-click based) commands for interacting with the files and folders and processes on a computer. It also turns

---

[1] "Pure-code" software does not come with a handy installer program to handle installation for us.

out that even installing the packages we need, and having our code behave the way we want, often involves rooting around in the operating system itself via the command line.

You may be wondering why we use the command line for these tasks, given that the "rest of the world" has moved on to GUI-based interfaces. There are several answers to this question. For one, *writing* GUIs that allow users to interact with a program takes a long time—it is an entire extra layer of programming on top of the base code that actually performs some tasks, and is often much more verbose and complicated than the actual code itself. Further, a GUI inherently limits what you can ask the program to do—that is, you can only carry out tasks for which someone coded a button or entry field, etc. Thus, if we can write code in such a way that the user can supply inputs and make choices entirely via text or customizable input files, we can save ourselves a lot of programming time while creating a more flexible program. Speaking of time, once you learn how to navigate a system via the command line, many tasks are actually much faster and more efficient to carry out from the command line than they are from a GUI. We'll see examples of this below. In general, it is only worth the considerable extra coding effort to create a GUI when (a) interaction in a point/click/drag paradigm is a much more convenient way of working with the data, and (b) when the code is going to see widespread use. SAOImage DS9 is an example of a GUI program used by many astronomers.

2.1 Operating Systems

Because the command line gives us a tool for navigating the filesystem of a computer, it naturally matters which operating system we are using. The primary operating systems used in astronomy are, in order of prevalence, Linux, MacOS, and Windows. Partly for historical reasons, and partly because of its open source and customizable nature, Linux systems tend to be favored by astronomers, and essentially all supercomputers and other large-scale computing systems use some version of Linux. That said, your typical astronomer's personal computer is often, these days, a Mac of some kind. As it turns out, the command line interface used by MacOS is UNIX-based, just like current Linux systems. This means that commands to the terminal are very similar for the two platforms, even if the underlying hardware isn't. Finally, we have Windows. Windows differs from the other two because its default terminal, *command prompt*, uses entirely different syntax from UNIX-based systems, and doesn't play well with the type of astronomical software we tend to use. Luckily, you're not out of luck if you have Windows, and many astronomers do. Recent versions of Windows support the installation of Linux directly on the Windows machine via something called *Windows Subsystem Linux*. For all three systems, Chapter 3 has up-to-date (as of publication) instructions for setting up a computing environment on your computer that is compatible with writing and running astronomy code.

In this book, I'll be using MacOS as the default operating system presented, with notes provided on where a task may differ significantly between operating systems. There are parts of this chapter that will likely be timeless (part of UNIX's strength is its long-term stability), but other parts may quickly fall out of date as operating systems evolve and methods of packaging and installing software changes. I recommend finding up-to-date guides by astronomers on installing a basic software stack for astronomy when starting up a computer from scratch.

2.2 Anatomy of the Terminal

On whatever operating system we're on, let's go ahead and open up a terminal (also called a shell). On MacOS and Linux, there is an application called Terminal one can directly access. On Windows, you'll want the shell that was installed when you installed wsl, likely Ubuntu.

Upon opening a terminal, you should see something *similar* to the schematic presented in Figure 2.1, though there will be some differences. Because a terminal is where a programmer spends a large amount of their time, it is perhaps unsurprising they are nearly infinitely customize-able, and no two coders will have theirs set up precisely the same way. That said, there are a few elements that will be common to all terminals.

On the left-hand edge of the terminal is the *prompt*. The prompt is one of the tunable aspects of the terminal, and can be as simple as a single character (like a > or $), and can be as complex as showing your full location in a filesystem. In Figure 2.1, this particular prompt has been set up to show the current folder/directory, which is "home". For now, don't worry about the value of your prompt; once you are comfortable with the shell, numerous online resources can help you customize it to your heart's content.

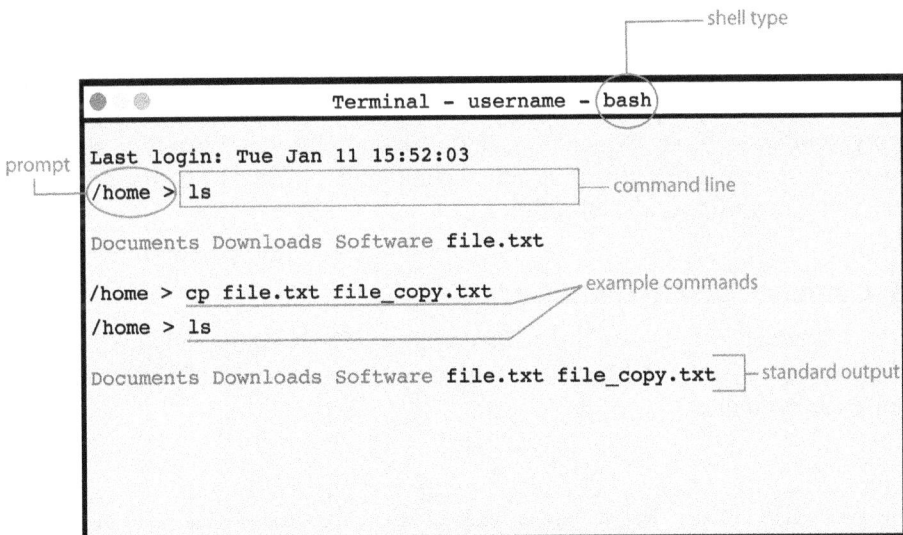

Figure 2.1. Schematic of a terminal (shell). Terminals on different systems will have slightly different presentations, and can be heavily modified, but common features include the prompt—whatever appears on the left before the cursor position, and the command line, which is where one types in commands. Commands that have outputs are printed to the screen after being run (an example of this, 1s, is shown).

To the right of the prompt is the *command line*. This is where we actually type commands and execute them (by hitting Enter). Generally, your terminal will show you the location of your cursor via a blinking underscore or box; this is also configurable in your terminal settings.

In Figure 2.1, I've shown a standard command which has been entered—in this case ls, which, when run with no arguments as I have here, lists the files and directories in the current location. Below that, we see that a set of words has been printed. This is the standard output, and in this case, shows us the contents of the home/directory.[2]

After the printout from the first ls command, a new line is available to type a new command. Here, I've used the cp command, which copies a file (and renames it). This function has no output, so the very next line was the command line again. Now, when we run ls again, we see the new file has been added to the listed contents of this directory. The terminal keeps a running log of such commands and outputs, almost like a receipt. So we can scroll up some amount and see what things were entered, and what, if anything, the computer output to the screen as a result.[3]

Also in Figure 2.1, you'll see that I've identified a shell type—in this case, bash. In the example provided, this is a bash shell, and we'd write any commands or scripts (collections of commands) in a language called bash. These shell languages can be quite powerful—many early astronomical analyses were carried out entirely using shell languages like bash—but they can also be quite obtuse to learn. We will be focusing, in this book, primarily on the shell commands we need to know to successfully install and use Python—but learning how to write scripts in a shell language like bash can be helpful in the longer-term as you advance in programming acumen.

Your default shell will likely depend on your system and is subject to change as operating systems change and upgrade over time. As our default for this text is MacOS (as of 2023), the default shell for commands in this book will be zsh, which is very similar to bash (which is still the default for Ubuntu). The differences between these very rarely affect the actual commands covered here, and arise mostly when you go to edit your shell profile.

2.3 Common UNIX Commands

In the following sections, we'll discuss common UNIX commands for navigating and using a file system. We'll use the following example "tree" of nested folders for all of these examples:[4]

---

[2] Note that if you try this command, you may not see directories highlighted in a different color than the files as shown—this is also a custom setting we can tune later.

[3] How far back this terminal history goes can be set in your terminal's setting; I recommend max-ing it out. The filesize of the history is small (it's just text), and being able to check what command you ran in the past is highly useful.

[4] Feel free to jump to the discussion of mkdir below and make a set of folders like this on your own system to follow along.

```
/root/physics/user/sally/documents/homework/python/week1/
```

If you are unfamiliar with the slash notation above, it means that inside the root folder, one created a folder called physics/, clicked into it, created a folder called user/, and so on. The path above refers to one specific nested subdirectory, but is convenient for explaining file system traversal below. If we have a file, file.txt inside week1/, you can think of the "true" file name as being /root/physics/user/sally/documents/homework/python/week1/ file.txt. The uniqueness implied by this full path name is what allows us to have other files with the name file.txt, so long as they are in a different directory than week1/ and hence their "full" name is different.

2.3.1 Printing the Current Directory

We can easily see at any time which folder our shell is in by using the

```
pwd
```

command. This will print[5] the full path of the "working directory." Depending on how you've set up your prompt, it may also display some of, or all of, the path to your working directory. Assuming we were in the week1/ folder indicated above, running pwd would show the full path shown above.

2.3.2 Changing Directories

There is a single command by which one can navigate the entire UNIX directory tree of any system, and as there are several subtleties to it, we will discuss it in some detail. The command in question is cd. The syntax cd is interpreted by the computer to mean "change directories." Clearly though, with just this command alone, it would be impossible for the computer to know where to change directories to. Because of this, the command cd takes what is called an *argument*. An argument is a part of the command necessary for it to function, but that is variable—the user specifies different values for the argument within a certain set of possibilities. In this example, the cd command takes as an argument a *path* (a.k.a., a location). For example, in the command

```
cd /root/physics/sally
```

/root/physics/sally is a path, and it serves as the argument to cd, informing cd the desired location in the file system to which we wish to navigate. The majority of commands in UNIX have arguments, although there are a few exceptions, and several have a default argument (for example, the current directory is often the default argument of file system commands). The standard syntax for UNIX commands is

---

[5] In computer parlance, print indicates showing information on the console screen, not printing on paper.

```
name flags arguments
```

where *flags* are modifiers that tell the function some extra information about how to carry out the command. Flags are distinguished from arguments by a prepended dash (e.g., `ls -a .`, where `ls` is the command, `-a` is the flag, and · is the argument). Flags usually also have a verbose version, in which the full name of the flag is prepended with two dashes.

So, how can we efficiently use the `cd` command to navigate between directories in UNIX? There are two types of path inputs to `cd`.

1. **The absolute path**: the full path (as seen above) is a unique locator of a folder on a file system, and can always be used to get to a specific folder. It's important to remember that this method involves writing out the path all the way from the root directory of the file system, which can be cumbersome. Note that the root directory of a file system need not be called "root."

2. **A relative path**: if, for example, you are in the directory `/root/physics/sally/` and want to `cd` into a folder called `homework/` that lives inside `sally/`, you can simply type

```
cd homework
```

This may seem confusing at first, because there is no "/" before homework. Essentially, the computer is interpreting your lack of a "/" to mean that the directory you are looking to `cd` into is *within* the one you are currently in. This is the reason we call it a relative path—you are specifying a directory relative to your current location. This concept can also be chained. If you want to `cd` from a current location to a directory two levels deeper in the nesting system, you can continue the relative syntax above to a longer path. For example, if you were in `root/physics/sally/` and wanted to get into not just `homework/` but all the way into an interior folder called `python/` you would type

```
cd homework/python
```

When working with relative paths, it is useful to be able to "back out" of your current directory and then dive into another. We can accomplish this without resorting to absolute paths. The command

```
cd ..
```

will take you out one directory; i.e., if you were in the homework directory of the sample tree, `cd ..` would take you to the `documents/` directory. This command can be strung together as well:

```
cd ../..
```

brings you out two directories, and so forth. From here, you can then move into other directories. You might be wondering, if the double-dot refers to the directory outside of the current directory, whether the single dot refers to the current directory. If so, you'd be right! The single dot as an argument isn't useful for cd, because we never want to change directories to the current directory. But it can be useful in other functions, for example, when copying or moving files.

Finally, if you use cd with *no* arguments, the command will assume a default directory—in almost all cases, this is your *home* directory. Any "user" on a system has a designated home directory, generally a location where the user has read and write privileges. To give a concrete example, the home directory on a Mac is typically

```
/Users/username
```

where username is the account name you are logged into on that computer.

As a note, there is a "shortcut" symbol for the home directory on most systems: the tilde (~). So, if you had a file myfile.txt in your home directory, and you were somewhere deep in the filesystem nowhere near home, you could still access that file with the path

```
~/myfile.txt
```

and copy, view, or otherwise interact with it from your current location. Note that not all programs accept the tilde, so sometimes you will have to supply a full path regardless.

2.3.3 Viewing Files and Directories

Once we are inside a given directory, we need to be able to visualize the contents of that directory. We saw this in the example shown in Figure 2.1. Typing:

```
ls
```

into the terminal will print a list of the files and folders in the current directory in which you are operating. The ls command has many useful flags for various situations, which are listed in the appendix under the UNIX guide. Additionally, typing

```
man(ls)
```

will bring up the manual for it (or for any UNIX command), right in the terminal.

Two of the most useful flags for ls are

```
ls -a
```

which lists *all* files in a directory, including those hidden by default,[6] and

---

[6] In most operating systems, there is a way to "hide" files from showing up in GUI folder views and ls commands, usually by prepending a period (e.g., .file.ext. Certain, usually auxiliary, files are often named in this manner.

```
ls -ltr
```

which prints an ordered list of files and directories that includes the owner of the files, the permissions of the files, the file sizes, and the date or time the files were last modified. This command also prints the files in order of modification or creation, meaning you can very quickly see the most recently modified files. Note that above, "-ltr" is not a specific flag, but rather a shorthand for combining three flags, "-l -t -r" would be equivalent.

2.3.4 Making Directories

Now that we know how to view the contents in our directories, it becomes important for us to know how to create and delete files and directories as well. In order to create a directory, use the command,

```
mkdir desired_name
```

which you may notice is a shortened version of the phrase "make directory." If we use ls we will see that this new directory is included in the contents listed.

You may have noticed that the folder name I used above, which had more than one word, connects the words with underscores. This is because spaces are special on a command line; they separate commands from flags, flags from arguments, and arguments from other arguments. If you attempt to make a directory with a space in the name, the output will not be what you expect (try it out)! There *is* actually a way to name/use files and folders with spaces within their names—it requires using the special character backslash. For example, to create a folder in sally/ called "my folder", we'd type

```
mkdir my\ folder
```

and it would have to be queried this way every time as well (e.g., when using cd to get into it). It is thus recommended that you not use spaces, and instead use underscores as shown here, or some other spaceless format, like camel case (e.g., desiredName).

Note that when creating a directory, the argument is technically a path, meaning you can make folders "down the tree" so long as you know the path to get there. For example, if we were in /root/physics/sally/, I could run

```
mkdir documents/homework/desired_name
```

to make a new folder inside the homework directory (inside the documents directory) called desired_name.

We can also make multiple sub-directories at once using the –p flag. This flag will create the outermost folder first (if it does not exist) then the folders within, e.g.,

```
mkdir -p new_docs/new_directory
```

Since new_docs wasn't a folder inside sally/, the above command wouldn't work without the –p flag. We can actually make multiple tracks at once, as multiple arguments can be passed (separated by spaces):

```
mkdir -p new_docs/new_directory new_docs/new_directory2
```

On shared-network file systems, having an organized system of directories that make logical sense is very helpful, both for keeping yourself organized and for allowing you to direct others to specific files and folders more easily. We will also find that well organized directory structures make interacting with files with Python (or any programming language) much more efficient and automatable.

Exercise 2.1:
Create a directory in your home directory called Software. This is actually useful generally—while you likely have an Applications folder already, in this book we'll learn how to install astronomical software downloaded from, e.g., Github. It is useful to have an easily accessible directory to keep these organized.

2.3.5 Deleting Files and Directories

Now that we know how to create directories, move between them, and look at the files inside, the next step is to learn how to delete things.[7]

Removing files is a relatively easy task in UNIX. If you are in the directory where the file to be deleted is stored, simply type

```
rm filename.ext
```

to delete it. To remove a directory (along with the files in it) we need to make use of the "recursive" flag:

```
rm -r directory_name
```

which will go into a directory, delete the files within, and then delete the directory itself. The example here illustrates the syntax for using flags/options in general (with a dash preceding the flag).

There is a secondary way of deleting directories,

```
rmdir directory_name
```

which will also delete the directory in question, but not if it contains files.

---

[7] UNIX is not like a Windows or Mac where files are sent to a trash bin. When you hit delete, things are gone forever.

2.3.6 Moving/Copying Files and Directories

The last major skill needed for operating in UNIX file systems is moving and copying files and directories from one place to another. Moving is done using the "move" command:

```
mv filename.ext new_location/
```

(This assumes you are in the directory with the file to be moved. Depending on where you are moving the file to, the new_location could be as simple as ".." (to move the file out one directory) or as complex as a full path name to another directory tree).

The command mv also gives you the option of changing the name of a file as you move it, for example:

```
mv file_name.ext new_location/new_name.ext
```

would move file_name to new_location, changing its name to new_name along the way. Interestingly, because of this functionality, mv serves as the "rename" command as well. To rename a file, "move" it to a new name without specifying a new location to send it (which will default to the current directory). If you want to copy a file instead of moving it, use:

```
cp filename.ext new_location
```

which will create a copy and put it in new_location. The command cp also has the ability to rename files in transit, by the same syntax as mv.

2.3.7 The Wildcard

Denoted by a * symbol, wildcards can stand for any character, or any number of characters. The strategic use of wildcards can save you a lot of time when working with large numbers of files. A few examples should make clear how wildcards are used:

1. **Deleting many files**: say for example you wanted to delete all files in a certain directory that were of the type .doc. If you entered

```
rm *.doc
```

The wildcard would feed rm every file with any combination of characters that ended in .doc for deletion. In a similar vein, if you have a group of research files that all started with simulation_run (where an example filename might be simulation_run10004.dat, simulation_run10005.dat, etc)

```
rm simulation_run*
```

would delete all of those files, as rm doesn't care what comes after the "n" in "run" anymore. As a note, a wildcard can be inserted anywhere; here we used it at the beginning and end of a term, but we could also insert into the middle. rm simulation_run1*.dat would remove simulation files but only those starting with the 1 (so files in the hypothetical 00 000 or 200 000 series would be left alone).

2. **Copying files**: this is somewhat of a trivial expansion, but it is useful to note that more often than not you are going to be copying and moving large numbers of files rather than deleting them (archiving data for later is safer than losing it). It becomes clear now why many research processes that output many files have a very regular system for naming: it allows for the easy extraction of subsets or all files within UNIX systems. Wildcards also work within names, for example:

```
cp simulation*.dat new_location
```

would copy all files starting with "simulation" and ending with ".dat" to a new location. This can be handy if your software also outputs files with the same prefix but different file endings, and you only want the .dat files.

This might be a good time to point out perhaps the single most dangerous combination of characters one could run on a UNIX system: from the root directory, one should **never, ever, type**

```
rm -r *
```

Can you determine why this is a bad idea? Now, most file system administrators have put failsafes in place against such commands, but on your own system, you may not be so lucky.

2.4 Cancelling Commands

Sometimes we begin a process we didn't mean to, or a process hangs without returning indefinitely. Some UNIX commands actually proceed indefinitely until we stop them, and sometimes we begin a process that will take a while (say, copying many files), and discover we picked the wrong destination. In all of these cases, we need to know how to kill a process from the command line.

Thankfully, the method is simple: type ctrl+C. Sometimes, when a process is hanging, the first ctrl+C doesn't end the hang, a chain of processes hanging. In this case, we usually spam the cancel command a few more times to see if that helps.

In addition to killing the *current* process from the command line, we can kill *any* particular process if we know its process ID. One can find the process ID of an offending code or other process that is hanging using the `top` command, which displays a live view of running processes and their IDs.[8] Once you know the process number, simply type

```
kill 259302
```

(where I've made up a process ID here).

As a final note, the same `ctrl+C` command is how we kill a hanging or incorrect process within the Python interpreter (which we'll learn about soon)!

2.5 Tab Complete

A lot of the commands we've covered so far are short and succinct—the point of the command line is to increase efficiency. Commands like `ls` take very little time to type and can easily be used to view a file system. On the other hand, certain commands, like copying one full path to another full path, can be quite verbose. There are two main ways of decreasing the amount of time you spend typing unnecessary information: **tab complete** and **aliases**. We'll cover aliases later in this chapter.

Tab complete is a feature of the terminal that allows you to quickly finish commands or filenames as you type them, *if* they are unique for a given directory. For example, let's say I have three files in a directory,

- `testrun1234876545635624.dat`,
- `testrun49232450238472034.dat`, and
- `testrun95432859234502598.dat`

It would be extremely annoying to type these all out in something like a copy command. Notice, however, that after the word testrun, all three have a different character, which allows them to be differentiated by only the word testrun and the first number (i.e., if I asked you to give me the file starting with "testrun9" there is only one you could choose). To quickly copy testrun9(...) we would simply type

```
cp testrun9<tab>
```

and when we pressed tab, it would automatically complete the rest of the filename, letting us move on to typing in the new location. In fact, tab complete works on typing in locations as well—if you are typing in a long path name, you can tab complete each directory name as you type it, as soon as it's the only one with those letters/numbers in its name.[9]

---

[8] Fancier versions of `top` exist and can be installed for a more granular look at your system processes.

[9] By default this is case sensitive, but there are shell settings to make it insensitive, and also display all currently non-unique entries when you press tab, if tab complete is not yet available.

2.6 Intermediate Shell Commands

The following shell commands come up less frequently than those above, but are still worth learning about. However, they are not generally required in order to get up and running with Python, so these sub-sections can be safely skipped for the time being if time is of the essence.

2.6.1 Touch/File Creation

A relatively common task relates updating the timestamp of a file, or creating it if it doesn't exist. If the file needs *contents*, it is usually best to simply open a code editor (e.g., vim or emacs in the shell) to add that content and save it. But if the file can be empty,[10] the touch command can come in handy. To create an empty file with a given filename, we can run

```
touch myfile.py
```

As a note, we can create any number of files at once using touch; every argument after touch will be treated as another file to touch.

Like nearly every shell command, there are many optional flags one can add to touch to change its behavior. The -a flag will update the access time of the file, meaning ls queries by accessed time will have it at the top/bottom; similarly, the -m file will change the last-modified time.

2.6.2 Previewing File Contents

When we are navigating around with the shell, we often would like to preview the contents of ASCII text files quickly, without needing to open them in a full editor like vim or VSCode. UNIX has several commands that can accomplish this. As a note, ASCII text files need not have file extensions of .txt or .ascii—astronomers in fact tend to assign arbitrary file extensions to what are all ultimately text files, to differentiate, e.g., different outputs of a code. So run104.sfr might contain the star formation rates from a galaxy fit, while run104.mass might contain the measured mass. The filenames here were descriptive both of which run the file is part of and which measurement it contains, though run104_mass.txt would be equally valid. The point is, regardless of the extension, if the file contains ASCII text, it is viewable by the following programs.

If we just need see the first few, or last few lines of the file, the head and tail command can handle this for us. Here I'll run it on a data file containing some flux data about star forming regions.

---

[10] This happens more often than you would think.

```
head relano2016_m33_phot.txt
```

ID	logFFUV	logFFUV_e	logFNUV	logFNUV_e	logFHa	logFHa_e	logF3.4	logF3.4_e
1	16.18	0.00	16.15	0.00	14.22	0.07	14.77	0.02
2	15.66	0.01	15.58	0.01	13.67	0.07	14.58	0.02
3	16.13	0.00	16.01	0.00	14.11	0.07	14.66	0.02
4	15.92	0.01	15.83	0.01	14.11	0.07	14.57	0.02
5	15.92	0.00	15.82	0.00	13.70	0.07	14.35	0.02
6	16.22	0.00	16.09	0.00	14.23	0.07	14.38	0.05
7	16.19	0.00	16.14	0.00	14.65	0.07	14.44	0.03
8	15.04	0.02	14.92	0.02	13.47	0.07	13.87	0.05
9	15.47	0.02	15.45	0.02	13.65	0.07	14.30	0.04

```
tail relano2016_m33_phot.txt
```

110	15.81	0.01	15.62	0.01	13.54	0.07	14.23	0.03
111	15.71	0.00	15.59	0.00	13.72	0.07	13.97	0.05
112	15.67	0.01	15.53	0.01	12.95	0.07	14.23	0.02
113	15.43	0.00	15.31	0.01	13.48	0.07	14.19	0.03
114	14.52	0.02	14.39	0.02	12.19	0.07	99.99	9.99
115	14.88	0.01	14.68	0.03	13.16	0.07	99.99	9.99
116	15.24	0.01	15.13	0.01	13.19	0.07	14.02	0.03
117	16.09	0.00	15.92	0.01	14.07	0.07	14.25	0.04
118	15.53	0.01	15.43	0.01	13.62	0.07	14.31	0.02
119	14.73	0.02	14.56	0.04	13.07	0.07	14.48	0.02

By default, these commands will output to the shell the first ten (or last ten) lines of the file, which can be great for quickly looking at a header or information stored at the top of a file, or seeing how it ends (in this case, we learn very quickly how many objects are in this file—119). Flags for these functions allow you to easily specify a different number of lines.

If we needed to dive a little deeper into the file, we would want to use the more or cat commands. These both allow you to print an entire file's contents to the shell. The cat command will do this all at once. The more command, meanwhile, is actually interactive. When you more a file, it will show an output similar to the head command—but if you start pressing the <Enter> key, it will begin scrolling down through the file line by line (showing you what percentage of the file you have seen in the bottom corner). You can also jump through the file faster by pressing the space bar. As with other shell commands, there are plenty of flags and options for further specialized navigation within more, but the basic functionality is usually sufficient for our purposes. Note: to exit these programs, press the 'q' key, which will return you to your shell interface.

2.6.3 Setting Permissions

When working on your own computer, you generally don't run into issues of file ownership: your user account should automatically have all access to all files on the system. In contrast, when you are on a shared network, like a department server, you may be able to navigate via cd from your own home directory to that of your colleague —but you will likely not have the ability to modify (or even in some cases read) files that are stored there. This is sensible for a shared network; you wouldn't want other users having the ability to accidentally or maliciously delete or steal your files.

But what if we *did* want to make a file accessible to a colleague of ours, so they could copy it or open it? For that, we need the chmod (change mode) command. There are three primary states to affect for a file: read (r), write (w) and execute (x). We can set the mode of a file (whether it can be read or not, written or not (edited), or executed or not) for three *groups* of people: user (the owner of the file), Group, and Others. Who exactly is in these two groups depends on the system; your colleague will likely be considered in the same group as you on the system, but might be categorized as other.

When you print the permissions for a file, (e.g., via ls -ltr), the output will look like this: rwxr-x--, in which there are three groups of three dashes, in order rwxrwxrwx for the three groups. This example has full permissions for the user, read and execute permissions (but not write) for the group, and no permissions for other.

To set the permissions of a file, there are several shortcuts one can use, which are a bit confusing at first, so I recommend using the symbolic mode at first. To set our access in this example, we would use

```
chmod u=rwx,g=rx,o=r myfile
```

This would give us full permissions, group the read and execute but not write, and others only read access.

As a final note, the one time we do sometimes have to use chmod on our own owned files is when we create a shell script. If, instead of writing a command in the shell, we made a file run.sh and added it there, we should be able to run it in our shell by typing

```
./run.sh
```

However, by default when you save a text file (as the run.sh file would be), it isn't given execute permissions. So we would want to run chmod to give ourselves all three permissions on the file. We could then proceed to run it.

2.6.4 Piping Outputs

When I introduced the shell, I described it as a "one-command-at-a-time" environment, and for most shell operations, this is true. But many shell-wizards would take umbridge at the characterization, because the pipe command (used via a | symbol allows you to pass the *output* of one function to the *input* of another. This allows functions to be chained together, which can be convenient when we want to accomplish this task without saving shell variables to hold intermediate values. For example, if we were to run

```
ls -l | wc -l
```

the ls -l command (which prints all contents of the current directory line by line) is "being piped into" the "wc" comand, which is the word-count command. The "-l" flag for wc tells it to count lines, meaning that the combination of ls and wc here results in a quick way to count the number of files/folders in a directory.

2.6.5 File and Directory Archives

When we need to send a large file, or large number of files in a directory, it is often useful to combine and compress the file(s), such that we can transport a single, somewhat-smaller file. Various algorithms exist for this task, and various programs exist to execute those algorithms (e.g., .zip files are common).

On UNIX-based systems, the default shell command for compressing a file or set of files is tar, which originally stood for "tape archive." To use it, we provide some flags to the tar command, the name of the archive we want to create, and then the directory or file to compress. To use a previous example, if we had a directory with many files of the form simulation_run1000.dat, in a folder called sim_output, we could tar that all into one file via

```
tar -cvf simulation_archive.tar.gz sim_output/
```

(assuming we were in the directory containing sim_output). Here, the flags I've chosen are −c (which tells tar we are creating an archive), −v to verbosely print what is going into the archive, and −f, which is needed in most cases (otherwise tar will attempt to use a system variable for the archive to be read/written).

By default, a tar file (also called a *tarball*), which we could create by specifying the filename as simulation_archive.tar is combined into one file, but is not compressed. The .gz suffix here informs tar to use the compression algorithm gzip and compress the file.

Once we have created our archive, we can move it around between systems as desired, etc. Once we are ready to open it back up into a regular directory, we will re-run tar with the create flag replaced by −x, for extract, and provide the location to extract to. Let's extract our archive into the current directory:

```
tar -xvf simulation_archive.tar.gz .
```

2.7 SSH and Servers

An extremely important aspect of working with the command line is ssh-ing into servers to work remotely. A *server* is a computer or system of computers that store files and contain programs that are often accessed remotely. Almost any computer can be converted into a server, though generally speaking servers are set up on computers with a lot of memory and free space. Astronomers use servers frequently because they allow for the storage of large data sets, or the use of many processors.[11] Additionally, servers allow us to log in and work on our research from any computer with an internet connection, without needing all the data and programs installed on our personal machines. Finally, with multiple users on the same server, it becomes easy to share data, code, and any other file with collaborators, instead of having to email or otherwise transfer things to their computers.

2.7.1 Logging into a Server

The standard method for accessing a server is to log in via **SSH**. SSH is a terminal command standing for "secure shell host". When you run a command like

```
ssh username@servername.address
```

in the terminal, your computer reaches out to the server and establishes a connection (assuming you have an account on the server). To give a concrete example, say you have an account under the name "sjohnson" on a server called "vega" on a "university" astronomy department network. You would type

```
ssh sjohnson@vega.astro.university.edu
```

to log in to the server. The first time you try to SSH to a new server, you will be asked whether to trust the RSA key and add it to your trusted list, (just hit "y" and enter). The server will then ask you for a password. The admin for the server will have made one for you when they created your account; once you log in you can generally change this to something of your choosing using, e.g., the `passwd` command (but this varies by system). Note that when you are typing in your passwords, nothing will appear on the screen—that's normal, just type the password and hit enter.

Now that you are in the server, everything works just like you are in a terminal on your own computer. You can `ls`, `cd`, and otherwise work with the files and programs installed on the computer you are ssh'd into. One extra step that's worth mentioning is that if you want to open programs with display windows (for example, `matplotlib`, which we will cover later), you will need to use an additional flag—currently, the "-Y" flag is preferred; that is,

```
ssh -Y sjohnson@vega.astro.university.edu
```

This will allow the windows to open on your computer (other common flags include -X and -L). For this to work, you will need something called X11 forwarding. On a Mac, this involves installing something called "XQuartz" (easily Googled), and on a PC it involves installing something called "Xming" and "Putty" which have X11 options.

---

[11] A typical astronomy department cluster might have computers with 16 to 32 cores, whereas a laptop has only 4–8. Those computers are also set up to best facilitate multi-processing.

2.7.2 Copying Files from a Local Directory to a Server Using SCP

Often we have the need to move files between the server we are working on and our own personal computers (or between two servers). The default command for this is "scp," which stands for "secure copy." To move a file called "test.txt" from a certain computer to, for example, a user directory on a remote server, the syntax is

```
scp test.txt username@server.address:/home/user
```

assuming, of course, you are currently in the directory with the file and that /home/user exists in the server's root directory. To give a concrete example using the same name as above,

```
scp test.txt sjohnson@vega.astro.university.edu:/home/users/sjohnson
```

would move the file to that location on the server after prompting for sjohnson's password.

2.7.3 Copying a File from a Server to a Local Directory

Pulling a file from a remote server uses the same structure as the section above, but switches the two arguments. For example, to pull the file above back to our own computer, we would use

```
scp sjohnson@vega.astro.university.edu:/home/users/sjohnson/file.txt .
```

where we specify any directory we want on the current computer (in the example, I chose the current directory).

These are the primary ways to copy files (to copy multiple files we could tar them into one file and move that, or use a wildcard). If you are *on* a server, and trying to transfer from there to a specific computer, it can be slightly trickier and involves looking up the hostname and IP address of the computer in question; for something like a laptop this is nontrivial.

2.7.4 Rsync

The rsync command is similar to scp; it facilitates file transfers between servers. However, it has several advantages over scp when dealing with many (or large) files. Unlike scp, which is a pure copy-to-destination protocol, rsync attempts to *sync* files between a local directory and a remote one. In practice, if one folder is empty and the other is full, the result will be copying the files from the full to the empty folder. However, rsync checks that files at source are not *already* at the destination, and if they are, it does not attempt to copy/overwrite those files.

A classic example which exemplifies the usefulness of this is the process of copying images from a remote server at an observatory to one's local computer *during an observing night*. We need to do this because quick reductions and analyses of the incoming data can impact our decision-making at the telescope, e.g., how long to remain at a given object.

Every time a new exposure, or several exposures, have been saved to disk at the telescope, we want to pull them over to our own computer. But we don't want to copy all the data files over every time, only the new ones. Here, rsync shines, checking both directories and only transferring the new files.

The usage of rsync mirrors scp. There are a few flags you might like to use, such as −v for verbose output, −z for compression of files during transfer (which reduces file sizes and speeds up the transfer), −a for archive, which indicates that we want to include everything (recursively) in our selection, and −P, which combines the partial flag, for deleting partially transferred files if the transfer is interrupted, and progress, which prints the progress of the transfer. These flags can be combined, so a typical rsync command might look like

```
rsync -azvP remote_location/ local_location/ .
```

2.7.5 Screen and Backgrounding Processes

An unfortunate fact of reality when it comes to remote servers is that when we start processes on the remote server, that process tends to care if the connection (via ssh) is maintained. If our computer falls asleep, or the internet drops momentarily, we may lose our connection to the server—this is accompanied by an oft-hated message of a Broken Pipe. It is hated because the login shell was running the process, so when the pipe breaks, the process being run by that shell (i.e., any code we started running) would be killed. Thankfully, we are not out of luck. There are multiple ways to ensure a process you start on a remote server continues to run even if the connection drops. One of the easiest is via a program called screen, which is installed on most servers. The usage is relatively simple. When we log in to a remote server using the ssh syntax presented above, before we run our code, we do the following:

```
screen -S description
```

in which I've used a flag to give the screen a name (in this case, description but you should choose something more... descriptive).[12] This will clear your terminal and essentially starts a new subprocess on the server not linked to your ssh login. Now, you can run your code as normal. Unlike normal, though, is if you disconnect (on purpose or due to a broken pipe), the code will keep running. When you ssh back into the server, you can reconnect to the screen you created. If you have multiple going,

---

[12] The −S flag and argument are optional, but if you exclude them, the screen will be known only by an ID number, which if you have multiple running, can be a pain to tell apart.

```
screen -l
```

will list them. Each will have a name (if given) and a unique number). Find the one you want to rejoin, and then use, e.g.,

```
screen -r 46741
```

to join it, and you'll be in the exact terminal you left. Ideally, your code will have finished running and you can evaluate the output to see if things went well.

2.8 Profiles

The shell profile is the configuration file that defines the behavior of the shell. By default, it is empty, but you can add plenty of useful things to a profile (and other configuration files) to make your terminal experience more seamless.

There are, annoyingly, several *different* profiles involved on any given system, some of which run only on startup, some of which run whenever a new shell is opened.

2.8.1 Aliasing

If we have an often-used command that is long to type, or wish to overwrite a program name without affecting the program (say, launch vim when vi is typed), the easiest way to accomplish this is using an *alias*. We set these in our shell's profile. Here is a real world example, in that it is an alias I created for an actual compute system in the last several days. The command in question was

```
docker exec pi-server-1 dfcore expose --camera 0 --duration 0.25
  --diable_overscan true
```

This command, on a particular computer, tells a webserver running in a docker container to run a program called dfcore in order to take an exposure with a telescope, with some arguments. This command has a lot of text to type out, and I was going to have to run it manually several dozen times with these particular flags. So instead, I popped into my profile and added the line

```
alias 'dockerexpose'='docker exec pi-server-1 dfcore expose --camera 0
  --duration 0.25 --diable_overscan true'
```

Once the file is saved, we can open a new terminal, or *source* our profile by typing

```
source .zshrc
```

in the terminal. Sourcing a profile executes the commands within to set any new aliases or environment variables in the current shell. You can use aliases for all sorts of commands, e.g., to create shortcuts to certain directories on your computer, or to carry out certain creation or deletion commands. Which profile (of several) you should update depends primarily on which shell you are using (e.g., bash, zsh).

One step above aliases is shell *functions*, which allow you to have adjustable inputs to your calls. It is also possible to set up (not via aliases) ssh commands that bypass entering your password each time. Both are topics worth exploring once you feel you have a familiarity with your shell environment.

2.8.2 Environment Variables

The other primary use for the bash profile is to set environment variables. An environment variable is a value, stored in a variable name, that can be accessed from the terminal. For example, if I were to type

```
export VAR=''value''
```

into my terminal, I could then access this variable. Let's print it using the terminal's print command, echo:

```
echo $VAR
 value
```

(note the dollar sign when accessing an environment variable). The environment variable we created in the terminal above is transitory—when we close or restart our terminal, it will be gone. But because our bash profile gets executed every time we spin up a new terminal, we can add our export statement there instead. If we do, then $VAR will be accessible in our terminal all the time.

We ourselves might not need to make much use of this feature early on, as we are not coding primarily in bash. But there will be astronomical software you install that may ask you to set an environment variable for it to function properly. Usually that environment variable is something simple, for example, the code fsps (a stellar population synthesis code) asks you to set an environment variable SPS_HOME that is just a path to the location where the fsps source code files are sitting.

Exercise 2.2: Easy cd.
Create an environment variable in your profile (i.e., ~/.bash_profile or ~/.zprofile) called SOFTWARE that contains the path to your Software directory. Source your profile, and show that running cd $SOFTWARE takes you to this directory using pwd. You could also add an alias, e.g., soft, which executes the command cd $SOFTWARE for you.

2.9 Summary

In this chapter, we took a whirlwind tour of the file systems of our computers via the shell, a command line, text-based interface for interacting with a computer. While lacking some of the intuitiveness of a graphical user interface, the shell is an essential tool for installing and working in astronomical software, as well as a convenient and efficient way to carry out certain tasks, like copying or transferring large swaths of particular data. Astronomers work in the shell every day, and feeling comfortable with this environment will make all of the research tasks we must carry out (even within Python) easier. Luckily, the UNIX shell is one of the best-documented programs in the world, with dozens of textbooks and websites dedicated to its use.

Astronomical Python
An introduction to modern scientific programming
Imad Pasha

Chapter 3

Installing Python and the Astronomy Stack

Perhaps one of the more challenging initial aspects of carrying out astrophysics research in Python is actually *installing* all of the necessary tools on one's computer —a task which is greatly aided by UNIX and shell familiarity one might not yet have. It is also one of the most challenging aspects to describe in a text, as the process changes considerably as computers, operating systems, and the managers for handling Python installations change or become outdated.

Before continuing, then, it is perhaps worth noting that the *fastest* way to access Python (with the tools needed for some astronomy research) is to not install it at all! An ever-growing list of services are offering access to virtual machines, which are accessible over the web and have Python and many core packages already installed. As of writing, perhaps the simplest (and most free) is Google's Colaboratory. One can open a Colaboratory notebook, in browser, directly from Google Drive. There are limits to this tool (single, short sessions at a time), which I will discuss later, but I have found that for many students, this is a convenient way to get going with some Python without spending half a day on installation woes.

Ultimately, everyone trying to set up a computer for scientific computing would benefit from sitting down with someone who has done it (graduate students everywhere are often friendly and might help!) and having them walk you through it. The more comfortable and familiar you are with computers and the shell/terminal, the easier things will be. Google Colaboratory, and other similar services, at present are not a permanent solution to computing needs for astronomy, so at *some point* you will need to get Python and the full stack of tools needed installed on your system. But if this chapter's instructions seem daunting, feel free to jump ahead and use whatever free online Python resource you can find to get started.

3.1 Prerequisites

Before we can actually get Python installed, there is a stack of dependencies we need to sort out first. These vary by system and these instructions may be out of date

3-1

relatively quickly, so it is always best to check online for new guides for installing science stack environments.

3.1.1 MacOS

To set up a Mac computer for scientific programming, you'll want to start by installing X-code developer command line tools. The easiest way to get this is actually to attempt to install the *next* tool, homebrew, which will prompt you to install developer command line tools during the process. Homebrew is a software manager that has several packages you'll need on your system. You can check if you already have these installed by opening the Terminal app and typing brew. If the command is not found, you'll need to install it. To do so, go to https://brew.shthis site and copy the link, which will look like

```
/bin/bash -c "$(curl -fsSL https://raw.githubusercontent.com/
Homebrew/install/HEAD/install.sh)"
```

You can run this in your open terminal to install brew and Xcode command line interface tools.

If you have an Intel chip mac, this step is now done, otherwise (on M1+ Macs), you'll need to add the location of the brew files to your PATH.

Your PATH is a set of folders on your computer that the shell/terminal knows to look in when searching for executable code to run. When you type a program name into your shell (like brew), the files associated with running brew must be in your PATH in order to be found and run. *Most of the time*, software we install will automatically add itself to our path, or will install in a location that is in our PATH. So we normally don't have to worry about this. For some reason, the current miniconda distribution does not appropriately add brew to your PATH.

To add the brew install path to your PATH, run the following two lines in your shell:

```
echo 'eval "$(/opt/homebrew/bin/brew shellenv)"' >> ~/.zprofile
eval "$(/opt/homebrew/bin/brew shellenv)"
```

Here, we are using the shell commands echo (which just prints the string after it) and >>, which appends and writes that to a file provided after, which in this case is your shell's profile, stored in your home directory (which ~ is shorthand for). Note that here I assume your default shell is *zshell*, which for M1 and M2 Macbooks should be the case.

After this, you should be able to restart your terminal, and when you enter brew, it should not warn about the command not being found.

Next we need to install Conda, the package and environment manager we will use to actually install Python and the various extra libraries we need to fully leverage Python for astrophysical research. Now that we have brew and CLI tools, we need to install Python, with a package manager. Conda is not the only package manager in the Python ecosystem, but it is the one used by most in the astrophysics community. To get conda, the current recommended method is to install it via the miniconda distribution. We'll do that here as well, as that will then translate to other software installation guides you come across. To install it, head over to the Miniconda Website and choose the installer for your operating system. In the past, I have found that installing via the pkg installer places files in a different, more annoying location. I recommend instead downloading the bash version for your OS/computer type. Then, you can use the shell cd command to, e.g., cd ~/Downloads, then run, e.g.,

```
bash Miniconda3-py310_23.3.1-0-MacOSX-x86_64.sh
```

and follow the instructions. Note that even if your shell is zshell, you can use the bash command to execute the installation via bash as shown above. I recommend accepting all of the suggested locations and defaults. Once this is done, we now have a working set of command line tools, and conda to help us manage Python and its environments. We'll cover environments in the next section.

3.1.2 Windows

Windows itself is not particularly conducive to scientific computing, and you will struggle trying to use a Windows OS for research purposes. Luckily, Windows now supports something called WSL, or Windows Subsystem Linux, which allows you to install a linux distribution directly on your Windows machine. When using WSL, you will essentially be using Linux, for which most scientific programs are written, and will be able to mirror all the conda environment and coding paths needed for your research.

To get it installed, simply open a Windows Powershell and type:

```
wsl --install
```

which will install the latest distribution of Ubuntu, which is fine to use as a Linux distribution for our purposes.

Once you have installed WSL, you will need to create a user account and password for your newly installed Linux distribution—it is basically like a second mini operating system, with its own username/password and file system. Follow the instructions and select a username and password (these need not be the same as for your PC).

When you are setting (or entering) passwords in terminals, here included, generally nothing shows up as you type. This is for security, just be careful while typing and hit enter to submit.

Once you have wsl installed, you should have a program on your computer which is an Ubuntu Shell. When you open it, you'll have a terminal, but this is not a windows powershell, it is in fact UNIX-like and is accessing your new WSL distribution. It is recommended to start out by upgrading all default installed packages via

```
sudo apt update && sudo apt upgrade
```

After that, you're ready to install miniconda, which is the Python + package distribution manager we'll be using.

There are several out there, but in the astrophysics community, the majority currently use the miniconda distribution to manage Python and Python environments. We'll do that here as well, as that will then translate to other software installation guides you come across. To install it, open up your Ubuntu shell and type

```
curl -sL "https://repo.anaconda.com/
miniconda/Miniconda3-latest-Linux-x86_64.sh" > "Miniconda3.sh"
```

Followed by

```
bash Miniconda3.sh
```

I would accept all the default suggestions.

3.2 Python Environments

We now have the core set of tools needed to get started with Python. You will likely need many more tools (software packages) as we proceed, but we should be able to install *those* using one of the core tools we have just installed—brew, conda, or pip.

Before we talk about Python, we need to discuss *environments*. Environments are a critical element of coding infrastructure that will allow us to use and write code that runs without conflicts.

Why is this? Primarily, it is because software is always evolving, and both the language (Python) and packages (Python codes) have new versions released nearly-continuously. Because of this, it is important to track, for a given code you are writing, which versions of Python and associated packages are compatible with your code, so that in the future, you know you can create such an environment again and your code will work. It is also critical for distributing code to others and ensuring *they* can run it. Developers releasing new code often try to make changes that are backwards compatible with previous versions, but over enough time, a change to a dependency (library you are importing and using in your own code) will break things in your code. Moreover, you might be creating different codes (or using different codes) that have conflicting version requirements for the same dependencies. Environments give us a system for handling these cases.

So what *is* an environment?

When we install typical software (say, Microsoft Office or Apple Keynote) on a consumer computer, that software has a version. It may even prompt you about updating the software when new versions come out. We typically don't concern ourselves with anything that looks like an environment. This is because Microsoft Office and Apple Keynote are *standalone* programs, internally consistent and not dependent on any other particular software being installed on your computer. Everything they need, they have internally, and the update-tree is managed by Microsoft or Apple, respectively.

Unfortunately, open source code does not quite work this way. Let's say we have three code packages—package foo, package bar, and package spam. Both foo and bar *use* spam in order to function. But, for various reasons, foo requires spam >v0.1.2 (that is, some version of spam released after v0.1.2), while bar *requires* spam v0.1.1. If we simply install spam globally, by default, it will be the latest release version (let's say that's v0.5.4). Presumably, this would be fine for running package foo, but bar is going to complain.

Environments allow us to create isolated "silos" in which any version of Python, and any number of dependencies of any version requirements, can be installed. It will not interfere with other environments on the system. Thus, if we have legacy code written in Python 2.x, we could create a Python 2 environment to run it, while also having a Python 3.x environment for newer code. In our example above, we could create an environment just for running package bar, into which we specifically install spam v0.1.1 (and, if needed, older versions of any other codes bar depends on that are incompatible with older versions of spam).

If that sounds like a nightmare, it can be. Slightly assuaging this headache is the fact that our package and environment manager, which for this text will be *conda* (as for most scientific applications you will run into use it), will try its best to resolve dependency conflicts and install a set of packages that are all compatible with each other.[1]

3.2.1 When to Create an Environment

In the following section, we'll discuss step by step how to make new environments. But when should you? In general, I recommend having a default environment that is more or less kept up to date with a simple set of packages that you use every day. Beyond that, you should take the time to make a new, separate environment:

---

[1] You may also be wondering what happens if you need both foo and bar in the same code. The answer: you cry, quite a lot. There is no way, currently, to have two versions of the same package in the same environment. You would have to try to split tasks requiring foo into one script, bar in to another, and run them in separate environments, combining their output results as needed. Thankfully this doesn't happen often.

- Whenever you start a new research project,
- Whenever you are installing a code with many dependencies and versioned-dependencies—often such codes, in their installation guides, will actively recommend a separate environment for their package,
- And whenever you know you need a specific, downgraded version of some code.

If you make new environments at these times, you will be most likely to avoid headaches later on.

3.2.2 Creating and Managing Environments

There are a few methods by which we can create new conda environments. We can create an "empty" environment, which will just install the core packages needed to control the environment itself. From our shell (e.g., bash or zsh), we can type

```
conda create -n NAME
```

where NAME is the name you want to give the environment (this is the name with which you will activate and deactivate the environment). We can also specify certain elements of the environment, e.g., the version of Python and some basic packages, right at the creation stage:

```
conda create -n NAME python=3.10 numpy scipy matplotlib astropy pandas
```

In this example, we created a new environment, and specified for Python 3.10 to be installed, along with several core packages. Because we didn't specify which version number of those packages to install, conda will attempt to install the most recent versions of those packages which are co-compatible with one another.

Once the environment is created and set up, we can enter and exit the environment using

```
conda activate NAME
```

and

```
conda deactivate
```

I previously described environments as silos—in that analogy, activating and deactivating an environment is much like then stepping inside each of these silos and taking advantage of the different equipment inside. Once *inside* an environment, we can proceed to install more packages if we wish to. There are two main avenues for this: conda install, and pip install. The pip program will download and install programs which have been uploaded to the Python Package Index, PyPI. Meanwhile, the `conda install` command will pull from the online `conda` index of packages. Often, code developers will upload their code to both of these locations, meaning you can install that code from either source. A third way to install software is directly from a hosted repository on, e.g., Github. There, you generally download the code, `cd` into the code directory, and run `pip install .` to install.

In general, it is best to install a package from `conda` if it is available there (the installation instructions for a package should indicate if so). If it is not, then installing from `pip` *inside your environment* is the next course of action. This is because when you install from `conda`, the environment manager that is part of `conda` can do a better job determining versions and dependency requirements.

An additional means by which we can create an environment is that of an `environment.yml` file.[2] Using the YAML format,[3] this simple text file can specify a name for an environment and set of packages, something like this:

```
name: NAME
channels:
  - conda-forge
  - defaults
dependencies:
  - python<=3.11
  - pip
  - astropy
  - matplotlib
  - numpy
  - pandas
  - photutils
  - sep
  - scipy
  - pip:
      - tqdm
```

Here, we've specified a few dependencies. The primary list are those which can be installed with `conda`. If you know there is a package only available via `pip`, you can specify those as shown at the bottom. Version numbers can also be set in this file— one could create a so-called "locked" environment file in which every version of every dependency is set explicitly. With the environment file in hand, creating the new environment becomes as simple as:

---

[2] Requirements.txt files work similarly, but `yml` is increasingly the preferred format.
[3] YAML files are text, but have rules about what dashes, colons, and indents mean that allow the files to be parsed by other programs.

```
conda env create -f environment.yml
```

After completing, you should be able to activate and deactivate this new environment by the name established at the top of the environment file. Finally, if you'd like to save an environment file representing your current environment's state (e.g., to setup on a second computer, or share with a colleague), you can do so via

```
conda env export > filename.yml
```

which will save every package installed in your current active environment to a file (with their current versions).

3.2.3 An Environment for this Textbook

In the spirit of the above discussion on environments and portability, I have created an `environment.yml` file for this textbook, available in the Supplementary Materials (which can be accessed from https://doi.org/10.1088/2514-3433/acfa9a). Once you have `conda` successfully installed, you can create an environment using it (it will create an environment called `astropython`; you can change this name within the file before using it if you wish). Once created and activated, you should have all of the packages we use in this textbook installed and ready to use.

3.3 Editors

Once we have our computing environment set up, with Python installed into one (or more) environments along with at least the core set of packages we need, we can actually turn our attention to writing and running Python code. There are several different ways to do this. In the next chapter, I will introduce the iPython interpreter: a program you open in the shell which allows you to type Python commands one at a time. Essentially, it behaves like a shell, but for Python commands, rather than the language of your shell (e.g., bash, zsh).

In the long run, this isn't where we will do most of our coding. That will happen in an *editor* of some kind—a specialized program that lets us write many lines of code into a document, and then run it all at once. In the following sections, I'll lay out some of the different options available to you for editing code. Some of these programs actually have access to the shell, and iPython, meaning you can run the code you've written from right inside the program. Ultimately, you'll need to find a workflow that works right for you. Try out all of these options, and see what you like! When you move on to Chapter 4, you can follow along, running commands in the interpreter, or, you can set up scripts in your editor of choice and run things there.

3.3.1 Terminal Editors

In the old days, before the tools I'll discuss below were developed, one opened, edited, saved, closed, and then ran code files all from within the shell itself. Almost any shell you log into (your own, or a remote server) will have one (or both) of the following installed: vi (or vim) or emacs. They can be used by typing their name into the shell, and usually if you provide a filename after, e.g., vim myfile.py, it will open that file if it is in the current directory or create a file with that name if it is not.

While some still use these as their primary coding environment, the majority of code development now occurs in more modern, GUI based programs. That said, both vi and emacs are efficient, fast editors, and those who know all the keyboard shortcuts associated with them can navigate and edit a file with ruthless efficiency.

I recommend *familiarizing* yourself with at least one of these two (at least enough to open a file, make some minor changes, and close it). This is very useful when working via ssh on remote servers, where you won't (by definition) have access to the text editors or IDEs you're used to.[4]

3.3.2 Text Editors

One step up from the terminal editors are text editors. These, as their name implies, allow you to open text files (of any kind, including Python code), and edit it. They typically have what is called "syntax highlighting" which helps you recognize certain patterns unique to a given language (i.e., opening a .py file, functions will be highlighted a different color than variables). They also usually do small things for you (such as auto-indentation and tab-completing variable names) to make your coding faster. They typically have built in search/replace and regular expression support, and are generally a clean, suitable way to develop code scripts or larger programs.

Technically, even non "coding" oriented text editors (like Microsoft Notepad) can be used to edit code. What *cannot* be used are word processors (such as Microsoft Word or Apple Pages). That is because these save files not in plaintext, but in proprietary formats which are not compatible with code.

Popular examples of text editors (as of writing) include atom, Sublime Text, nano, and VSCode. Some of these have numerous community-developed plug-ins to make for even more efficient coding—VSCode in particular has numerous quality of life features and extensions. I generally recommend students start with a text editor such as those listed above when working on the basics. I myself use VSCode.

3.3.3 Jupyter Notebooks

An alternative, hybrid integration of permanent/saved code and interactive execution is the Jupyter Notebook. Notebooks are separated into different "cells", which can be either "code" cells or "markdown" cells.[5]

Jupyter notebooks are becoming extremely widespread in both academia and industry, and like anything, have their pros and cons (and subtleties). One thing to emphasize is that, broadly speaking, something we would term "Software" is not

---

[4] Although, these days, it is possible to open remote connections within a locally-running instance of VSCode, and I suspect this will become more common in the future.

[5] Markdown is a mark-up syntax for text, allowing you to add headers and some light formatting with very little effort.

(and will not ever be) written in notebooks, though algorithms for software may be developed partially in them. For the purposes of following along with this textbook, it does not matter whether you code in a notebook or save scripts which you run from the terminal. Within this chapter, I'll make some distinctions and notes that may be relevant when running code in a notebook vs from a script.[6]

There are two main ways to open a Jupyter notebook. You can run the command Jupyter lab from your command line (within an environment with it installed), which will open a web browser window with the interface (Jupyter itself is a GUI). From there, you will have access to a file browser, the ability to open or create notebooks, etc. Make sure to leave the terminal from which you launched Jupyter open, as that is where the actual Python kernel is being controlled from. When you are done, you can hit <ctrl>+C in that terminal window to shutdown the running kernels.

Alternatively, an editor like VSCode can simply open and create notebooks and files. If you have it installed, an empty file saved with the .ipynb format will auto convert to a notebook. Unlike the Jupyter lab implementation, when you open a notebook in VSCode, it doesn't automatically start and connect to a kernel. This is useful for, say, glancing at the contents of a notebook when you don't want to run anything. It will also disconnect automatically after a period of time. As soon as you run an individual code cell (which you can do with <Shift>+<Enter>), the kernel will connect, which takes a few seconds, and then run.

One major thing to watch out for with notebooks is that cells can be run out of order. This can cause problems when a variable is being overwritten or modified later in a notebook, but then earlier cells using it, if re-run, will be dealing with the new value. My advice with notebooks is always to run the cells in order, or at least, chunks of the cells in order. If your variables are ever behaving strangely in a notebook in ways you can't explain otherwise, restarting the kernel (the connection of the notebook to an instantiation of Python under the hood), then running through the cells in order, is a good troubleshooting technique. Similarly, when you delete a cell after it has been run (or delete the text within it), the variable (or anything else) that was run in that cell will still be "active" in your environment. This can cause issues as well, if that (now deleted) variable is sprinkled throughout the rest of your cells. Once again, restarting the kernel and re-running cells in order will inform you if this has happened, because instances of that variable will cause errors.

I generally recommend students to learn the fundamentals of programming *without* using notebooks first—this will cut the number of errors you are dealing with down, and leave you only with those introduced by your code, and not the medium of notebooks. Once you are comfortable with basic coding and understand exactly *how* notebooks can cause issues, it is then fine to switch to notebooks. In any case, even if you do use notebooks for prototyping and testing things on the fly, I recommend having both a notebook and a regular Python file open. As you experiment and vet

---

[6] Also note that some text editors, such as VScode, can open and edit notebooks as well.

elements of code, moving them over to the .py script (preferably in a function) is a good workflow for ensuring stability and reproducibility over time.

As a note, this textbook was written in Jupyter notebooks. So the cells of Python code you see reflect cells in a notebook—but you can easily transcribe them into .py scripts to run as well. Certain shortcuts (like a bare variable name as the last line of a cell, which prints it) won't work in script form; I have tried to use print statements such that code is copy-paste-able to scripts, but you may find one or two cases where slight modification is needed.

3.3.4 IDE (Integrated Development Environment)

Finally, the most complex/feature rich option for developing code is the IDE, or integrated development environment. Here, the program has not only the features of a text editor, but can run the files through specialized software to find bugs (debuggers), determine bottlenecks (find places where your code slows down), and more. They are typically large programs and are most useful when developing expansive pieces of software with many interconnected parts. Many argue that an IDE is not a necessity for Python development, compared to other languages, and I tend to agree with this assessment. Popular examples include PyCharm and IDLE.

3.4 Summary

In this chapter, we used our knowledge of navigating the shell to install the Python Astronomy Stack, a set of core software needed to get up and running with data analysis in Python. I'll re-emphasize that the exact packages, distributions, and order of operations for installation change as operating systems update (along with the softwares themselves), so it is always worth checking in with a tech-savvy astronomer near you, or online resources, for the current norms. As another final note, I'll also remind you that some resources online will allow you to open and access a (usually temporary) Python environment with most of everything you need pre-installed (or easily installable). This can be a reasonable stop gap, if installing on your own machine runs into issues.

Astronomical Python
An introduction to modern scientific programming
Imad Pasha

Chapter 4

Introduction to Python

I mentioned in the Introduction that Python (as used by astronomers) could in some ways be thought of as a glorified calculator. Let's put this to use now.

To use Python, we have to access a special interpreter, which understands how to read Python syntax, and translate it to the computer, before returning the answer. There are three primary ways we can access this interpreter:

1. By calling the standard Python interpreter from the command line,
2. By calling the interactive iPython interpreter from the command line
3. By using a an integrated development environment (IDE)—a program, which accesses Python "under the hood"—examples include VSCode and IDLE, and
4. By using a `Jupyter` environment—Jupyter also accesses a Python kernel under the hood, but is organized into "cells" of code that can be run independently. Some IDEs can open and run these so-called `Jupyter notebooks` alongside shells which can execute Python scripts or individual lines of code.

For our purposes in this text, we will not concern ourselves with option number 1. The iPython version of the interpreter is more convenient, has more quality of life features, and provides no drawbacks compared to the standard interpreter[1]. That's not to say we never use option 1; I'll return to this when we discuss standalone scripts in a later chapter.

We will also discuss Jupyter notebooks in depth below. But for now, let's use iPython. To start, you'll want to have open a shell (aka a terminal). On Mac and Linux systems, the standard terminal is fine. On Windows, you specifically want to *not* be using the windows command prompt, and instead open the Anaconda prompt, which is a program installed when you install the Anaconda distribution of Python on Windows. Alternatively, if you are not using Anaconda, you will need to

---

[1] If you install the environment for this book, you will have iPython. In any environment without it, iPython can be installed via 'pip install ipython'.

make sure your Windows system is configured to have an Ubuntu/UNIX style prompt available (if you installed ws1, this will be a Ubuntu terminal), and that your installed version of Python is accessible from this terminal.

Once you are in your terminal, launching iPython is as simple as typing

```
ipython
```

When you do so, you should notice your terminal change. You will have some summary information about the version of iPython being used, and you should see the prompt change from your terminal's default to a bracket with a "line number" followed by a colon. An example from my system is reproduced below:

```
Python 3.11.6 | packaged by conda-forge | (main, Oct  3 2023, 10:37:07)⊔
  ↪[Clang 15.0.7 ]
Type 'copyright', 'credits' or 'license' for more information
IPython 8.16.1 -- An enhanced Interactive Python. Type '?' for help.
```

```
In [1]:
```

Once inside iPython we can begin using it as a calculator. The basic operations all work out of the box:

```
1+1
```

```
2
```

```
2*3
```

```
6
```

```
5/2
```

```
2.5
```

```
2**4
```

```
16
```

You may notice that we haven't attempted, e.g., trigonometric functions like sin(), cos() or tan(). These will require the use of external libraries which are not automatically loaded when Python is opened.

4.1 Variables

The basic operations we performed above could easily be carried out on a hand calculator. With Python, however, we can actually *store* the output of calculations. To do this, we define a ***variable*** and set it equal to our calculation. This variable then refers to a location in system memory, from where the quantity can be retrieved. Strictly speaking, because of the way things are actually stored, Python variables are not formally variables or pointers (e.g., in the same way as in C++) but are rather *names*. This distinction is not particularly important to us for now.

4.1.1 Defining Variables

```
my_variable = (5+6) * (20/6.5) / 45.
```

```
print(my_variable)
```

0.7521367521367521

As we can see, the output of the above calculation was stored into the variable my_variable, and we were able to print the value back out to the screen. It is also worth establishing at this juncture the way a Python function works. Much like $\sin(x)$, Python functions (of which print() is a member)[2] are called by their name followed by open/close parenthesis. Arguments to the function are passed inside the parenthesis (above, I pass my_variable to print this way). This is in contrast to, e.g., bash, in which arguments are supplied after a space. We'll get much deeper into functions soon, but it's important to remember this particular syntax.

The ability to store variables gives us the ability to construct more complicated expressions by combining variables:

```
new_variable = my_variable**2 - 4.0
```

```
print(new_variable)
```

-3.434290306085178

When we are dealing with a physics expression that has complicated parts, it can be useful to separate the calculation into sub-parts which we store in separate variables, and then combine these results together. We can also overwrite variables that we've created, even within a calculation that involves a variable's own value, as follows:

```
new_variable = new_variable * 2
```

A shorthand method for performing the above (and other similar modifications) is

```
new_variable*=2
new_variable+=1
new_variable-=10
```

---

[2] Throughout this text, I will generally refer to custom functions by adding, e.g., func() parentheses after to distinguish functions from variables. However, I will generally not do this for common functions such as print.

and so on. When we do this, the right hand side is evaluated first (using the currently stored value of the variable), and then the result is stored (in this case, overwriting the variable's older value).[3]

4.1.2 Copies of Variables

There's one subtlety regarding variable declarations (namely, the *copying* of variables) that merits some discussion now, as it may cause your code to behave strangely. You may want to read ahead to the section on Data Types before reading this section, as we are jumping slightly ahead to cover it.

Let's say we want to make a copy of a variable we have in our code.

```
my_variable = 10
```

Now, we can try to make a copy of the value stored in my_variable by setting my_variable2 equal to my_variable.

```
my_variable2 = my_variable
```

Printing our two variables confirms that both now have a value of 10, as expected.

```
print(my_variable,my_variable2)
```

```
10 10
```

Is the 10 stored in my_variable2 independent from the one stored in my_variable? Let's find out, by modifying the first variable.

```
my_variable=15
```

Printing both again returns

```
print(my_variable,my_variable2)
```

```
15 10
```

Okay, so perhaps as your intuition might suggest, we now have 15 and 10. However, things are not quite as simple as this! We can illustrate this using a similar, but slightly different example. Let's make a list instead:

```
my_variable=[1,2,3]
my_variable2 = my_variable
```

---

[3] Under the hood, it's actually slightly more complicated than. But in practice, the above describe what happens on the user end.

Printing both our lists produces

```
print(my_variable,my_variable2)
```

[1, 2, 3] [1, 2, 3]

and we can once again perform the experiment of modifying one of the variables:

```
my_variable[2] = 10
print(my_variable,my_variable2)
```

[1, 2, 10] [1, 2, 10]

Uh oh! We got a *different* behavior this time. When we changed the second index (third element) of the first list, the *second* list *also* got changed. Our "copy" of my_variable turned out not to be a copy at all! Indeed, when setting the *name* variable2=my_variable, we told Python to essentially point this name to the same spot in system memory. When the list stored at that location got modified, our new name (still pointing to the same place) also returns the changed list.

"But wait," you might ask. "Why then, did changing the variable the first time, in the case of 10 and 15, not produce the same results?"

This is where the subtlety comes in. The difference between these two examples has not to do with the concept of copying. As I said above, the new name will point to the same location, and this remains true. The difference is that Python treats some data types (see more on data types below) as ***mutable*** (modifyable) and others as immutable, or unchangeable. Integers, floats, and tuples are examples of data types Python considers ***immutable***. When we changed my_variable from 10 to 15 in the first example, Python, under the hood, allocated new spots in memory for my_variable and my_variable2, setting the new value for my_variable to 15, and moving the 10 to a new spot allocated to my_variable. But with lists, which are mutable, it has no problem modifying it in place, and letting both names point to the same spot.

So how do we handle (and remember) this? You may have noticed that in the code snippets above, we used variable names liberally in equations being used to set the values of other equations. As it turns out, because that involved multiplying by integers and floats along the way (immutable), it was guaranteed that a new location, independent location in memory would be allocated for those variables. We can mimic that behavior here, e.g.,

```
my_variable2=my_variable*1
```

```
my_variable[0] = 3
print(my_variable,my_variable2)
```

[3, 2, 10] [1, 2, 10]

As you can see, this time, by setting my_variable2 to my variable1*1, changing a value within my_variable does not propagate to my_variable2. However, this is not a very elegant solution, and leaves some confusion for anyone reading the code as to what's going on. Instead, when we truly want to make a copy of an otherwise mutable variable, ensuring it is independent, we can use the copy library. We haven't gotten to importing libraries yet, so don't be too concerned with this solution yet—I included this aside here primarily because even before we learn about data types, libraries, etc, this behavior with variable declarations has the potential to rear its head and create confusing results for new programmers.

```
import copy

my_variable2 = copy.deepcopy(my_variable)

my_variable[1]=5
print(my_variable,my_variable2)
```

```
[3, 5, 10] [3, 2, 10]
```

We can see that by running this special "deep copy" command, we created a copy of the first list that gets its own spot in memory and is unaffected by further changes to the original variable.[4] Alternatively, if you know a list-like quantity should be immutable at the time of instantiation, you can use a tuple instead (parenthesis instead of square brackets—see Table 4.1).

Table 4.1. Python Data Types

Datatype	Type	Description
string (str)	text	contained in "quotes", usually for paths or names
integer (int)	numeric	counting numbers
float (float)	numeric	decimal-valued numbers
complex (complex)	numeric	complex numbers (e.g., 4i+2j)
list (list)	sequence	series of values, can be of any type, mutable
tuple (tuple)	sequence	series of values, any type, immutable
set (set)	set	like a list, but single valued (obeys set rules)
boolean (bool)	boolean	True or False
dictionary (dict)	mappable	key-value paired collection
byte/bytearray	binary	mutable array of values (0-256) in binary.

---

[4] There is also a copy.copy() function. I recommend always using deepcopy... the idea is just that copy only goes one layer down, and wouldn't handle lists within lists and similar cases very well—the inner objects would still be pointers to the original memory locations.

4.2 Importing External Libraries

The base Python environment when we launch, e.g., iPython or a notebook does not have much functionality beyond the basic mathematical operations, variable declaration, and the attributes and methods of the built-in data types (we'll cover these more extensively throughout this chapter). But as we've already seen with the copy library above, the practical usage of Python involves the *importing* of sets of libraries which have useful functions (e.g., copy.deepcopy above). These libraries can be divided into two categories: those which come *with* Python, and those which we install ourselves.

In Chapter 3, we installed a set of core external libraries, including Numpy, SciPy, and Astropy, into our Python environment. Python actually also comes with a set of very useful libraries, including functtools and collections. Within Python, we treat these the same way—we just didn't have to install them ourselves. In both cases, these libraries are not loaded into the global namespace of our coding environment by default.

Instead, we must explicitly import these libraries if we need to make use of their functions. If we had called the copy.deepcopy() function above without importing copy, we would have gotten an error.

There are a few ways we can import libraries and functions into our code. Above, I used the simplest means, which is simply typing import library. This imports the entire library (e.g., all functions within the copy library), and all of them are accessible under the name copy using *dot-notation*. Thus, to get to copy's deepcopy command, we type copy.deepcopy() in our code.

Very soon, we will begin using more libraries, especially Numpy. Because we use it so ubiquitously throughout scientific programming, it is convenient to import the module but give it an *alias* that is shorter and easier to type. For Numpy, everyone in the world, more or less, has agreed to the convention of using np as its shorthand. Thus, when we go to import Numpy, we assign it this alias right away:

```
import numpy as np
```

By importing Numpy *as* np, when we use that library in our code—for example, to take the sine of a number—we would type np.sin() instead of numpy.sin(). This may seem like a small difference now (three letters' worth), but it makes a huge difference when calling Numpy functions hundreds of times throughout a script. It also helps keep individual lines of code shorter.

Because we often need a *multitude* of functions from Numpy, it makes sense to read in the entire library (that gives us immediate access to hundreds of functions). On the other hand, if we only need a single function from some library (e.g., we *only* need the deepcopy function from copy), we can just import that one function:

```
from copy import deepcopy
```

Here, we use the `from` keyword to indicate that we should only import the functions from the chosen listed after the `import` call (you can list more than one, separated by commas).

When we import this way, we can call `deepcopy()` directly within our code with no prefix. But we also can't decide to call some other function from `copy`, either.

Finally, it is worth noting that the wildcard (*) can be used to bring all functions from a library into the global namespace with no organization. For example, if we did the following (not recommended):

```
from numpy import *
```

Here, we've told Python to import *every* function from `numpy`—but instead of being bound to the name `Numpy` or `np`, would just be immediately available. Typing `sin(1.4)` would use the `Numpy` library to carry out that computation.

You might already be able to see why this is generally a bad idea: now, when we see `sin(x)` in our code, we don't know explicitly which library is responsible for computing it. For example, there is also a `math` library with its own sine function. If we imported all functions from both the `math` and `Numpy` library, it would be ambiguous which had taken precedent—for many functions, the actual computation might vary slightly, so we need to know which libraries we're using. The long and short of it is, from external packages, it is best to import either the library name (possibly with an alias), or to import the set of individual functions needed.

4.3 Comments

Not formally a data type, comments are text within our code that are *ignored* by Python when it runs our code. Why would we want to add text that will be ignored? The primary reason is for *documentation*. Documentation, in general, is the process of labeling what our code does, how it works, and how it should be used properly. It is surprisingly easy to forget how a piece of code works, and the last thing we want to do is return to our code after a few weeks (or months, or years) and not be able to parse it. Documenting our code in various ways also makes it much easier to share code with others.

We have two primary ways to set text up as a comment: the hashtag/pound symbol, and triple quotes. Hashtags are used to define single-line comments, while triple quotes can enclose multiple lines:

```
"""
Script Summary:
This script defines some constants and carries out computations to␣
 ↪determine
the temperature of a cloud of gas for a given temperature and density.
"""
# define needed constants
gamma = 5/3. #adiabatic index
nH = 0.14 # 1/cm3
T = 1e4 # Kelvin
```

As we can see, single line comments can span a whole line, but can also be added *after* some valid code. Anything before the comment character in a line is executed, anything after it will be ignored. As we can see here, the comments can establish the purpose of some code, indicate what some lines of code are doing, and give further context to individual variables (e.g., indicating their units). I've also used the handy calculator-style 1eX format to indicate 10^4. A similar trick is that one can add underscores to a long number to help with readability, e.g., 1_000_000 would execute to the integer one million.

4.4 Data Types

In the above examples, we have dealt with three of Python's built-in data types: integers (int) and floats (float) and briefly, lists (list). When you assign a number in Python such as 42, *without a decimal*, the default assignation is that of an integer. We can force the issue as well, by defining int(42). Integers differ from floats primarily in how much memory they take up; for most calculations we carry out in basic data analysis, it is perfectly fine to simply cast all values as floats. As a note, any time an integer and at least one float are used in the same calculation (e.g., in the definition of my_variable above), the integers are all cast automatically as floats, and the resulting answer will be a float as well. Furthermore, from Python version >3, a calculation between two integers that does not result in an integer (e.g., 5/2) will return a float as well. There are times when integers are specifically useful—in particular, as "counting numbers" when indexing an array or looping over a range, two topics we will come to shortly.

Table 4.1 provides a summary of the most important native Python data types.

It is challenging to motivate the use of these different data types without examples. So rather than an item-by-item treatment, I'll proceed by establishing some astrophysically motivated examples, and use them to elucidate each data type as it becomes relevant throughout the book.

4.4.1 Strings

One data type that stands somewhat apart from the others discussed here are **strings**. Strings are defined by quotations, and allow us to input arbitrary text into our programs. Why do this? In the context of a data scientist, the main reason we need strings is that we will load files containing data from our computer by giving the file path as a string to some function. Generally the same applies for saving files—we'll define the save path as a string.

Another use case is displaying text to the user. If, for example, a certain point in our code will ask the user for input, we need a way to input the phrase we want displayed to them. Error messages (or simply informational messages) will also be stored as strings, and we will also use them (often) as keys to dictionaries (see the next section).

Additionally, we'll soon learn about `dictionaries`, which store data as key-value pairs. Keys to dictionaries need not be strings, but it is often convenient to use them for this purpose. For example, if I had a dictionary of stellar fluxes, I might set the key-value pair for each entry to be the name of the star, e.g., `'Deneb'` and the value to be its flux.

Finally, our data itself is sometimes in string form. As astronomers, this is *usually* not the case—we deal primarily in numbers. But you might have a table, for example, with a number of supernova light curves, for which one of the columns contains values like `TypeIa`, `TypeIIc`, etc, listing the type of supernova. We would want to read this column in strings, as they are "words" and do not fall into any of the other data type categories. In the chapter on libraries, we'll discuss in detail how to read every column of a data file into different data types.

4.4.1.1 F-strings

It is useful to be able to insert values from our code into a string. We can do this using f-strings, which is the currently preferred method, though there are a few different methods that work. To define a string as an f-string, we add the letter "f" to the front of the string, and then use single curly braces within the actual string to insert anything Python can evaluate (like a variable name or expression).

```
name = 'M31'
printout = f"The Galaxy to be observed is: {name}"
print(printout)
```

```
The Galaxy to be observed is: M31
```

Plugging a string into a string is fine, and this use case sometimes emerges when combining, say, a file path (location) with a set of file names. But our primary use case is going to be viewing or saving the output of a calculation. For example, if we have a measured galaxy flux and distance, the formula

$$F_g = \frac{L_g}{4\pi D_g^2}$$

can be used to derive the intrinsic luminosity of the Galaxy:

```
import astropy.units as u
import numpy as np
galaxy_distance = 10 * u.Mpc
galaxy_flux = 2.5e-16 * u.erg / u.s/ u.cm**2
# Thus
galaxy_luminosity = galaxy_flux * 4 * np.pi * galaxy_distance**2

print(f'The Galaxy Luminosity is {galaxy_luminosity.to(u.Lsun)}')
```

In this example, I've given us a preview of the Numpy and Astropy libraries that we'll be learning in detail later. Given a distance and flux, I use the formula to determine a luminosity. I can then insert the *numerical* quantity into an f-string to seamlessly integrate it with the printout. I also convert the units to those I want, and Astropy is clever enough to print those units when an Astropy quantity is printed. There is, perhaps, one issue with the above. We have many digits of our answer, beyond the relevant significant figures. While retaining this "full" decimal float for further calculations may be useful, it's not ideal when printing. Luckily, string formatting will allow us to round off our answer:

```
print(f'The Galaxy Luminosity is {galaxy_luminosity.to(u.Lsun):.2f}')
```

The Galaxy Luminosity is 781.41 solLum

By adding a colon after the inserted value in the f-string, and then specifying .2f, we told Python to round to the second place after the decimal when printing the answer.

F-strings have plenty more formatting options than this simple exploration, but we'll discuss these as they become relevant when carrying out data analysis later in this text. For now, we can think of the f-string formatting as a way to simply insert things into strings.

4.4.2 Collections of Data: Lists and Dictionaries

As mentioned in the Introduction, individual calculations with ints and floats aren't all that interesting—they are doable by hand, or on a calculator. The more interesting situation arises when we want to deal with a *collection* of values (usually floats). In the Introduction, I used a specialized library called Numpy to carry out a calculation on many values at once. We'll come to this solution soon—it's the fastest (computationally) way of performing simple math operations on many numbers at once. But even sticking with the built-in Python data types, we can learn about collections and how to operate on them.

For practical purposes, lists and dictionaries are the two most useful built-in collection structures.[5] Lists are defined by square brackets, and permit almost anything to be made an entry. For example, I could define a simple list as

```
new_list = [1,2,3,4,5]
```

or

```
new_list = [1.0,2.2,3.5,4.2,5.1]
```

if we were dealing with floats. Alternatively, I could fill a list with all sorts of data types:

```
new_list = [1.0,'hello',3.5,int(4),['another',' ','list']]
```

That is to say that a list is a very permissible container—you can dump almost anything in it, and it won't complain. That's great for a lot of use cases, but we'll also find that for mathematical calculations, uniformity of the data being represented can help speed up calculations.[6] We'll come to the process of *extracting* entries in a list momentarily (we call this slicing or indexing), but first, let's look at the setup for a dictionary.

```
my_dict = {'key1':'value1',
           'key2':10,
           45:[1,2,3,4]}
```

A dictionary is defined by curly brackets, (where a list uses square brackets). Instead of simply having entries, dictionaries have *keys* and *values*. The keys can be any basic type—strings, integers, floats—so long as they are *unique*. Strings are by far the most common key type. The values can also be pretty much anything, much like a list, including other collections. You can nest dictionaries within dictionaries, for example. To extract values from a dictionary, we use the associated key. It is a form of indexing, which we address shortly, but to give a preview:

```
print(my_dict['key2'])
```

```
10
```

As we can see, square brackets *after* the variable name we created tells Python we are going to extract something, and by feeding in a valid key, it returned the

---

[5] Sets are the unique elements of a collection—which can be useful, but data often have repeated values. Tuples are like lists, but are immutable; we cannot change values in them after the fact, making them less commonly useful for our applications.

[6] I want to take this juncture to point out that we've defined some lists here by hand, but very often in our code, some abstracted code we've written will, e.g., parse a file or perform calculations and then *create* these lists for us.

associated value. This turns out to be the syntax for indexing/slicing/extracting data from any collection of things in Python—but what goes *inside* the square brackets depends on the particular data type. If we had put a dictionary inside my_dict, we would chain the indexing. Let's say that interior dictionary had a key called 'key5'. I would use

```
print(my_dict['key2']['key5'])
```

At any time, we can see the keys and values of our dictionary by querying my_dict.keys() or my_dict.values(), which are methods that will retrieve the keys or values, respectively.

4.5 Indexing

Dictionaries are mapped by their keys, meaning extracting values is a straightforward prospect as above (as long as the keys are sensible). Other collections, like lists, sets, and as we'll see soon, Numpy arrays, don't have keys to "tag" each value. So how do we extract elements from these collections?

To do so, we use the *position* of the element in question. In Python, positions are counted starting from 0, so in our final new_list established above, "1.0" would be index 0, "hello" would be index 1, and so forth (see Figure 4.1). We can plug the desired index into square brackets after the variable name of the collection (or directly after the collection), e.g.,

```
new_list[0]
```

```
1.0
```

These indices can be chained together—as shown in Figure 4.1, we have examples of strings and other lists (containing strings) within new_list. To get at these *nested* values, we can use bracketed indices one after the other, e.g.,

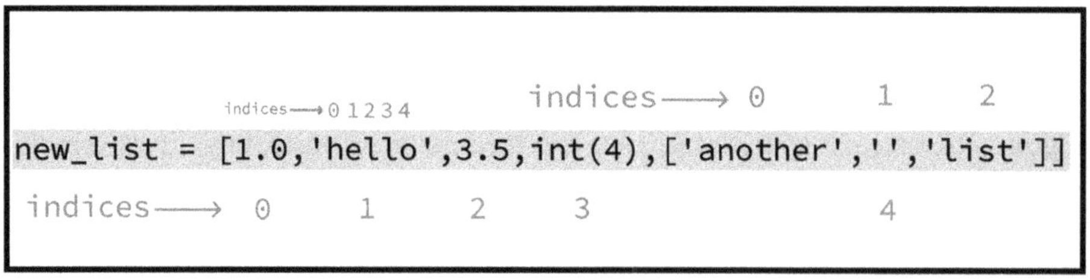

Figure 4.1. Some data types in Python, including lists and strings, can be *indexed* by position, starting at 0. Here, I show what the index is for each entry in new_list, as well as for some of the entries themselves (which are themselves index-able, also known as being *iterable*). This means that I can index the letter "e" in "hello" by indexing new_list[1][1].

```
new_list[1][4]
```

```
'o'
```

which gives us the last letter of the string `hello`. This indexing is carried out left to right—so the entity `new_list[1]` is constructed first (now the string hello) and then that is indexed at position 4.

Anything that can be indexed in this manner is known as an *iterable*. Lists and strings, as shown above, are examples of iterables—but integers, for example, are not. If I had the integer `x=10503` and I wanted to index individual digits in this number, I'd want to turn it into a string first. I can then index it—and even turn it back into an integer. We can do that like this:

```
print(int(str(x)[2]))
```

```
5
```

You'll notice that I used the `str()` function to turn the integer into a string, and then the output of the index position 2 of that string, which is '2', was turned back into an integer via the `int()` function. These built-in functions are useful for the above purpose, but you can't turn *every* data type into every *other* data type.

Finally, we can also index iterables using *negative indices*. When we do this, we start at the *end* of the iterable and work backwards. Above, I indexed the final letter of "hello" by indexing the first index of `new_list` and the 4th index of 'hello'. I could get the same result by typing

```
new_list[1][-1]
```

```
'o'
```

The second to last character would be obtained with index [–2], and so on.

As a final note, we saw above that `dictionaries` are indexed by *key*, rather than by index number—and indeed, dictionaries are not stored in a particular order. But we *can* actually iterate through a dictionary if we put our minds to it. Dictionaries have two helper methods—functions that we can run on any dictionary via dot-notation after a given dictionary's name (i.e., `.keys()` and `.values()`)— which allow us to construct lists containing either the keys or the values of a dictionary.

```
list(my_dict.keys())
```

```
['key1', 'key2', 45]
```

```
list(my_dict.values())
```

```
['value1', 10, [1, 2, 3, 4]]
```

As we can see, we can extract the keys or values of our dictionary this way, but we could also use this to turn around and index our dictionary using them. This will become a little more useful later when we learn about looping.

4.6 Slicing

In addition to querying individual entries of iterable collections, we can also pull out *sections* of those collections, e.g., from index 1 to 5. This process is known as *slicing*.

We use the same square bracket indexing notation to extract slices of iterables, but add a colon and specify the start and stop of where we want to index on either side.

```
new_list[1:3]

['hello', 3.5]
```

Immediately we find something rather interesting. We know that `hello` was the 1st index of our list, but 3.5 is not the third index, it's only the 2nd. This isn't an error: it takes a while to get used to, but Python uses a closed-open indexing scheme, meaning that the left-hand index is included in the index, while the right-hand index input is *not*. If you want the first through third index of a list, you would need to index, e.g., `new_list[1:4]`. You can read a slice as "retrieve the elements of X from index 1 up to (but not including) index 4."

Sometimes we want to slice from a certain index through the end of an iterable. We can do this by selecting the start index and adding the colon, but not specifying any end index:

```
new_list[1:]

['hello', 3.5, 4, ['another', '', 'list']]

new_list[-3:]

[3.5, 4, ['another', '', 'list']]
```

Notice again that we can use positive or negative indices as desired. We can do a similar trick to index up through a specific index from the beginning of the iterable:

```
new_list[:2]

[1.0, 'hello']
```

Later on in this book, we'll return to the concept of slicing when discussing multi-dimensional arrays, which can be sliced in multiple dimensions at once.

4.7 Operations

Finally, you may have noticed that we have been using operations (such as the "+" sign) when adding numbers. In a more general sense, every data type in Python are abstractions we call "objects". These objects have rules assigned to them which define what it means for various operators to be applied to them. In the case of integer or float numbers, the "+" sign does exactly what we expect: it adds the numbers. But what if we use the add symbol between lists? Try it out:

```
[1,2,3] + [4,5,6]
```

You should see that there is a particular behavior here. That behavior is defined for the object type "lists", and thus all lists can be operated on in this way. Now try multiplying a list by an integer. What happens? What happens if you try subtracting a number from a list?

At this point, you should get an error — the developer of the Python "lists" decided there was not an obvious meaning to the subtraction of a number from a list object. So the subtraction operator between these two datatypes has not been defined.

The same kind of rules apply for transforming one datatype to another. Some are so obvious and seamless (like turning an integer into a float) that we almost never have to do it ourselves. But if you try the function int("123"), you should find that Python is able to convert this string into a real integer number. That only works because the value within the string looks like an integer; int("hello") would obviously not work.

Taken as a whole, this means we can in general perform operations (when defined) on our variables based on their object type (list, string, etc.), or indeed turn one object type into another assuming the the transformation is defined and makes sense. Which operations and transformations are defined? The best way to learn this is by trial and error — most are intuitive; some are not. As we move through this text, I encourage you to try turning one datatype into another, or try adding, multiplying, subtracting, or dividing between datatypes to see if such operations are defined and what they actually do.

4.8 Reserved Words

Python reserves some words in order to facilitate the language—these words cannot be used as variable names, and it is worth being aware of them, though nearly all editors with syntax highlighters will highlight these reserved word in a specific color to indicate them as such.

You've seen some of the reserved words all ready; there are thirty-five in total.

- To handle imports: import, as, from,
- To handle truthiness: True, False, None,
- To handle control flow: if, else, elif,

- To handle iterating: for, while, break, continue,
- To handle operators: and, or, not, in, is,
- To handle function/class creation: def, class, with, as, pass, lambda,
- To handle returns: return, yield,
- To handle exceptions: raise, try, except, finally, assert
- To handle asynchronous tasks: async, await, and
- To handle variables/scope: del, global, nonlocal.

We'll be using many of these throughout this text, but for now, they are worth mentioning here for both completeness and to be sure you recognize them as scaffolding for the language's logic. Thankfully, these reserved words don't tend to lend themselves to being variable names anyways, and with our editor warning us, it is unlikely to accidentally attempt to use one of these words incorrectly.

They are also worth mentioning in the context of operations. In the next few sections, for example we will use conditionals, which check whether an expression evaluates to True or False. When we employ these reserved words in a conditional check, e.g., `a is True`, the result will be `True` if a is *truthy*, and `False` if a is *falsy*. A Python variable is only falsy if it is the reserved keyword `False`, a value that is numerically zero, an empty string, the reserved keyword `None`, or some special cases of class objects. You can check if a given variable is truthy by passing it to the built-in function `bool`.

In practice, you don't generally need to consider the truthiness of expressions—the conditionals we employ read much like logical English, and if the condition is true, the value of the expression will be truthy. There are a few (rare) gotchas, and some interesting cases of leveraging truthiness, but those are not within the scope of this text.

4.9 Filtering and Masking

Filtering is the process of indexing and retrieving a subset of some set of data based on one or more conditions. For example, in a list of star magnitudes, we may wish to retrieve only those with apparent magnitudes less than some value. We can easily mask the list and retrieve those values as follows:

```
subset = starlist[starlist<25]
# note that the mask itself is just
mask = starlist < 25
```

In the first line above, we've combined the process of masking and slicing. The *mask* is simply the statement `starlist<25`. Running this by itself returns a boolean array, with one True or False for each value of `starlist`, depending on if the condition evaluated to True for that item. If we want to use that mask to index the same array from which the mask was derived, it is common practice to place this expression directly into the indexing brackets (as shown in the first line). A programmer would read this line of code as *"subset equals starlist indexed where starlist is less than 25"*.[7]

Python considers the order of operations, so first the expression `starlist<25` is run, producing the boolean array, and when a boolean array is provided as the slice for an iterable, only the items for which the corresponding location in the boolean array is True are indexed. We also see in the second example that we can define the mask explicitly if we wish, which can be helpful if we wanted to index multiple different lists using the same mask (i.e., based on some condition applied to a single array).

To be clear on terminology, a mask is a boolean array, and filtering is the process of using a mask as a slice to extract a subset of data. This distinction can matter, because later we will learn about *masked arrays*, which are a special Numpy object in which data and mask are kept together, (and there is often also a version of the data with masked values set to NaN), but the data has in that case not been filtered or indexed.

Notice that in the above usage of masks, we never know (or care) about the actual *indices* where the condition was true. We don't know if the stars which met our magnitude cut were the first, third, and fourth in the list, or the second, sixth, and ninth. In many cases, we *do* need to know both the indices where our condition is true, and also want to index the array for just those values. In that case, we will use a Numpy function, `np.where()`, which is detailed in the section on Numpy below.

4.9.1 Multiple Conditions

If we have multiple conditions, our syntax changes slightly in two ways:

- First, we need parentheses around each condition, and
- Second, we need bitwise operators & and | to define the logic across our conditions.

Let's continue with the starlist example from above. If instead of a magnitude cutoff, I wanted all star magnitudes *between* 20 and 25, then my mask-filtering now looks like this:

```
subset = starlist[(starlist<25)&(starlist>20)]
```

---

[7] Bear in mind, then, that the subset may be a different length than starlist, but mask's length will always match that of starlist.

Notice the new parentheses, and we now use & to indicate that *both* conditions must be true for a given magnitude to be included. We can similarly use the bar ("or") to, e.g., *exclude* stars with magnitudes in the 20-25 range:

```
subset = starlist[(starlist>25)|(starlist<20)]
```

4.10 Conditional Statements

Conditionals are the first major way that we can use code to not only execute multiple commands, but to actually automate tasks that involve decision making. When we insert a conditional into our code, we assess the validity of a conditional statement (i.e., we ask if the statement is true), and then have space to take different actions depending on the outcome of this check.[8] This turns our code into something like a decision tree, with forks that guide calculations through different channels depending on (usually) the state of the data.

Conditionals in Python are created using the if, else, and elif statements. Here is a simple example of an if-statement:

```
star_magnitude = 15.5
```

```
if star_magnitude > 6.5:
    print('Star is undetectable by eye.')
else:
    print('Star is visible to human eye.')
```

Star is undetectable by eye.

An if-statement is defined using the special Python word if, followed by something which Python can evaluate to either by True or False. It ends in a colon, and anything which you want to run if the particular statement is true needs to be indented using four spaces.[9]

In this case, we see that Python assessed the validity of the statement star_magnitude>6.5. To do so, it inserted the mathematical value of star_magnitude in for its variable name and compared the two numbers directly. Because the statement evaluated to True, the first print() statement executed, while the second was never run. If the first statement turned out to be False, the block of code contained within the else clause would execute.[10]

What if we want more than a simple yes-no boundary line? For example, what if we want to check if a star is visible to the naked eye, visible to a 5 meter telescope on the ground, or visible to the JWST telescope in space? One way of writing that is as follows:

---

[8] Technically conditionals return an expression and checks involve the truthiness of the type, but the end result works the way we would logically expect.

[9] Tabbing in technically works (as long as you are consistent, as tabs and spaces can't be mixed), and some editors can be configured to turn tabs into four spaces to be consistent with the Python style guide.

[10] Of course, this only works if star_magnitude *can* be compared to 6.5, meaning it must be a float or integer.

```
star_magnitude = 4
```

```
if star_magnitude < 6.5:
    print('Star is detectable by eye.')
if star_magnitude<28:
    print('Star detectable by ground telescopes.')
if star_magnitude<34:
    print('Star detectable by space telescope.')
else:
    print('Star is undetectable by all.')
```

```
Star is detectable by eye.
Star detectable by ground telescopes.
Star detectable by space telescope.
```

When we place multiple if-statements in a row, all of them have the opportunity to be True. In this case, the first three statements all printed, because a 4th magnitude star is visible to all of these telescopes.[11] The else-statement here, by definition, will only execute if the star magnitude is larger than 34.

So far, we have used if and else, but what is elif? This special word is short for "else-if". Situations often arise in which we want to check one condition, but if the condition is false, we then want to chose between two (or more) *other* options. In standard Python code, this would involve a lot of indentations:

```
if condition1:
    do_something
else:
    if condition2:
        do_something_else
    else:
        if condition3:
            do_next_thing
        else:
            do_final_thing
```

Here, we use the standard if and else, but for each new additional condition we want to try, we need more indents. By using elif, we can keep all these options aligned at the outer indent level, which keeps our code cleaner and easier to read and debug:

---

[11] Though, ironically, a 4th magnitude star is actually too *bright* to point these telescopes at, without saturating the detectors and potentially damaging the optics.

```
if condition1:
    do_something
elif condition2:
    do_something_else
elif condition3:
    do_next_thing
else:
    do_final_thing
```

In short, if you find yourself writing an if-else statement in which there is another if-else statement buried inside the `else` of the first, you can re-write it using `elif`.

4.10.1 Multiple Simultaneous Conditions

Very often, we want to ask if two (or more) statements are true simultaneously before triggering a certain block of code. For example, we might have a catalog of galaxy fluxes, and while one column contains the flux, another column contains a flag, either 0 or 1, which indicates the photometry is either usable or suspect. As we work through the entries in the catalog (more on that below, in the section on loops), we might want to perform a calculation only if a flux is above a certain value and *also* the flag for that galaxy is 1, not 0. We can combine multiple conditions in one conditional statement by separating our conditions into parentheticals and using the `and` keyword to combine them:

```
flux = 14
flag = 1

if (flux>10) and (flag==1):
    calculation = -2.5 * np.log(flux)
print(calculation)
```

```
-6.597643324038146
```

In some cases, the parenthesis around each individual condition is not needed, but it is useful for keeping code readable. In the above example, our calculation only occurred because both conditions were evaluated to be `True`. You'll notice that I have been using several comparative operators thus far, such as > (greater than), < (less than), and == (equals to). Table 4.2 shows these different options for Python. Similarly, we can choose to evaluate a block of code if *any* of several possible conditions are true. This is handled using the `or` operator.

Exercise 4.1:
What would the `print(calculation)` line have produced if the flux had been less than 10? Try it out and see if the result matches your expectation, and explain the difference if there is one.

Table 4.2. Comparison Operators in Python

Symbol	Meaning
<	less than
>	greater than
>=	greater than or equal to
<=	less than or equal to
==	equal to
!=	not equal to

Statements using these symbols evaluate to True or False (barring any other errors).

4.10.2 Equality versus Identity

So far, we have been using the comparison operators (such as == or !=, to check whether the value of the thing on one side of the operator is equal or not equal to the value of the thing on the other. But remember that in Python, variables represent actual places in memory. Two separately created variables, with different locations in memory (i.e., different objects), can have the same *value* without *being the same thing*. This distinction matters when we consider the special Python word is. We can use is to check if two things are actually the same. For example,

```
a = 1.0
b = 1.0
```

```
a == b
```

```
True
```

```
a is b
```

```
False
```

In this case, a==b evaluates to True because they are both numerically equivalent. But they are not the same object and point to different locations, and thus, a is not b. In contrast, if we had set b=a above, then both the equality comparison and the identity comparison would be true.

You don't need to worry about this often—a general rule of thumb is when performing any mathematical comparison, the mathematical operator (==) is appropriate, and only when specifically checking the identity of objects should you use is instead (or when comparing to a built-in, e.g., the booleans True/False, or None). We'll see examples of where this might be useful later on.

4.11 Loops and Iterators

Looping is the process of taking a block of code and repeating it many times with different (or the same) conditions. For example, if we have a set of files in a directory, `spec1.dat`, `spec2.dat`, and so on up through `spec100.dat`, we might want to load these files into a large array, where each column is a spectrum from one of these files. However, typing out something like `np.loadtxt(spec1.dat)` over and over again to load each file would mean typing one hundred lines of code (and take a considerable amount of time). Using loops, we can make that process only three lines.

It is worth noting that loops (and *particularly* nested loops) are often the culprit of the most considerable slowdowns to a code. Many lines (or blocks of lines) of code can execute in fractions of a second—but if a given block takes one second to run, and we have a loop with 100 iterations, our code will now take more than 100 seconds to run. In general, then, we want to avoid loops whenever possible. Once data is ingested into Python (particularly into arrays), this is often feasible. In some cases (such as the case of loading 100 files), there is no avoiding the loop.

4.11.1 For-Loops

For-loops are the most commonly used loop in Python. We have some iterable object (such as a string, a list, an array, etc), and we want to extract each item from that iterable object in turn, and perform some action with it. Some objects might look iterable (like a long integer, e.g., 1 234 566) but are not—we'd need to make that integer a string to iterate over it, turning each value back to an integer of its own (if needed).

Let's say your advisor has sent you a list of names—targets for an upcoming observing run you'll be helping with. The list was provided in an ASCII text file, with one name per row, e.g.,

```
NGC 1054
NGC 4344
BD+75d325
IC 1157
GD 153
```

... and so on.

At the moment, this list is a mix of galaxies (those starting with NGC, IC, or UGC) and standard stars for calibrations (everything else). Your advisor has asked you to split the list into two separate files, one with science targets, and one with standard stars. Additionally, they've asked you to grab the RA and DEC coordinates of these objects, such that the output files look like this:

```
...
IC 1157 16:00:56.2577 +15:31:35.2359 J2000.0
...
```

Now, we could open the file, manually look up the coordinates of each object on Simbad[12], and copy and paste it into new files in this format. But this gets tedious, especially when the list is tens of objects long, and your advisor keeps sending updates with new objects sprinkled throughout. With Python, we can approach this problem using a loop.

Let's assume we have the data (names) in a list or Numpy array. What we want to do is

1. Iterate over each name in the list,
2. Retrieve its coordinates somehow,
3. IF it is a star, *append* to one list, otherwise append to another,
4. Finally, write those two lists to new files.

Here is a sample of code that accomplishes this.

```
stars = []
galaxies = []
for i in names:
    coords = SkyCoord.from_name(i).to_string('hmsdms')
    output_string = i + ' ' + coords + ' ' + 'J2000.0'
    if i.startswith('NGC') or i.startswith('UGC') or i.startswith('IC'):
        galaxies.append(output_string)
    else:
        stars.append(output_string)
```

Let's unpack what just happened, line by line. We create two empty output lists, to store the stars and galaxies separately. Then, we use the key phrase for XXX in YYY:, which tells Python to iterate over names (names, of course, must be iterable). When looping over an iterable, we select a *temporary* variable name to hold each individual item from that iterable (these are overwritten each time the loop runs). It doesn't matter what we choose for this, though i, j, and k are commonly chosen (in that order) for loops because they are unlikely to be the name of any other variable, and loops often resemble a mathematical sum or similar operation, for which those letters are also used. However, writing for item in names: would've worked just fine, assuming we used item in our indented block instead of i.

So now we have the indented block (just like with if/else statements), which will run as many times as there are items in names, and will store the *current* value from names in the variable i. We first use the astropy.SkyCoord module's from_name() method to create SkyCoord objects from the names—this is handy, as it will look up and retrieve the coordinates for us from an online server.[13] SkyCoord objects will be covered in depth in Chapter 8, but for now, I'll just assert that these objects have a handy to_string() method with some options for outputting the coordinates into a readable string.

---

[12] Simbad is an online catalog reference for looking up measured properties of astronomical objects.
[13] So this won't work if your advisor misspelled/badly copied a name.

Next we take that coordinate, the name (i), some empty strings with spaces in them, and the ' 'J2000.0' we need, and use string concatenation to make the final 'line" that needs to go in the file.

Finally, we use an if-statement to determine which of our lists to append to—in the case of names which start with NGC, UGC, or IC we will append to the galaxy list, while in all other cases we will assume the object is a star. This might be a bad assumption; this code might need to grow more complex to catch more types of name conventions being passed through it.

Actually saving these lists to files is trivial (and can be done several ways), but as we are focusing on loops here, I'll leave that for later.

4.11.2 While-Loops

While-loops define a block of code which will continue running over and over *while* some condition remains True. If the condition has become False, then at the next iteration, the loop ends, and the next subsequent code after the loop begins executing. By default, if the condition becomes False midway through the loop, the block still finishes, as the check for truth occurs at the start of each repetition. Below, we will see how to break out of a loop at any moment.

As a note, one can easily construct a while-loop which acts (functionally) as a for-loop. To do so, you would create some counter variable outside the loop (set to 0), and at the end of the indented block increment it by one (counter+=1). Your while condition would then be while counter < = len(iterable):, meaning once the counter reaches the length of the iterable, the while condition will become False and the loop will stop. Note that in this case there is no automatic iterator variable, but our counter would serve as an index—in our names example above, in the loop, we'd need names[counter] to extract each name.

Exercise 4.2:
Rewrite the galaxy and star sorting loop above as a while-loop instead.

You may be wondering when it is most appropriate to use a for-loop versus a while-loop. The answer is that a while-loop should only be used when it is not known a priori how many iterations of the loop will be necessary before it terminates, and there is not an iterable with a set length involved. Any time we are "looping over" an iterable (like a list or array), we do know the scope of the loop, and thus should use a for-loop. This is in part for safety—a poorly constructed while-loop can end up in an infinite loop which hangs forever if the stopping condition is not met. One can use ctrl+c to attempt to interrupt one's code and stop execution, but if, for example, each loop allocates new computational resources, such a while-loop could very quickly overclock a computer processor.

An example of a situation in which a while-loop is required over a for-loop is one of user-input. Python has methods which allow a program to prompt the user to input text via the terminal. If one deigned to write a code which checked that the user's input met some condition (like, say, requirements that a password have symbols and numbers in them), and re-prompted if it did not, a while-loop would be the appropriate tool to use. Pseudocode for this might look like:

1. textVerified = False
2. while not textVerified:
3.   userInput = input("enter password")
4.   if userInput matches conditions:
5.   textVerified = True

Note that I've used the shorthand `while not textVerified`, which is equivalent to writing `while textVerified is False`. In the above example, if the user's input passes all conditions (we might check this by calling some other function), textVerifed gets set to `True`, and the next time the loop runs, the while-statement no longer evaluates to True and thus the code moves on to whatever comes after the full block of the loop.

Exercise 4.3:
Turn the above pseudo-code into functioning code, and write a loop which demands a password be entered with

1. at least 8 characters
2. at least one capital letter
3. at least one lower case letter.

Your loop should keep running if bad passwords are entered (perhaps add a print statement to tell the user what they did wrong?), and then complete and print a final success statement after the loop if a proper password is entered. *Hint: you may find the string methods* `.upper()` *and* `.lower()` *useful.*

4.11.3 Continuing Through and Breaking Out of Loops

While looping, it is sometimes required to either break out of the loop, or jump to the next iteration and continue. An example of a break condition might be looping through a list of star names until one begins with a certain set of letters—if we only need one, we do not need to finish the loop. In our `names` example above, this might look like

```
for i in names:
    if i.startswith('BD'):
        break
print(i)
```

In this case, we begin looping over the names as usual, but the first time a name does in fact start with the string "BD", the code will break out of the loop immediately (skipping any lines below the break within the loop) and then, in this case, print that particular name.[14] If that item was the second in a list of 100, the other 97 would *not* be checked.

On the other hand, we might want to not stop looping entirely, but simply "give up" on the current index of the iterable and move to the next one. For example, if our list of names had some which began with SKIP: :, added by our advisor, we could simply continue our loop whenever one cropped up:

```
for i in names:
    if i.startswith('SKIP'):
        continue
```

In this case, the code jumps back to the first (indented) line within the loop, increments to the next item in the iterable, and continues. This allows us to loop over some iterable, but skip any computations on elements which for some reason are not of interest to use.

4.11.4 Generators

A generator in Python is something which behaves like an iterable, and can thus be used in a loop or other iteration operation... but which is not, in fact, an iterable. Let's examine this:

```
print(range(5))
print(np.arange(5))
```

```
range(0, 5)
[0 1 2 3 4]
```

We can see that range(5) and np.arange(5) have different representations—the Numpy version is a full array, but the range object is not. Yet, we can use either in a loop:

```
for i in range(3):
    print(i)
for i in np.arange(3):
    print(i)
```

```
0
1
2
0
1
2
```

---

[14] Note that there are faster Numpy based methods to perform this particular check.

And we can index either of them equivalently:

```python
print(range(5)[0])
print(np.arange(5)[0])
```

```
0
0
```

So what is the difference? Primarily, the difference is that when we initialize the Numpy array, we place that whole array in memory. At a length of 5, this is no issue. But if we needed to iterate over ten million values, having that whole array in memory just as our "loop counter" makes little sense (and may not even fit in memory). The generator, meanwhile, only knows how to compute the requested value — so in the loop, it will continue providing the correct next value, without ever making a full list in memory.

In practice, this difference doesn't matter too much in most cases — but there are scenarios when it is useful to use (or indeed, create) a generator. We won't cover that in this text, as it is less likely to come up early in your coding practice.

As a final note on loops, Python has several built in functions that can facilitate more complex loops. Two commonly used ones are enumerate() and zip(), which handle cases where

1. We need to know which index of an iterable we are on as well as the value of the iterable at that index during each loop, and
2. We need to loop over multiple (same-length) iterables at once,

respectively.

4.12 Cancelling Code Execution

You may have discovered during that section on loops that you on occasion would start a Python code, and then immediately realize something was wrong. It is useful to cancel these executions so that we don't waste time waiting for a loop with a bug to finish, or wait indefinitely while something is causing our code to hang. Thankfully, just like in the shell, hitting <ctrl+>+C a few times usually does the trick to cancel the code's execution. This may not work in Jupyter notebooks; thankfully, the interface for notebooks whether in browser or an editor like VSCode will have a button to press for "Interrupt Kernel" and "Restart Kernel". The first is like a <ctrl>+C, while the second amounts to killing the whole Python shell and restarting. If things get *really* bad with a Python process, you can try killing the whole Python process from the command line, but this isn't usually necessary.

4.13 Shell and Shell-like Commands in Python

Early in this text, we spent time learning about how to navigate and interact with the shell environment. This time was well spent; practically speaking, it is challenging or impossible to install and use Python in a scientific setting without understanding some elements of shell operation.

On occasion, we would also like to be able to use some of our shell commands from *within* a Python program. We may need to list the files in a directory, move files around, or even send an arbitrary shell command from Python.[15]

We've seen several methods thus far illustrating how to load a file into Python, almost always by providing some function a filepath as a string. What if we do not explicitly know the filenames we need within a Python script, but do know how to choose files in a directory based on some conditions?

For that, we need to be able to read the contents of a directory—a task we would accomplish with `ls` in the shell. We have a few ways to mimic this behavior within Python.

```
import os

os.listdir('../../BookDatasets/imaging/')
['hst_f814W_crop.fits',
 'hlsp_ceers_jwst_nircam_nircam6_f277w_dr0.5_i2d.fits',
 'hlsp_ceers_jwst_nircam_nircam6_f444w_dr0.5_i2d.fits',
 'jwst_f444W_crop.fits',
 'hst_f606W_crop.fits',
 'egs_all_acs_wfc_f814w_030mas_v1.9_nircam6_mef.fits',
 'jwst_f150W_crop.fits',
 'm81-narrowband.fits',
 'm33-continuum.fits',
 'm33-halpha.fits',
 '.ipynb_checkpoints',
 'egs_all_acs_wfc_f606w_030mas_v1.9_nircam6_mef.fits',
 'hlsp_ceers_jwst_nircam_nircam6_f150w_dr0.5_i2d.fits',
 'jwst_f277W_crop.fits']
```

As we can see, the `os` submodule's `listdir` command returns a Python list of the contents of a given directly. We have not specified any conditions in our call, we are also seeing other directories and generally ignorable files, e.g., `.ipynb_checkpoints`. While we could filter these out of our list with some extra code, it turns out there is a different library and function that will allow us to use the wildcard syntax we are familiar with from the shell to, e.g., specify only FITS files.

---

[15] As a note, this final task, of executing arbitrary shell commands via Python, can be quite unsafe from a systems security standpoint. It is discouraged in general, especially for code that is part of a package meant to be used widely.

```
from glob import glob
glob('../../BookDatasets/imaging/*.fits')
```

```
['../../BookDatasets/imaging/hst_f814W_crop.fits',
 '../../BookDatasets/imaging/hlsp_ceers_jwst_nircam_nircam6_f277w_dr0.
 5_i2d.fits
 ',
 '../../BookDatasets/imaging/hlsp_ceers_jwst_nircam_nircam6_f444w_dr0.
 5_i2d.fits
 ',
 '../../BookDatasets/imaging/jwst_f444W_crop.fits',
 '../../BookDatasets/imaging/hst_f606W_crop.fits',
 '../../BookDatasets/imaging/egs_all_acs_wfc_f814w_030mas_v1.
 9_nircam6_mef.fits',
 '../../BookDatasets/imaging/jwst_f150W_crop.fits',
 '../../BookDatasets/imaging/m81-narrowband.fits',
 '../../BookDatasets/imaging/m33-continuum.fits',
 '../../BookDatasets/imaging/m33-halpha.fits',
 '../../BookDatasets/imaging/egs_all_acs_wfc_f606w_030mas_v1.
 9_nircam6_mef.fits',
 '../../BookDatasets/imaging/hlsp_ceers_jwst_nircam_nircam6_f150w_dr0.
 5_i2d.fits
 ',
 '../../BookDatasets/imaging/jwst_f277W_crop.fits']
```

You should notice two behavioral changes here: first, we can now use our wildcard in order to subselect only the files that end with .fits, and second, that glob returns the *full relative path* between the current working directory and the files of interest, while os.listdir() returns only the file names (almost as if cd's into the directory first).

Thus, when it comes to listing directory contents, glob is generally more versatile. However, this doesn't mean the os module is not useful. In particular, there are a set of file system operations for which it is extremely useful:

Consider the scenario in which we need to construct a path to a specific file. One could use pure string concatenation:

```
filepath = 'home/data' + '/' + 'subdir/' + 'filename.extension'
filepath
```

```
'home/data/subdir/filename.extension'
```

However, using string concatenation within a code for this purpose can be a bit tricky, leading to bugs in which a slash may be missing and produce an extension like

```
basePath ='home/data'
subdir = 'subdir_name'
fname = 'filename.extension'
filepath = basePath + subdir + fname
filepath
```

```
'home/datasubdir_namefilename.extension'
```

Notice that we neglected to enforce a slash at the end of the basePath or the start of the subdir, or at the start of the file name. While in this canned example it is easy to see where we need to add slashes, if we were constructing those paths from information in our code, it might be more difficult to catch. We can instead use the `os.path` module to handle the creation of this path:

```
filepath = os.path.join(basePath,subdir,fname)
filepath
```

```
'home/data/subdir_name/filename.extension'
```

This module has taken the individual names of each path component and joined them with slashes appropriately. In fact, even if we have some trailing slashes (and not others), it will appropriately add or not add them as needed.

```
basePath ='home/data'
subdir = 'subdir_name/'
fname = 'filename.extension'
filepath = os.path.join(basePath,subdir,fname)
filepath
```

```
'home/data/subdir_name/filename.extension'
```

The only quirk to be aware of is that any *leading* slashes in one of the sub-paths will be treated as a root path—anything prior will thus be discarded.

Assuming we have created a filepath, we can also check if it exists or not:

```
os.path.exists(filepath)
```

```
False
```

This can be very useful when writing automated pipelines which read, or create, many files, allowing us to make decisions about how to proceed based on the existence (or not) of a given file. There are numerous other similar commends; some useful ones may be `os.path.isdir()` (returns whether a path is a directory) and `os.path.isfile()` (returns whether a path is a file).

The opposite is also possible: we can take full paths and split them into the part that is a file and the part that is the directory containing that file:

```
os.path.split(filepath)
```

```
('home/data/subdir_name', 'filename.extension')
```

We can also do more than check the contents or existence of certain directories, we can also *create* them. As one might expect, the commands for this are `os.mkdir()` and `os.makedirs()`, (with the complementary being `os.rmdir()` and `os.removedirs()`).

As you begin working with code that is opening and closing many files, these interactions with the filesystem become increasingly important.

As a final note: it is possible to execute arbitrary shell commands from within a Python program. Often, we don't need to do this—for example, the `os` module means we don't need to pass the `ls` command to the shell and read back the results. Sometimes, however, we need to run certain commands in the shell; this is often a piece of astronomical software which can only be run from the shell, such as `source extractor`. The `os` module has the ability to pass commands to the shell, but as with `glob` there is a library with additional functionality: `subprocess`. In particular, `subprocess` gives us the ability to *capture* the output of a given shell command and pull it back into Python, something `os.system` cannot do. When we are writing our own personal code, it is fine to use this module. Caution should be used when employing it, because as mentioned, code which runs shell commands has the power to, potentially, perform almost any dangerous activity that we would typically refer to as hacking. Examples of this might include installing foreign software from somewhere online, logging information about the system and sending that information to an external source, etc. It is especially important to be vigilant when installing software, and ensuring it is not running these types of commands on our systems.

4.14 Interpreting Error Messages

Python has a variety of errors that will appear when your code cannot be run successfully. It can be very intimidating at first to see the wall of text that flies by when Python has an error, but most of this text is not important to us—there are a few key lines of the output we can use to investigate what is going wrong. In particular, if we have a single error that didn't come from a cascade of other errors, the final line of the error output (which is known, aptly as a *traceback*) is the most helpful. The rest of the traceback is a "stepping stone" of the errors that occurred under-the-hood, and are usually difficult to parse.

Over the years, the degree of information in Python tracebacks has improved dramatically. In earlier versions of Python, (<3.10), a syntax error would tell you the *line* on which an error occurred, though sometimes it would be the line before the indicated line causing the problem. As of Python version 3.11, however, syntax errors will actually indicate *which word/command in a line of code* is causing the issue.

The following is a list of exceptions (Pythonic word for error) you are likely to run into as you start learning Python, and what they mean.

- **SyntaxError**. A syntax error is raised when you've failed somewhere to speak Python properly. Somewhere there is a typo of some kind, and Python cannot run your code. When you try to run a script, all syntax is checked first. These are generally easy to solve because Python will show us the offending line or character, which we simply need to examine.

- **NameError**. Name errors are raised when you reference a variable, function, or other object name that hasn't been defined. Often this is *also* a typo, like referencing the variable `vega_mag` when, earlier in the code, you defined it as `vegaMag = 15`. Usually also fairly easy to debug. If you think the name you've referenced *should* exist, try finding where it is defined. Is it defined in a conditional (like an if-statement)? If so, perhaps that condition is not being triggered.

- **TypeError**. Type errors are raised when you try to operate on incompatible types. For example, if we try to use the addition operator between a string and a list (`[1,2,3]+'hello'`), we will get a TypeError. The error itself will tell us which two incompatible types were being used, and which incompatible operation was attempted. Usually, we know we want the same or compatible types being combined, and which they should be, so one is usually wrong for some reason, and we have to follow that variable back through its history to see why it is not the type we expect.

- **IndexError**. Raised when we try to index an iterable (anything with items in it) with an index outside of the range of applicable values (e.g., indexing a number higher than the length of a list). We should make sure we are matching our indexing scheme to the length of the iterable, if necessary.

- **ZeroDivisionError**. Occurs when we divide by 0.

- **ValueError**. Value errors are raised when we supply an inappropriate value to a function. For example, the `int()` casting function can convert numbers within strings to int, e.g., `int('3')`, but the value within the string must be a number to be cast that way. If we input `int('three')`, we will get a ValueError.

- **AttributeError**. If we attempt to query an attribute that doesn't exist, we'll raise this error. For example, we know that `Numpy` arrays have a `shape` attribute. They do *not* have a `birthday` attribute. For an array `arr`, attempting `arr.birthday` will raise an AttributeError.

- **KeyError**. Similar to an AttributeError, but raised when we provide a key for a dictionary-style object that doesn't exist.

As you practice coding in Python, you will become very familiar with the standard built-in exceptions, and most will become easy to debug. However, anyone can define their own exceptions, and raise them at will. Such "custom" exceptions will be common in external packages and libraries you are using, and when triggered, if they are not very informative, they can become trickier to debug. In the past, the go-to avenue for learning how to debug an exception you cannot figure out or understand was a website called *Stack Overflow*. There, thousands of programmers asked and answered questions daily.

I still recommend Stack Overflow as a source of useful debugging knowledge. There is a strategy to looking up errors. First, the exception thrown by your code may include values and numbers in the error message that are specific to the code you tried to run. You generally want to strip these out, and then Google search the remaining, "generic" part of the error message.

Often, this search will produce several Stack Overflow forum pages in which someone was trying to accomplish something very similar to you, with answers guiding toward a solution.

Alternatively, it has become increasingly apparent that large language models (LLMs) are a valuable tool in the writing and debugging of code. With a conversational input and the ability to read, assess, and describe code, LLMs (many of which were *trained* on websites like Stack Overflow) have the advantage of helping you debug your specific code. In some cases, this will still fail; e.g., when the exception is from a package the LLM hasn't seen and cannot look up. Additionally, LLMs, at least as of writing, can produce answers which are incorrect.[16]

Ultimately, when trying to understand errors in your code to fix them, a mix of the resources at your disposal, from Stack Overflow, to an LLM, to your friends and colleagues and advisors, is the best way to get your problem solved.

4.15 Handling Exceptions

In our for-loop example of galaxy and star names last chapter, we were relying on the SkyCoord.from_name() method to get coordinates, and this only works if the names we insert are both spelled correctly and are recognizable names tagged to the object in question. If one of the names in the list *fails* to return a valid SkyCoord object, our code will fail at that point and return an error of some kind, and we won't be able to continue.

Usually, when our code runs into an error, this behavior is desirable; the code gives us a (hopefully informative) error message and we resolve the bug. But there are some cases (such as a long list in which we know bad values are present which would break our code) where the desired behavior is something akin to *stick the bad value somewhere else for further inspection, but continue on to the rest of the list.*

We can accomplish this behavior using a try/except statement. This works similarly to if/else, if the logic were *IF this block of code executes with no errors, continue, ELSE if an error occurs run this block instead.*

Here's an example of a try and except block as implemented in our example of names from before:

---

[16] To be fair, so can people on Stack Overflow. But there, answers are up-voted by the community if they are valid.

```
stars = []
galaxies = []
bad_names = []
for i in names:
    try:
        coords = SkyCoord.from_name(i).to_string('hmsdms')
    except:
        # name must have been bad
        bad_names.append(i)
        continue

    output_string = i + ' ' + coords + ' ' + 'J2000.0'
    if i.startswith('NGC') or i.startswith('UGC') or i.startswith('IC'):
        galaxies.append(i)
    else:
        stars.append(i)
```

As we can see, the first thing we now do is try the line in which the coords are retrieved based on the name. If this does not raise an exception, the code skips the except statement and jumps to the line beginning with ouput_string = ..., leaving the created coords variable accessible. However, if the attempt to run the SkyCoord line *does* raise an exception, the except block will run. In this example, we append that name to a new third list called bad_names for further inspection, and then use the continue command to tell Python to simply skip the rest of the indented block and start again with the next item in the iterable.

In the above example, except: will catch *all* exceptions that are raised by an offending line in the try: block. This is a bit dangerous, as it means we can't be sure that the issue raised was that the name was bad. For example, if the formatting of the string output had a typo in it, that would also cause the coords = ... line to fail, and would thus mark *every* item in the list as bad, and move them to the bad_names list.

Thankfully, Python allows us to be more specific with our exception catching. In the example below, I show the *exact* exception that should be raised if a name is not resolvable by the CDS server:

```
SkyCoord.from_name('fake_galaxy')
```

```
    NameResolveError: Unable to find coordinates for name 'fake_galaxy'
→using http://cdsweb.u-strasbg.fr/cgi-bin/nph-sesame/A?fake_galaxy
```

(I have truncated the exception for clarity). So the exact exception SkyCoord raises when the input name is not resolvable is a custom-defined error called NameResolveError (nice, this is well-named and captures one specific offending behavior).

Let's update our own code to only catch these types of errors. First I'll need to find where `Astropy` defines their exception they raise (don't worry about this for now), and then I add that exception to our code:

```
from astropy.coordinates.name_resolve import NameResolveError
```

```
stars = []
galaxies = []
bad_names = []
for i in names:
    try:
        coords = SkyCoord.from_name(i).to_string('hmsdms')
    except NameResolveError:
        # name must have been bad
        bad_names.append(i)
        continue

    output_string = i + ' ' + coords + ' ' + 'J2000.0'
    if i.startswith('NGC') or i.startswith('UGC') or i.startswith('IC'):
        galaxies.append(i)
    else:
        stars.append(i)
```

This code is more robust than our previous iteration, because if the error we expect is possible (name resolving errors) occurs, we have the appropriate catch for it (appending to a new list and continuing), but if any other exception occurs that we are not anticipating and do not have specific code set aside for handling, it will still halt our code and inform us what happened. As a note, we are allowed to add multiple `except` blocks after a given `try` block, meaning if we are aware of (and want to specifically handle) several different known ways a given line (or lines) of code can fail, we can do so—and if our code absolutely needs to continue and not fail at this step, a final empty `except` block will catch anything else not caught by our previous, named exception catches. Here, we must be responsible to log or note that an unknown exception of some kind occurred (and ideally note for which items it occurred). As a note, we can also *define* our own exceptions and raise them—see Chapter 14.4.

4.16 Summary

This chapter in some ways is the core of this text. Armed with the basic Python datatypes and Numpy arrays and functions, we can, by combining variables, control flow, and loops, accomplish the vast majority of simple research tasks. As we move beyond this chapter, we will cover, primarily, the set of libraries that extend the capabilities covered here for the very specialized applications in astronomy. But from a learning and logic perspective, this chapter is likely the most important to feel you have in grasp of any in this book.

Part II

Core Research Libraries

Introduction

Armed with the basic syntax and operational usage of Python, it is now time to dive into the core set of libraries that form the bedrock of data science in astronomy. Functionally *no* research in astrophysics and astronomy occurring today proceeds without using at least one of these packages. They are:

1. *Matplotlib*: a library for constructing visualizations. We use this library both to investigate and interpret trends in our data, as well as to produce figures for publications in journals. Several other libraries exist, some of which build on `Matplotlib`, and it is a foundational and highly extensible library.
2. *Numpy*: numerical Python, a library for handling vectorized calculations on arrays of data, containing hundreds of critical mathematical and statistical functions that can operate on large amounts of data very quickly.
3. *Scipy*: scientific Python, a library building in part on `Numpy` which contains more complex algorithms useful in science, from numerical integration to signal processing.
4. *Astropy*: a library which leverages all of the above, along with significant additional contributions, to make working with astronomical data of various kinds more tractable—e.g., unit handling, cosmological calculations, astronomical coordinates and the world coordinate system, and calculations therein.

By the time you begin carrying out novel research in astrophysics, you will almost certainly be using these packages in nearly every code you write, along with other, more specialized packages which usually involve the use of one of these under the hood.

All four libraries are extensive enough that whole textbooks exist to document their use. Here we will introduce each and tackle one or two aspects of the library that is commonly used, showing that usage in a realistic context. From there, it will be up to you to seek our documentation and other texts to deepen your knowledge of these libraries or determine how to use them to help you solve a particular research task.

Astronomical Python
An introduction to modern scientific programming
Imad Pasha

Chapter 5

Visualization with Matplotlib

5.1 Introduction

The visualization of data is a critical aspect of any data science or data analysis workflow, as well as the primary means by which results are communicated to others. The art of visualization can be split into two equally important, but separate categories: visualization theory, and visualization construction. Visualization theory —which investigates how we communicate and receive visual data—is a rich field. Examples include color theory (how does our choice of color affect the way a viewer interprets a figure), or the selection of the *type* of visualization for a given scenario. We make such choices all the time, and are influenced by them, but that will not be the topic of this text—as this is primarily a practical text, we will be focused on the visualization *construction* side of things.

Because visualization is so important, there are multiple libraries (now $N > 10$) which all exist in Python to build visualizations. They each have pros and cons, and many target different use cases. For the purposes of figures (also known colloquially as plots) that are to be included in a publication of an astronomical journal (e.g., The Astrophysical Journal, the Monthly Notices of the Royal Astronomical Society, or Astronomy & Astrophysics), the *primary* library used for figure construction is Matplotlib (Hunter 2007).[1] It is by no means your only option, especially for more atypical figure types (for example, interactive plots for a webpage). Other libraries growing in popularity include seaborn (built on Matplotlib), Altair and plotly. Many of these libraries are oriented toward a particular application (such as data science and machine learning oriented visualizations). By placing a lot of logic under the hood, they make it possible to create a complex plot in only one or two lines of code (where Matplotlib might take 40), usually at the expense of full customization. All of these libraries are useful—but for the purposes of this

---

[1] For context, the name is short for Matlab plotting library, as the original impetus was to allow for matlab-style plots to be made in Python.

text, we will focus on the basics of Matplotlib as the place most astrophysical visualizations begin.

One could (and several have) written entire textbooks about Matplotlib alone. In general, Matplotlib is one of the most flexible and versatile packages in the Python ecosystem, and with the right usage, one can make *almost any* visualization of two-dimensional (and even many three-dimensional) data. However, this flexibility comes at the cost of increased verbosity, and occasionally lack of consistency across the library. We won't be able to cover anything close to the entirety of visualizations with Matplotlib (much less the many other libraries!) If you finish this chapter and want to read more, I recommend Nicolas P. Rougier's *Scientific Visualization: Python & Matplotlib* (Rougier 2021). Once you have a strong handle on Matplotlib, learning other plotting libraries is primarily a matter of translation.

In this chapter, we will cover the basics of how a plot is constructed with Matplotlib, and discuss several of the ways we can take the basic, one line call for a plot and use tweaks to move it towards something you would see in a published paper in astronomy. Much like Python itself, this library has its quirks, and generally the best way to learn is by practicing making different plots of various kinds, and looking up the commands you need along the way and experimenting with them.

You may be wondering why we are diving into visualization before we cover any of the actual *analysis* related libraries. The reason, quite simply, is that visualization is critical to even the most basic research operations. When we discuss the Numpy, SciPy, and Astropy libraries below, *nearly every example* will require us to be making plots along the way to interpret our progress. It is thus valuable to have a working handle on Matplotlib before using it in those applications. In this section, we will be using fake data generated with some simple code. If you wish to follow along, you can copy those data and focus on the plotting aspects.

5.2 A Simple Plot

Matplotlib has two methods of creating a plot. The most robust way uses Object-Oriented Programming (OOP) principles, creating figure and axis objects you can manipulate directly to a high degree of specificity. This is the method we should use, but there is also another method, included for convenience and for quick scratch visualizations, that lets you plot some data in very little code. In this example, we'll use some randomly spread data between zero and one. Once we have an x and y array to plot on the horizontal (abscissa) and vertical (ordinate) axes, we can simply run the plt.plot() command, after calling import matplotlib.pyplot as plt:[2]

---

[2] It is standard to use plt for this import, though it is not *as* universal as np is for Numpy. You may see pl or mpl used.

```
x = np.linspace(0,10,100)
y = np.random.random(size=len(x))
plt.plot(x,y);
```

We can see in Figure 5.1 that with little to no work, Matplotlib created us a plot, and sure enough, our x-data is on the *x* axis and y-data is on the *y* axis. As a note, if you are using a Python script, rather than a notebook, you will need an extra line, plt.show() as the last line of your plot commands to make the window with the plot appear. I created this data using the Numpy library; we will dive much deeper into Numpy in Chapter 6. You can feel free to focus on the plotting commands here, though the Numpy functions used should be fairly explanatory.[3]

Under the hood, Matplotlib actually created a *figure* object and *axis* object, made some assumptions about what should get plotted where, and assumed some defaults about how to represent the data. We can't actually modify some of those choices when we use Matplotlib this way, though there are a few things we *can* set right in the plot command. For example, our data isn't inherently "connected" like a time series. Visualization theory tells us that we thus shouldn't connect the points together (from left to right) as this implies some sort of temporal connection. Let's use separated points instead. Again in the spirit of "quick plotting," the third argument of plt.plot() sets the marker, and if a marker is chosen, the new default behavior will be to not connect the points (Figure 5.2).

```
plt.plot(x,y,'.');
```

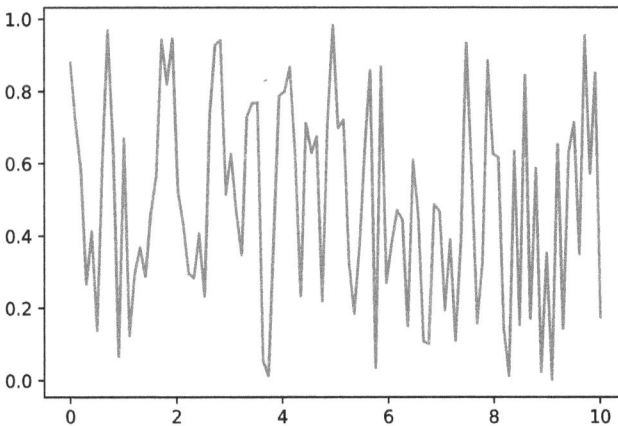

Figure 5.1. Random data generated (between 0 and 1), plotted using Matplotlib's defaults. Note the axis ratio, line color and width, and the fact that sequential points in the array are connected by lines are all defaults which can be set with arguments, which we will now proceed to explore.

---

[3] If you are having trouble getting a plot window to appear, try Googling "change Matplotlib backend" plus your operating system.

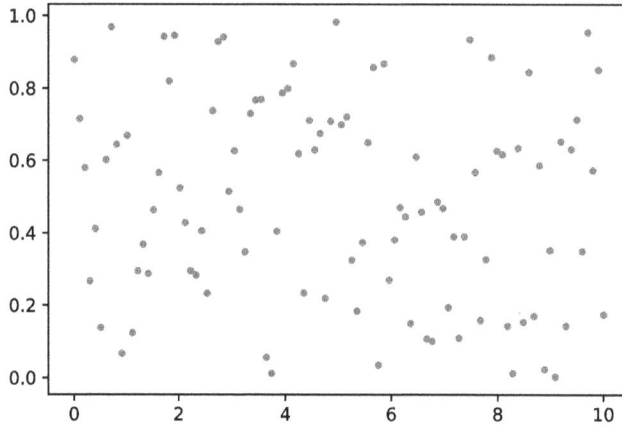

Figure 5.2. By specifying a marker, we *disable* the automatic connecting of the points. Note that it is possible to have both markers and a connecting line, if desired.

We can now see that our points are displayed without the connecting line. If we were done, we could save this figure to our computer using the command, e.g., `plt.savefig('myfig.pdf')`, and as long as this line is below our plot command, it will retrieve the most "recent" plot and save it. But to make further tweaks and progress, we will want to use the (`Matplotlib`-recommended) OOP syntax, to give us full control over our plots. It is my recommendation to almost *never* use the functional `plt.plot()` method. As a note, the `Matplotlib` collaboration strongly encourages everyone to *always* use the object-oriented way of creating plots that we will see in the rest of this chapter—and in the rest of this book, you will see that I have tried my best to adhere to this suggestion.

5.3 Figures and Axes

When we ran `plt.plot()` above, `Matplotlib` actually generated two distinct entities needed: a figure, and a set of axes. Think of the axes as the thing actually being plotted in (the lines (or spines) defining the plot boundaries, the labels and ticks, the data, etc.), and the figure as the "canvas" behind it in the background that the axes sit on. Quite literally, if you painted the example plots above on a canvas, the canvas would be the domain of the `figure` object, and the "paint" on it (everything you drew) would belong to the `axis` object. We can add multiple axes to one figure object, for example, in arbitrary locations on the canvas created by our initial figure creation call. If you wanted to adjust, say, the plot limits, that would be an axes action, if you wanted to change the width and height of the overall final PDF in inches, that would be a figure command. Most of our work with modifying our figures will be done at the axis level; we usually only have one or two tweaks at most to make at the figure level.

5.4 Subplots

Subplots are a way of putting multiple axes objects within one figure object, allowing you to make multipanel plots. We're not there yet, but I mention this now because the single plot can be considered a "subplot" call with only one set of axes. If you define your plots this way, it becomes easy to learn how to do multiple plots later.

The way we get started then, is to set up figure and axes objects using the `plt.subplots()` command:

```
fig, ax = plt.subplots()
```

As seen in Figure 5.3,[4] `Matplotlib` has created an empty plot with a set of axes on a blank canvas. Using the `plt.subplots()` function is not the only way to generate one (or more) axes on a figure, but it is the most straightforward, and we will use it most of the time. Figure 5.3 shows what happens when we feed no arguments to `subplots()`—but just to hint at what comes later, we can also do the following:

```
fig2,ax2 = plt.subplots(2,sharex=True)
```

and easily create two subplots which are stacked vertically, and share axes labels (Figure 5.4). The `subplots()` function can return a list of any number of axes so long as they are a regular grid of some kind. We'll learn how to create non-regular grids in Section 5.9; for now, we'll return to our single plot above.

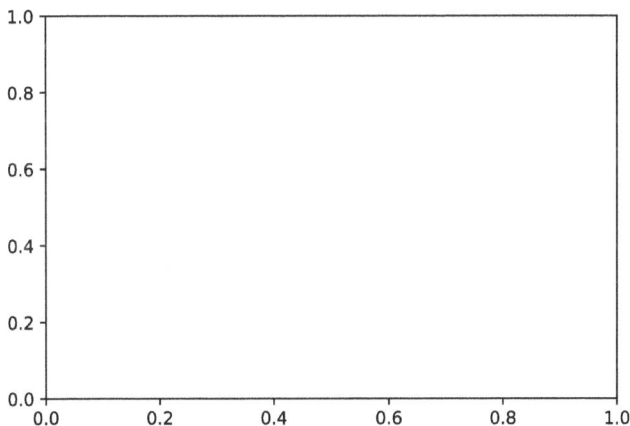

Figure 5.3. Creating explicit figure and axes objects is possible with `plt.subplots()`, which when run without arguments will show a default axis on a default canvas.

---

[4] I believe this figure earns the prize of "most boring figure included in the textbook" ... perhaps in *any* textbook.

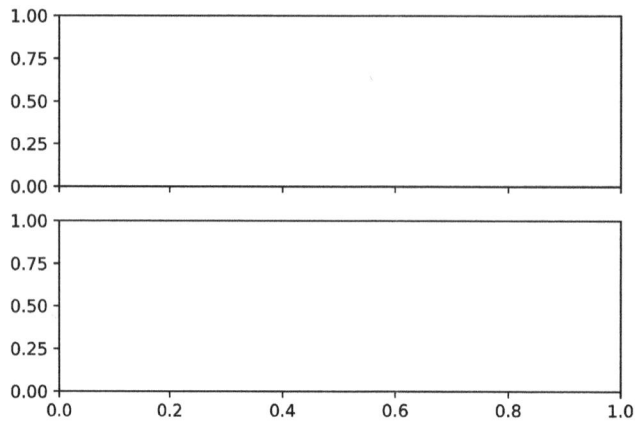

Figure 5.4. Flexing the more involved use of `subplots`, here we specify to create two subplots (by default stacked vertically; we could use the call `subplots(1,2,…)` to create horizontally placed plots). We also specify that they should share the x-axis, such that we only have axis labels on the bottom plot.

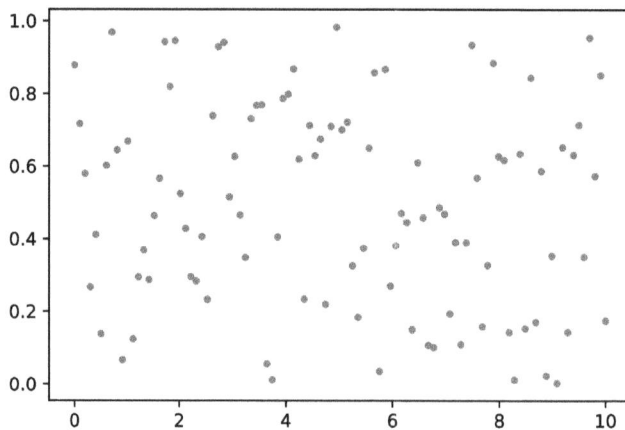

Figure 5.5. Using `ax.plot()` instead of `plt.plot()`, we now explicitly bind these data markers to the axis object `ax`.

We now want to get the data from above onto our plot. This has to do with the axes, so I'm going to call the `ax.plot()` command, the result of which is in Figure 5.5.

```
fig, ax = plt.subplots()
ax.plot(x,y,'.')
plt.show();
```

Why is this better than the original plot call, which took one fewer lines?

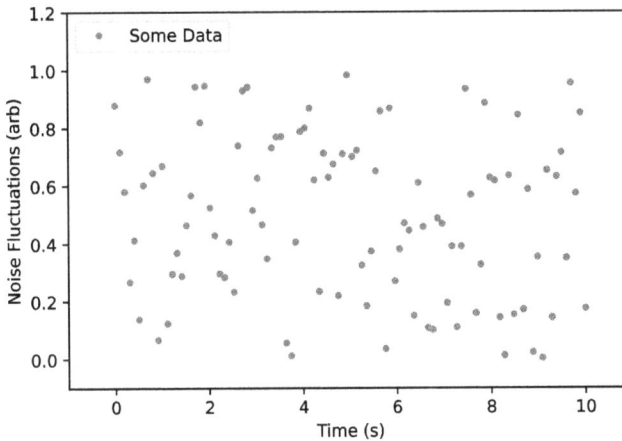

Figure 5.6. With a few extra plot calls, we can set a legend, axis labels, and specific boundaries for where our axes start and stop.

The reason is that we now have exposed figure and axes objects that define this plot (notice that `plt.plot()` does not return separate figure and axes objects).

Exposing these objects is useful because unlike for the `plt.plot()` technique, in which subsequent attempts to modify axes properties are limited to the set of options `Matplotlib` can infer how to handle, we can now use the methods of this object directly.

In order to set the axes limits (i.e., the boundaries of the plot), we will use the methods `ax.set_xlim()` and `ax.set_ylim()`; similarly to set the axes labels, we will use `ax.set_xlabel()` and `ax.set_ylabel()`. We can also add a legend via `ax.legend()`. In general, we use the `set_param` syntax when setting extant properties of the axis, while we use the direct `.plot()` or `.legend()` to create new elements on the axes.[5] All of these additions can be seen in Figure 5.6.

```
fig, ax = plt.subplots()
ax.plot(x,y,'.',label='Some Data')
ax.legend(loc=2)
ax.set_xlabel('Time (s)')
ax.set_ylabel('Noise Fluctuations (arb)')
ax.set_xlim(-1,11)
ax.set_ylim(-0.1,1.2)
plt.show();
```

5.5 Adjusting Marker Properties

Thus far, we have been using the "point" marker. There are numerous available markers available in `Matplotlib`, all of which have a symbol to select them. Popular

---

[5] This pattern doesn't always hold, e.g., a new axis is added to a figure via `fig.add_axes`.

options include circles (o), squares (s), triangles (∧ (upright), v (downward) < (left), or > (right)), stars (*), pixel (,), diamonds (d), and x or plus symbols (x and +).

Furthermore, we can adjust the color and size of these markers. The optional argument ms=... (markersize) for ax.plot(), or s=... (size) for ax.scatter() will set the size of our markers.[6]

To set colors, we can use the color keyword argument, specifying colors in many ways, including:

- a set of shortcuts ('b' (blue) "g" (green), "k" (black)) or words "purple",
- one of the colors in the default color cycle (our plots so far have used C0; we could select 'C1', 'C2', etc, and
- full hexadecimal colors (e.g., #5634d9).

Across these options, we can create markers of a huge variety of shapes, sizes, and colors.[7]

5.6 Adjusting Ticks

The next thing on our list might be adjusting how the ticks look. Scientifically, outward pointing ticks reduces the chance of overlapping with data, but many scientific instances also favor plotting inward ticks (even if just aesthetically.) For the sake of having further examples, let's change our ticks to face in, adjust their lengths, add minor ticks, and put all of them on all four spines of the plot. All of these tasks can be accomplished with the ax.tick_params() function (Figure 5.7).

```
fig, ax = plt.subplots(figsize=(6,6))
# old stuff
ax.plot(x,y,'.',label='Some Data')
ax.legend(loc=2)
ax.set_xlabel('Time (s)')
ax.set_ylabel('Noise Fluctuations (arb)')
ax.set_xlim(-1,11)
ax.set_ylim(-0.1,1.2)
# new stuff

ax.tick_params(axis='both',
               direction='in',
               length=6,
               top=True,
               right=True)
plt.show();
```

---

[6] Perhaps frustratingly, these sizes are not in the same units—the s for scatter plots denotes the size in units of markersize squared, while markersize is in linear units. So a ax.plot(x,y,ms=10) call is equal to ax.scatter(x,y,s=100).

[7] See https://Matplotlib.org/stable/api/markers_api.html to explore more options.

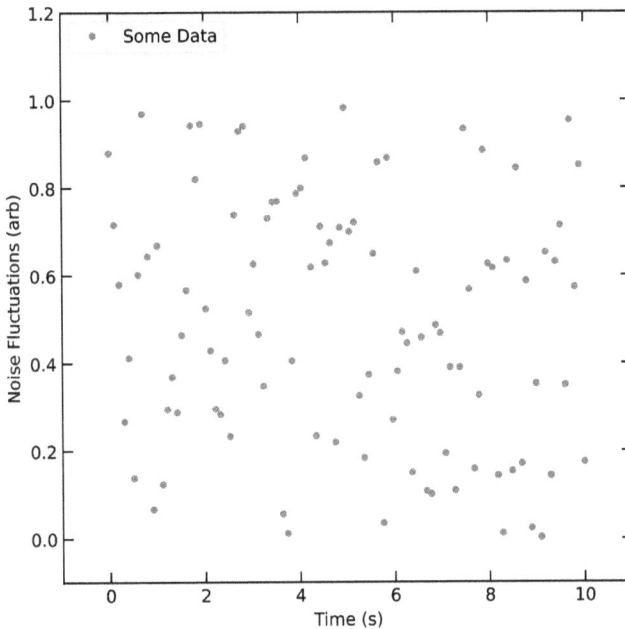

Figure 5.7. We have modified our figure further, moving to a square representation, and specifying ticks on all four axes, all pointed inward.

It is often desirable to have ticks on all four axes for astronomical data, as data may fill all areas of the plot and readers want to be able to "read off" values for a given data point fairly accurately.

Next, let's add minor ticks. For this, we need to "locate" them, and then "format" them onto the plot. Using the `MultipleLocator` class, we can set intervals of ticks, to choose appropriate minor tick intervals for each axis and then send them to the plot (Figure 5.8). We'll add a call to `tick_params()` to make them point inward as well. Because we want the major and minor ticks have different lengths, we need two separate calls to this function, in which we specify `which='major'` or `which='minor'`, but since most of the parameters are the same, we can create a simple loop with a parameter dictionary to save some lines:

```
fig, ax = plt.subplots(figsize=(6,6))
# old stuff
ax.plot(x,y,'.',label='Some Data')
ax.legend(loc=2)
ax.set_xlabel('Time (s)')
ax.set_ylabel('Noise Fluctuations (arb)')
ax.set_xlim(-1,11)
```

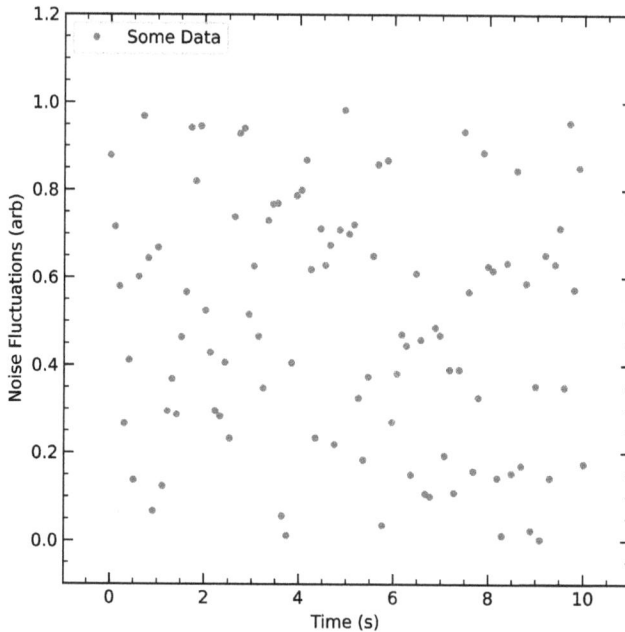

Figure 5.8. Addition of minor ticks to all four axes of the figure, spaced by dividing up the axis based on the number of small ticks per large tick we desire.

```
ax.set_ylim(-0.1,1.2)
for prop in [('major',6),('minor',3)]:
    ax.tick_params(which=prop[0],
                   axis='both',
                   direction='in',
                   length=prop[1],
                   top=True,
                   right=True)
XminorLocator = MultipleLocator(0.5)
YminorLocator = MultipleLocator(0.05)
ax.xaxis.set_minor_locator(XminorLocator)
ax.yaxis.set_minor_locator(YminorLocator)
plt.show();
```

5.7 Adjusting Fonts and Fontsizes

One of the major issues with Matplotlib's defaults is that, when compared with how plots are inserted into often 2-column papers, the fontsizes are just too small to work with. We can set the fontsize globally, or adjust it for a lot of the individual components. To set it globally, we'd use:

```
plt.rc("font", size=16,family='serif')
```

I've also switched our font family over to serif—this is just to demonstrate the functionality. Now if we replot using the same code as above, we get

```python
fig, ax = plt.subplots(figsize=(6,6))
# old stuff
ax.plot(x,y,'.',label='Some Data')
ax.legend(loc=2)
ax.set_xlabel('Time (s)')
ax.set_ylabel('Noise Fluctuations (arb)')
ax.set_xlim(-1,11)
ax.set_ylim(-0.1,1.2)
for prop in [('major',6),('minor',3)]:
    ax.tick_params(which=prop[0],
                   axis='both',
                   direction='in',
                   length=prop[1],
                   top=True,
                   right=True)
XminorLocator = MultipleLocator(0.5)
YminorLocator = MultipleLocator(0.05)
ax.xaxis.set_minor_locator(XminorLocator)
ax.yaxis.set_minor_locator(YminorLocator)
plt.show();
```

Figure 5.9 is much easier to read than previous iterations. You can also adjust the exact fonts being used in a similar manner.

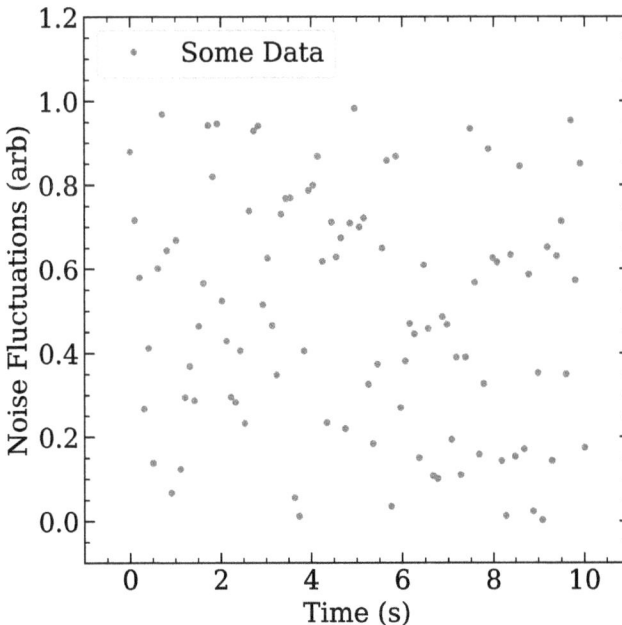

Figure 5.9. Modifying our figure to use larger font sizes and serif fonts.

Alternatively, we can set the fontsize of some individual components with calls that look like the following:

```
# Change the fontsize of an axis label
ax.set_ylabel('Time (s)',fontsize=15)
# Change the fontsize of the tick labels
ax.tick_params(labelsize=15)
# Change the fontsize in the legend
ax.legend(loc=1,prop={'size':15})
```

5.7.1 LaTeX in Labels

You can also insert LATEX expressions into your figures, in order to show units properly or display an equation. Within Matplotlib calls, you must prepend a string with the letter "r", and then use dollar signs within the string, e.g., ax.set_xlabel(r'Luminosity [$L_\odot$]'). I'll use this syntax several times in examples later in this book.

5.8 Multiple Subplots

Let's say we wanted to put two plots side by side, both of which were plotting noise fluctuations from 0 to 1 (like two trials of the same experiment. We could do that using plt.subplots(), as hinted at near the beginning of the chapter. I am going to loop over our axes, and over our major/minor parameters (mostly to save space).

```
new_y = np.random.random(size=len(x))
fig, ax = plt.subplots(1,2,sharey=True,figsize=(12,6))
# old stuff
ax[0].plot(x,y,'.',label='Trial 1')
ax[1].plot(x,new_y,'.',label='Trial 2')

XminorLocator = MultipleLocator(0.5)
YminorLocator = MultipleLocator(0.05)

for a in ax:
    a.set_xlabel('Time (s)')
    a.set_ylabel('Noise Fluctuations (arb)')
    a.set_xlim(-1,11)
    a.set_ylim(-0.1,1.2)
    for prop in [('major',6),('minor',3)]:
        a.xaxis.set_minor_locator(XminorLocator)
        a.yaxis.set_minor_locator(YminorLocator)
        a.legend(loc=1)
        a.tick_params(which=prop[0],
                      axis='both',
                      direction='in',
                      length=prop[1],
                      top=True,
                      right=True)

plt.show();
```

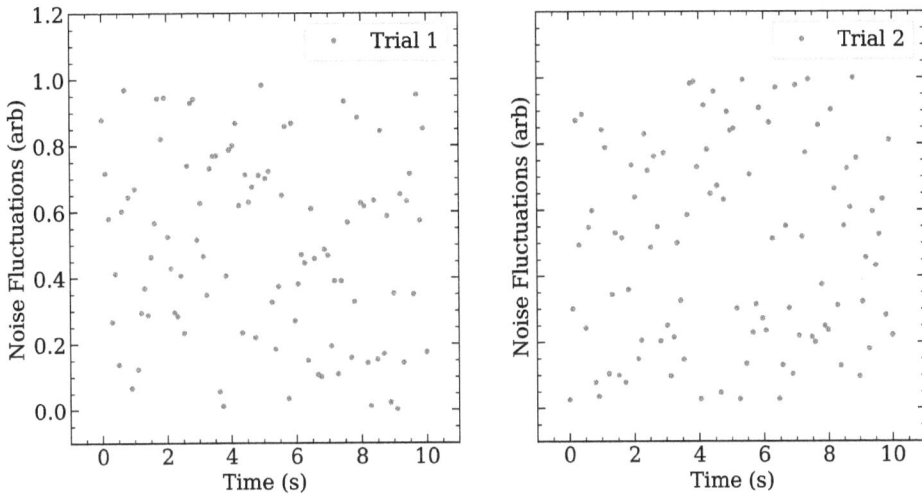

Figure 5.10. Using the `subplots()` function, we can create a grid of axes (here 2). When we do so, our `ax` variable (the second output of the function) becomes a **Numpy** array *containing* the axes objects.

Let's take a look at what we did to create Figure 5.10.

First, note that when you call **subplots** in a way that generates more than 1, **ax** becomes an array of **axes** objects, each of which can be indexed in order to be adjusted. In this case, because many of the axis settings were identical for the two, I looped over them to save repeating code. Importantly, when we create subplots that have more than one dimension (for example, if we had called `plt.subplots(2,2)`), by default, **Matplotlib**'s indexing will be two dimensional. In the case here, that would be values of [0,0], [0,1], etc. Since we often want to loop over our axes, this is a little cumbersome; thankfully, the `ax.flatten()` method allows us to, as it states, flatten this into a 1D array of axes we can then loop over.

Finally, we can take a look at bringing the two plots closer together, and removing the y-axis label of the second plot, as they share the same y axis. We can set our plots to share an axis using the `sharex` and `sharey` commands in `ax.subplots()`, which are booleans set to either **True** or **False** (default is **False**).

Below, I show the changes to the code above needed to get our axes close together, namely, the use of the `plt.subplots_adjust()` function to set the width-space (as opposed to `hspace` for vertical plots). The one other thing we'd need to do is remove the `ax[1].set_ylabel()` line (Figure 5.11).

```python
# Notice the sharey setting now added
fig, ax = plt.subplots(1,2,sharey=True,figsize=(12,6))
# All other plotting code here
# subplots adjust to bring the spacing in
plt.subplots_adjust(left=None, bottom=None, right=None, top=None,
                    wspace=0.04, hspace=None)

plt.show();
```

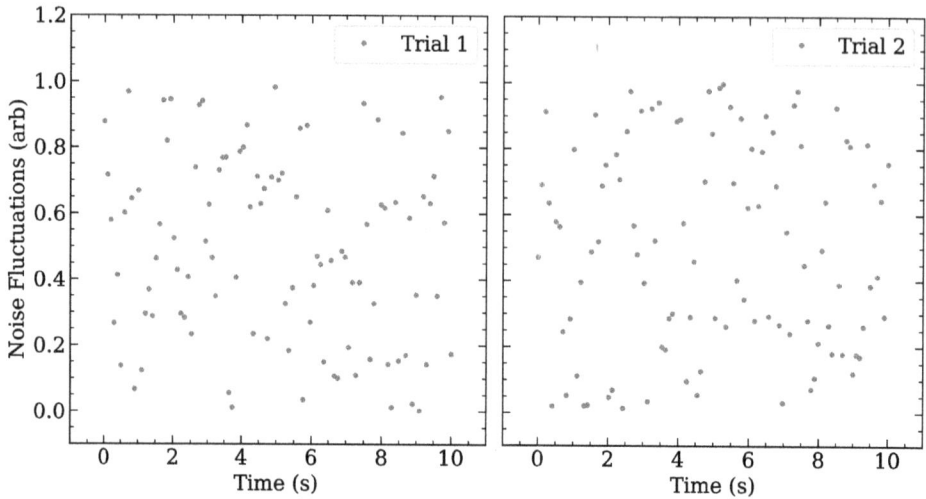

Figure 5.11. Here, we utilize the sharey argument to share an axis, and then use the 'plt.subplots_adjust()' function to adjust the spacing between the figures.

For now I only adjusted the "wspace" parameter, which controls the horizontal spacing between subplots in your figure. If they were vertically stacked, I would have used the hspace instead. Left, Right, Top, and Bottom adjust how far away you are from the edges of the overall figure.

5.9 Subplot Mosaic

If we need a more complex layout of figures than a simple grid, we can use the subplot_mosiac() function to generate a set of axes with different lengths, widths, or axis ratios relative to one another.

The function takes in a string specification that easily allows us to use letters to indicate our layout:

```
layout="""
AAABB
CCCCC
"""
fig, ax = plt.subplot_mosaic(layout,figsize=(12,7))
fig.save
```

As we can see in Figure 5.12, our axes now span multiple columns, in differing amounts, based on the layout inserted. We can also insert periods into our string to create empty space within the figure (Figure 5.13).

```
layout="""
AAABB
C..DD
"""
fig, ax = plt.subplot_mosaic(layout,figsize=(12,7))
```

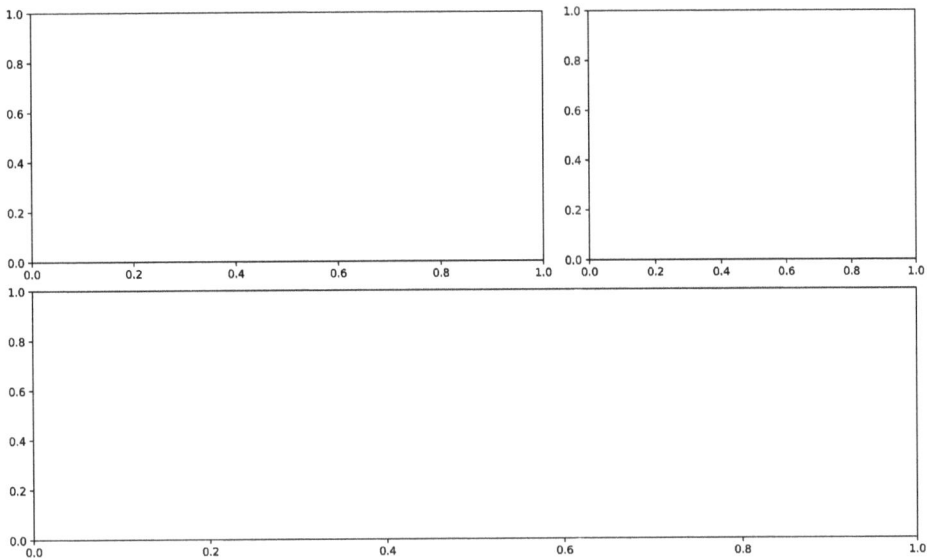

Figure 5.12. Axes created via `subplot_mosiac()`, demonstrating the ability to have axes span varying widths.

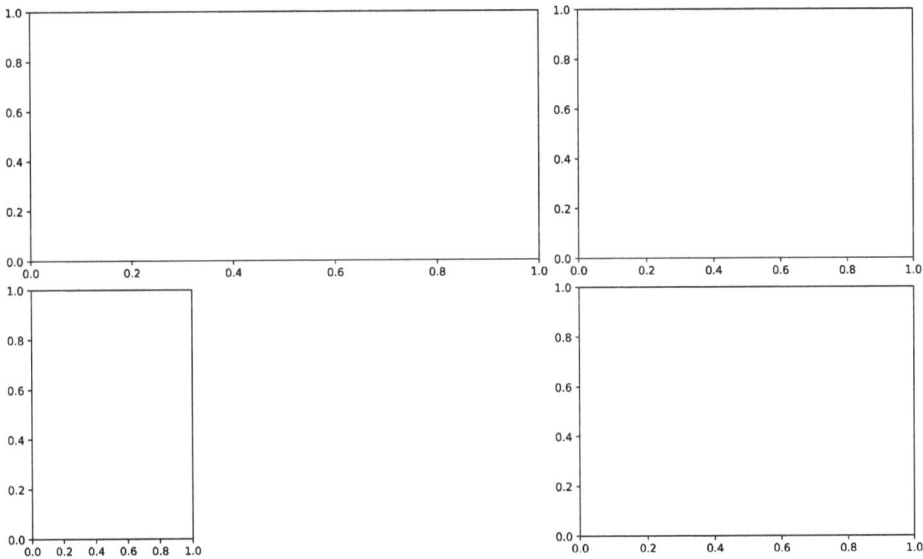

Figure 5.13. Subplot mosaic now featuring gaps of empty space in the grid, created by the insertion of periods into the relevant layout string.

As a last example: let's make a large panel (say, for a beautiful astronomical image), and a set of 5 small square panels in an "L" shape around it, which might be zoom ins on some of the objects in the larger image (Figure 5.14).

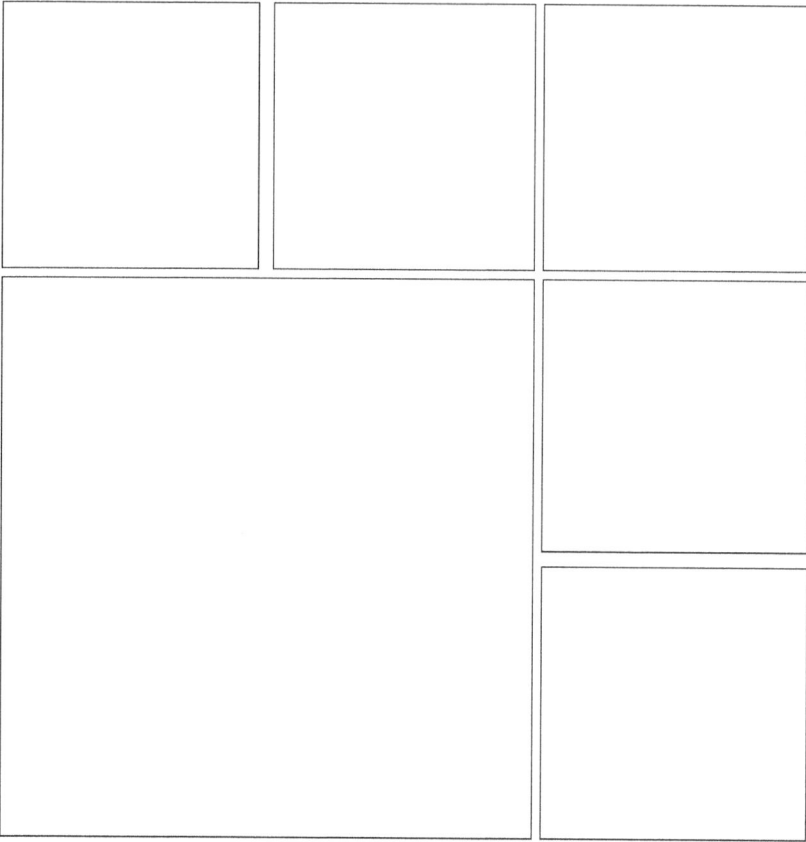

Figure 5.14. Layout which creates a large panel surrounded by smaller panels, perfect for displaying an astronomical image and several zoom in views.

```
layout="""
AABBCC
AABBCC
DDDDEE
DDDDEE
DDDDGG
DDDDGG
"""
fig, ax = plt.subplot_mosaic(layout,figsize=(12,12))
for i in ax.values():
    i.set_xticks([])
    i.set_yticks([])
```

One important thing to note with `subplot_mosaic()` is that unlike `plt.subplots()`, which returns an *array* of axes objects, this returns a *dictionary* of axes objects, using the letters we used for each axes as the keys. So in the above example, when I elected to turn off the ticks for this figure, I had to iterate over all the dictionary `.values()` explicitly.

The `subplot_mosaic()` tool is extremely useful when creating multipanel plots, and can be customized even further than is shown here.

5.10 Research Example: Displaying a Best Fit

Here's a quick example that makes use of the "alpha" (transparency) parameter, the "zorder" (order of plotting things on top of one another), and "fill_between" option, which lets you shade regions between two lines on a plot.

We'll start by inventing a data set. I show the code for creating it; the output is in Figure 5.15.

```python
import numpy as np
import matplotlib.pyplot as plt

#noise = np.random.normal(1,0.08,100) #mean 1, spread 0.5, 100 elements
data_x = np.arange(0,100)
data_y = np.linspace(25,36,100)*noise
```

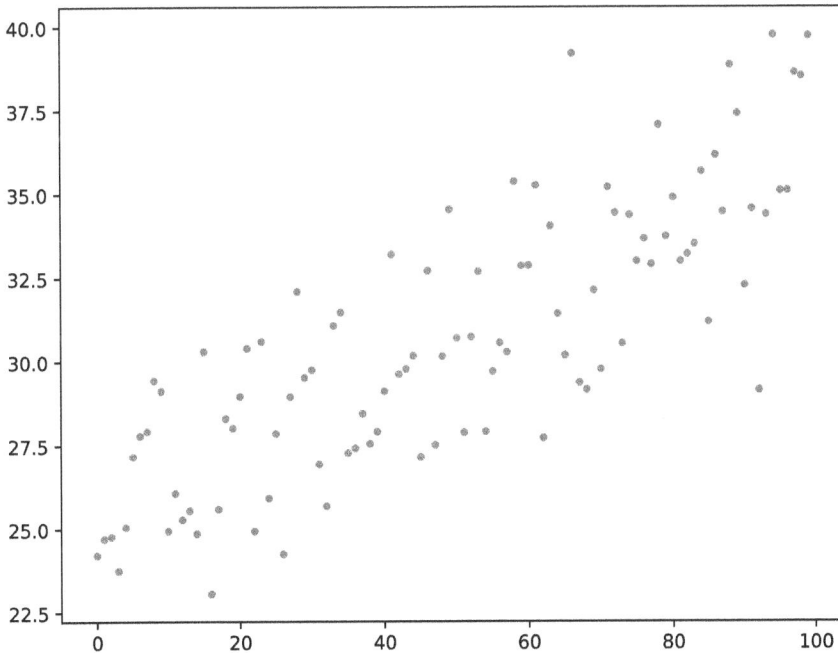

Figure 5.15. Mock data set to be used. We'll fit a line to this data and explore various plotting methods along the way to displaying that fit at the level required for insertion into an academic paper.

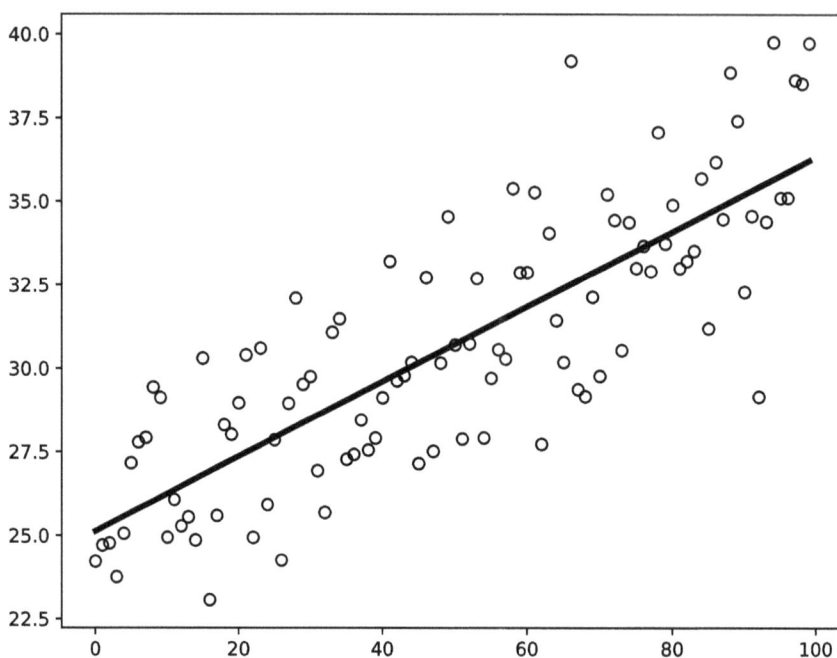

Figure 5.16. Fit to the data using the `np.polyfit()` function. We additionally change the data points to unfilled circles.

What we appear to have here is a data set that increases from around 25 to around 40, over an *x*-domain from zero to 100.

In the following example, I will use some Numpy functions we haven't talked about yet—you can jump ahead to Chapter 6 and Chapter 7 to reference these methods, or simply follow along.

We'll begin by using the `np.polyfit()` function to use linear least squares to fit a linear relation to these data (Figure 5.16).

```
fit = np.polyfit(data_x,data_y,1,cov=True) #linear
# first index takes fit params m,b,
# second specifies cov matrix
best_fit_y = fit[0][0]*data_x + fit[0][1]
plt.plot(data_x,data_y,'o',mec='k',color='None')
```

The `np.polyfit()` algorithm used linear least squares to determine a best fit to this data set. We can see that it (reasonably) passes through the points, which I have plotted via circles with edge colors set by the `mec` keyword and facecolors set to the string "None."

By setting the `cov=True` argument, Numpy will return a *covariance matrix*, which encodes some degree of the uncertainty on the fit parameters (*m,b*). The uncertainties can be extracted from this 2x2 matrix via

```
cov = fit[1]
unc = np.sqrt(np.diagonal(cov))
print(f'Slope: {fit[0][0]:.4f} | Intercept: {fit[0][1]:.4f}')
print(f'Slope unc: {unc[0]:.4f} | Intercept unc: {unc[1]:.4f}')
```

Note that the off-diagonal terms contain the co-variance, and fitting a line to these data certainly has a strong covariance between m and b—imagine for a moment choosing a steeper slope. It would have to have a smaller intercept value to still fit the data.

We can visualize the full covariance matrix by creating a 2D Gaussian[8] with values set by our covariance matrix. For this we'll get a quick preview of the `scipy.stats` module and the handy Numpy function `mgrid`:

```
from scipy.stats import multivariate_normal

fit_cov = multivariate_normal(mean=np.
    array([fit[0][0],fit[0][1]]),cov=cov)
```

```
x,y = np.mgrid[.09:.135:.005,24:26.5:.05]
pos = np.dstack((x,y))
```

```
fig, ax = plt.subplots()
ax.contourf(x,y,fit_cov.pdf(pos))
ax.set_xlabel(r'$m$',fontsize=20)
ax.set_ylabel(r'$b$',fontsize=20)
fig.savefig('contour-f.pdf')
```

We can see in Figure 5.17 the strong covariance between m and b. What actually happened above was the creation of a *contour* plot (in this case filled). We won't discuss contour plots further in this text, but they are one of the many visualization types you can create with `Matplotlib`.

We can also use this mulivariate normal to appropriately draw *samples* of m, b, and thus plot a variety of models over our data that represent the actual uncertainties (if they are indeed Gaussian) (Figure 5.18).

```
samples = fit_cov.rvs(size=50)
```

```
fig, ax = plt.subplots()
ax.plot(data_x,data_y,'o',mec='k',color='None')
ax.plot(data_x,best_fit_y,'k',lw=3)
for i in samples:
    l = data_x*i[0] + i[1]  #mx+b
```

---

[8] We assume a Gaussian because LLS assumes uncertainties are Gaussian.

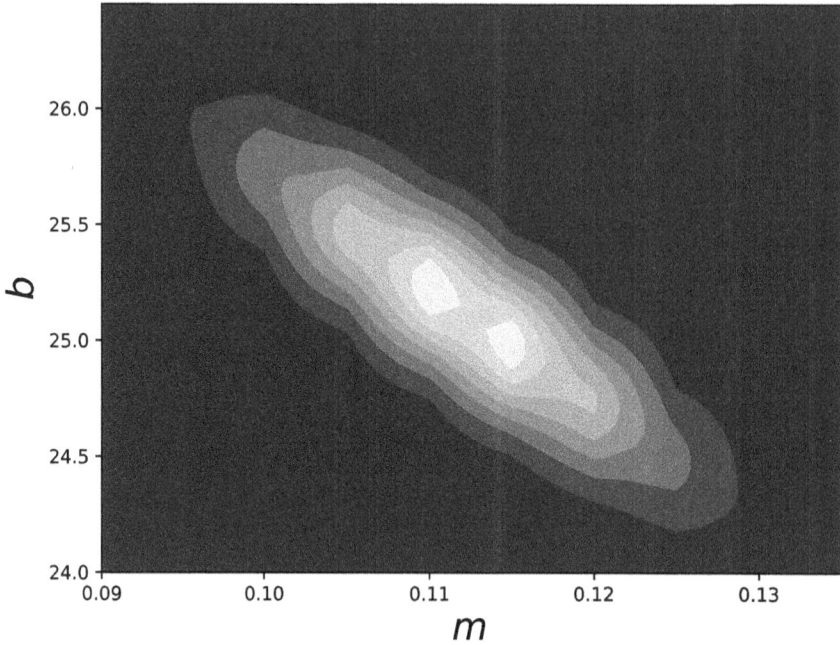

Figure 5.17. Filled-contour visualization of the covariance between slope and intercept. This distribution is created using our best-fit slope and intercept as its mean, and the covariance matrix output from `polyfit`.

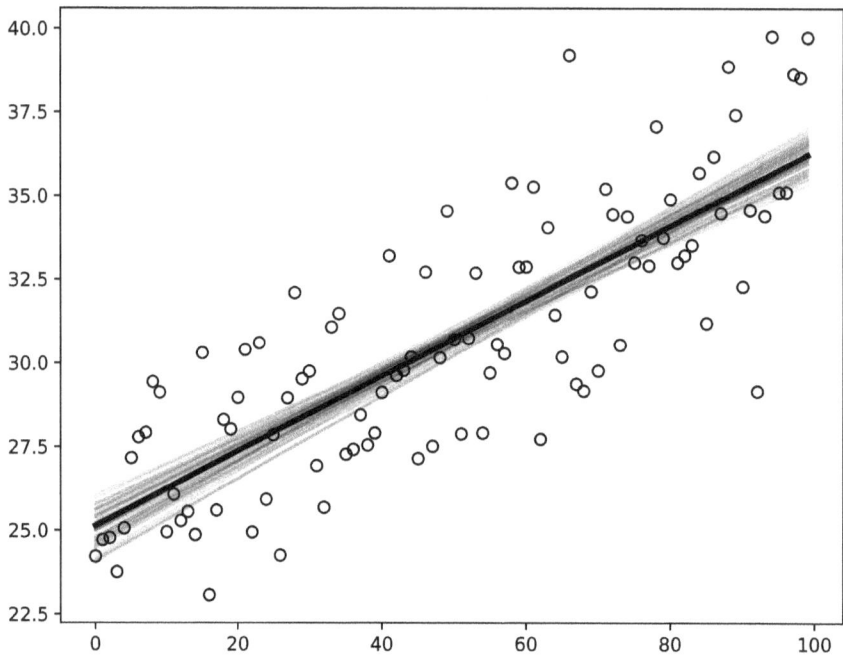

Figure 5.18. A set of lines created from a set of random (m, b) pairs sampled from the distribution defined by the best fit location and covariance. This sampling will most often produce slopes and intercepts near the center of the distribution.

```
ax.plot(data_x,l,alpha=0.2,color='k')
```

Using the `alpha` parameter to make the lines less opaque, we can see where most of the models lie, and the general envelope between which models within our parameter uncertainty lie. Instead of plotting many lines, we can instead compute the upper and lower edges of this envelope of models and use the `fill_between()` function to uniformly fill that region. I am going to resample the multivariate to get more models for this calculation. From these models, I'll calculate the spread in those models using percentiles which contain 1, 2, and 3σ variations.

Let's also leverage everything we've learned thus far about plotting to push this figure in the direction of being worthy of including in a paper (Figure 5.19).

```
from matplotlib.ticker import MultipleLocator

# Statistical calculations
samples = fit_cov.rvs(size=500)
models = np.array([data_x*i[0]+i[1] for i in samples]).reshape(500,100)
perc = np.percentile(models,[0.3,5,16,84,95,99.7],axis=0)
# Plotting
```

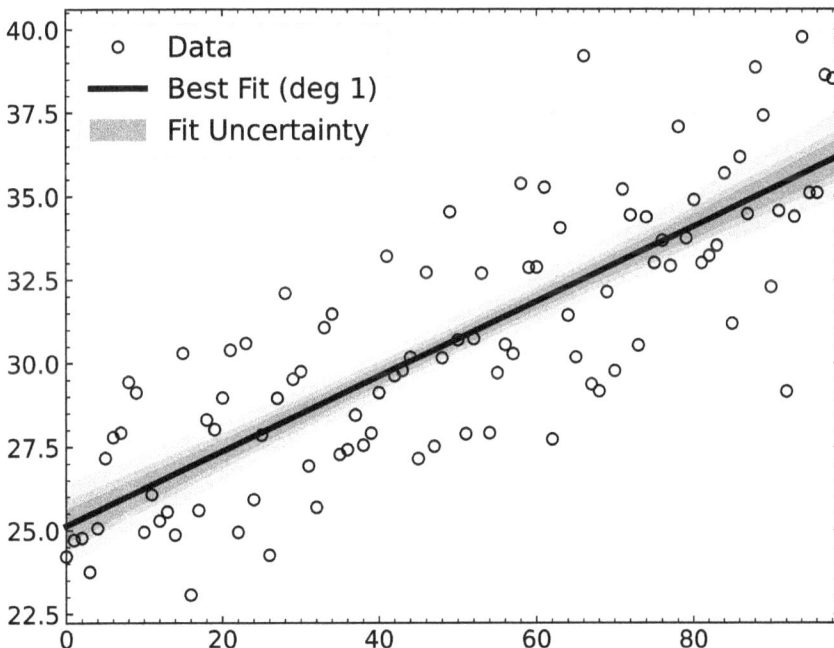

Figure 5.19. Publication-worthy plot of our fake data (minus the lack of axes labels). We have good visual clarity on our data points, clean axes with ticks, the fit line is visible, and the shaded regions around it convey the sense of the spread in uncertainty—and of course, we have a legend.

```
fig, ax = plt.subplots()
ax.plot(data_x,data_y,'o',mec='k',color='None',label='Data')
ax.plot(data_x,best_fit_y,'k',lw=3,label='Best Fit (deg 1)')
ax.fill_between(data_x,y1=perc[0],y2=perc[-1],color='gray',alpha=0.3)
ax.fill_between(data_x,y1=perc[1],y2=perc[-2],color='gray',alpha=0.5)
ax.
 fill_between(data_x,y1=perc[2],y2=perc[3],color='gray',alpha=1,label='Fit
 Uncertainty')
XminorLocator = MultipleLocator(2)
YminorLocator = MultipleLocator(0.5)
ax.xaxis.set_minor_locator(XminorLocator)
ax.yaxis.set_minor_locator(YminorLocator)
ax.tick_params(direction='in',
               which='both',
               top=True,
               right=True,
               labelsize=13)
ax.set_xlim(0,99)
ax.grid(alpha=0.2)
ax.legend(fontsize=15)
```

If we believe the uncertainties on *m* and *b*, then this filled in region provides a reasonable estimate of the uncertainty in our fit line. We've used the alpha parameter here to show the different spreads overlapping each other. Notice how many lines of code it took simply to nudge all of the parameters of this plot into the right directions — on the other hand, consider the degree of specificity with which we were able to determine every element of the figure's appearance.

5.11 Errorbars

What if we have uncertainty information about the individual points being plotted here? The traditional way to indicate this uncertainty would be via the inclusion of "error bars" on the data.[9] Matplotlib has us covered with an error-bar function we can use to add these to our plots (Figure 5.11).

Let's assume the y-uncertainties have a magnitude of 3 for each point (in whatever units our fake y-data has). Instead of replicating all the code, I'll show what you would add above to create errorbars (Figure 5.20).

```
ax.errorbar(data_x,data_y,yerr=3,fmt='o',color='k',alpha=0.4)
```

---

[9] We often use the words errors and uncertainties interchangeably but formally they are different quantities.

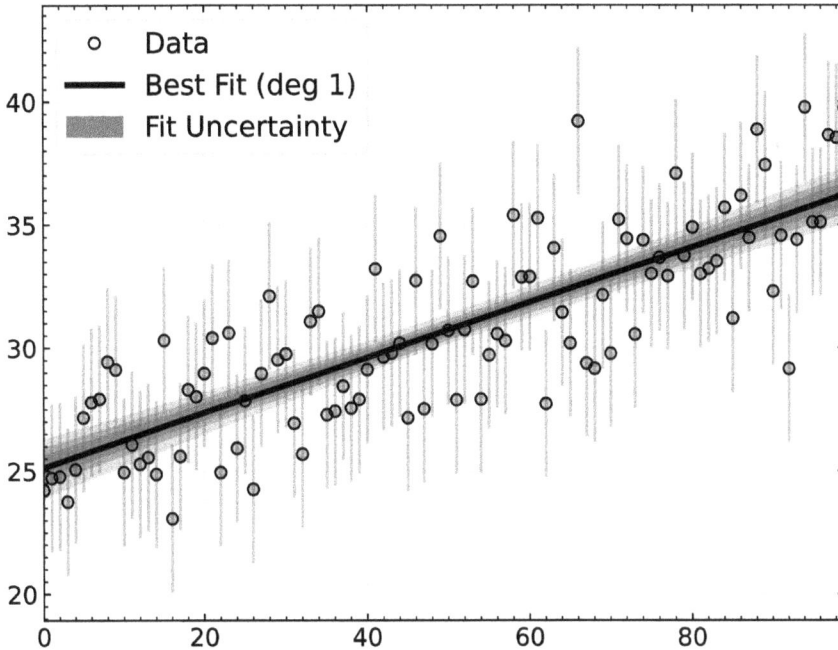

Figure 5.20. Fit, now showing some errorbars on each datum. With the figure more busy, I have changed the shaded uncertainty region to a blue color, though this may still need improvement if when printing to black-and-white the region is difficult to parse.

5.12 Plotting *N*-Dimensional Data

Thus far, all of our plots have been made in two dimensions (the abscissa and ordinate of the figure). There are several ways of adding additional dimensions to our data representation (though there is a limit to what we can accomplish when the final plot is still going to be 2D).

The first is to encode a third variable into either the marker size or marker color of our plots, which we can do fairly easily with optional arguments to the ax.plot() or ax.scatter() functions.

But what if we have something like an astronomical *image*? An image is also a form of three dimensional data—for every pixel (i.e., every array index), we have three pieces of information, (x_i, y_i, F_i) where F is the flux or brightness of that pixel. None of the plotting commands we have learned thus far in this chapter can handle this type of data.

There are two primary methods of plotting data of this sort: imshow(), and pcolormesh(). The primary difference between these two methods is that imshow assumes you have a regular grid of pixels (e.g., a 2D Numpy) array, while pcolormesh() can handle non-regular grids, or scenarios in which you have separated *x*, *y*, and *flux* arrays (I use flux here as a stand-in for whatever unit happens to be the *values* of our plot). As a rule of thumb, astronomical images can be plotted with imshow() because detectors generally have pixels all of the same size. Meanwhile, the rendering of simulation data will often require pcolormesh().

In Chapter 6 we will use some images from the TESS satellite to measure an exoplanet transit, and Chapter 8 we will explore astronomical images much more deeply in the context of some real images from JWST. So here, I will only briefly introduce the imshow() command. I will load up one of the images Chapter 6 for this example.

```
from astropy.io import fits
import numpy as np
import matplotlib.pyplot as plt

im_array = fits.getdata('../../BookDatasets/tess/HAT-P-11/0.fits')

fig, ax = plt.subplots(figsize=(7.5,6),constrained_layout=True)
ax.imshow(im_array)
```

When we feed a two dimensional array into ax.imshow() with no arguments, Matplotlib selects several defaults for us—several of which we wish to change. But we can see we have some sort of object in Figure 5.21 (in this case a star).

Take a look at the axes of this plot. You'll notice that there is no single origin—the 0th column starts on the left hand side, but the 0th row starts at the top of the

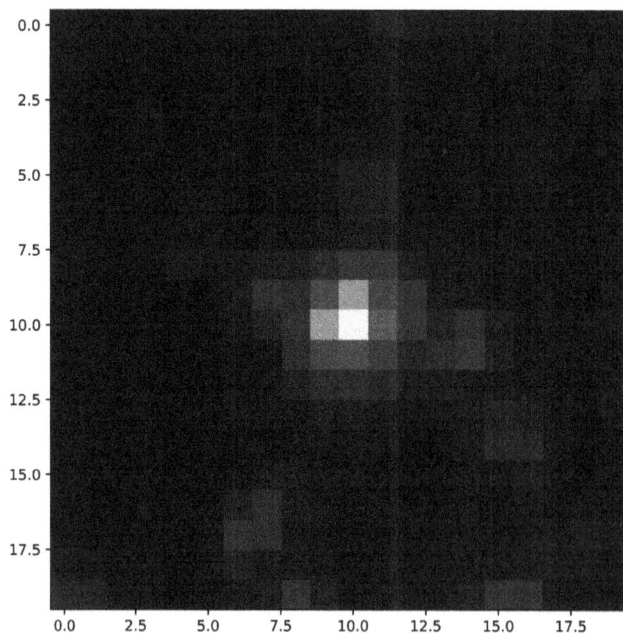

Figure 5.21. Default output of an imshow() command given a 2D array of data. Much of the dynamic range is not captured by the default vmin and vmax, and the origin of the rows is in the upper right hand column.

image. This is because we index arrays using this convention (you may not have thought about it in this way before).

For plotting purposes, we would rather have our origin in the lower lefthand corner, which we can accomplish by setting the `origin` keyword to "lower." Let's make our plot again—this time, flipping the origin, and for added clarity, I will take the log of the data array to make more of the image visible.

```
fig, ax = plt.subplots(figsize=(7.5,6),constrained_layout=True)
ax.imshow(np.log10(im_array),origin='lower')
```

Already in Figure 5.22 we have much improved our visualization: more of the image structure is visible now, and the origin is properly in the lower left hand corner. However, it may be the case that we want to *display* the data in this log form, without actually taking the log of the data *values* (e.g., when we add a colorbar, we want the actual values there, but in a log axis).

We can accomplish this using the `LogNorm()` class from `matplotlib.colors.LogNorm()`, to produce the same looking output.

We can also set the minimum and maximum values associated with our color scaling: `vmin` and `vmax`. When our color mapping is *linear*, which it is by default, these two parameters set the boundaries between which the gradient from black to white (or in the case of `viridis`, the default `cmap` used above, blue to yellow) occurs. Any value above `vmax` will be saturated at the color map's last color, and

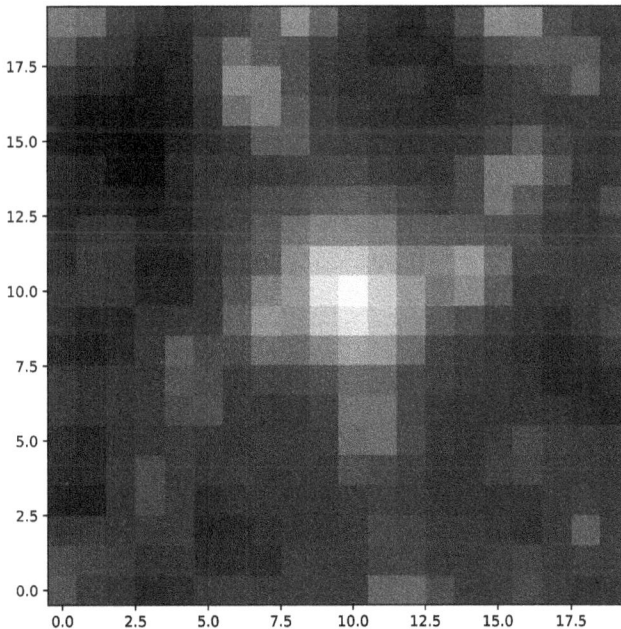

Figure 5.22. Visualization of the star, now with the origin in the lower lefthand corner and the data values logged so that more of the dynamic range of the image is visible.

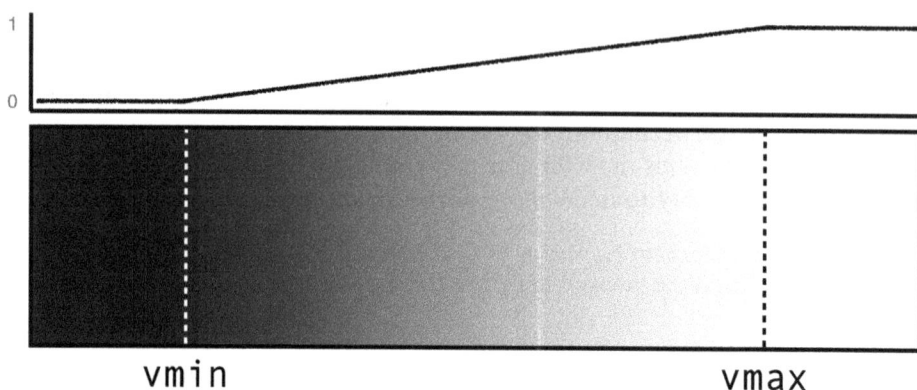

Figure 5.23. Schematic illustrating the function of vmin and vmax on a Matplotlib color map. For linear maps, the gradient smoothly varies between vmin and vmax, and outside of these bounds, the gradient is saturated at whatever colors define its endpoints. Generally, we wish to choose vmin and vmax such that the interesting dynamic range (the range where data values and differences between data values are meaningful) is captured between vmin and vmax.

any value below vmin will be saturated at the colormap's first color (see Figure 5.23 for a visual schematic of this). We'll see these arguments used extensively in Chapter 8, as astronomical images often require us to set very careful vmin and vmax values in order to see the objects in them.

As mentioned, we have not been setting a color map in our imshow() calls above—so Matplotlib is selecting the viridis colormap by default. It is a perfectly reasonable colormap to use in many cases, and was designed to be perceptually uniform (i.e., your eye sees linear gradients as linear). On the other hand, numerous other colormaps exist, which we can use for different purposes. Some are linear, like viridis, others are diverging (getting brighter or darker away from some central value), and some are categorical (fixed chunks of single colors). Additionally, most colormaps can be inverted simply by adding _r to the name of the colormap. You can find a full list of available colormaps on the Matplotlib documentation website; in Figure 5.24 I sample four colormaps from the library. As a note: it is often wise to check how a given visualization (and in particular, colormap) reads when the plot is rendered to grayscale or black and white, as often the papers you publish will be printed and read in that format.

As a final note, though we won't cover it here: you can create Matplotlib axes with three dimensions as well—this means picking a rotational angle to view from, but does allow for 3D information (such as star positions in 3D space) to be plotted. This means ultimately that four dimensions can be captured (if the size or color of those markers is used to denote an extra property, or, e.g., an arrow track is used to indicate time evolution). Recently, the adoption of other libraries that produce interactive (rotatable, zoomable) figures has opened up better avenues for visualizing this kind of data.

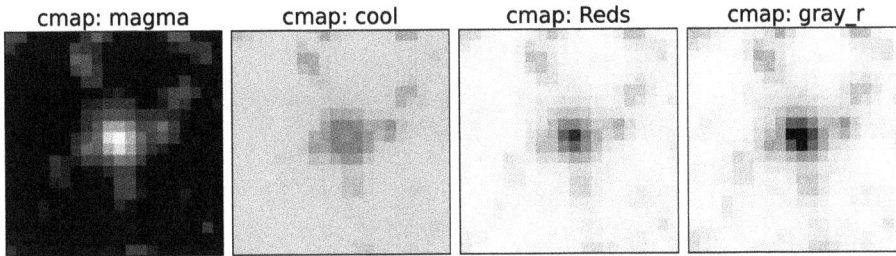

Figure 5.24. Four colormaps from `Matplotlib`. Though it may seem somewhat boring, it is often best to use the final option—gray, or inverted gray, which smoothly varies between white and black. It is simple to interpret, and will be guaranteed to work if someone prints your paper or figure in black and white.

5.13 Colorbars

When we plot using colors that are mapped from our data (e.g., a scatter plot with the c argument passed, or any `imshow()`), it is often important to display that mapping, so that those interpreting our plot know what physical value a given color or intensity of color corresponds to. In a sense, using color or intensity in our plots is a third dimensional axis, but one for which our eyes can detect relative differences, but not the absolute values. In that analogy, our color bar is a way of displaying that third axis (its range of values and units).

To add a colorbar to our figure, we need to specify an axis for it (known as a colorbar axis, or `cax`), and we need to provide it the mappable quantity. Thus far, when using `imshow()` or `scatter()`, we have not been saving the *output* of those function calls anywhere. We'll do so now, as these outputs are what the `plt.colorbar()` function requires.

There are several ways we could create the axes for our colorbar, including the direct `fig.add_axes([l,b,w,h])` function. But in this text, we'll use a helpful interrogative `Matplotlib` function to actually "steal" a bit of space from one of our current axes and create a colorbar axis there automatically.

Continuing with our star image, we can use the following technique to add a colorbar (Figure 5.25).

```python
fig, ax = plt.subplots(figsize=(6,6),constrained_layout=True)
mappable = ax.imshow(np.log10(im_array),origin='lower')

# Now for the colorbar
from mpl_toolkits.axes_grid1.axes_divider import make_axes_locatable

ax_divider = make_axes_locatable(ax)
cax = ax_divider.append_axes('right', size='7%', pad='2%')

plt.colorbar(mappable=mappable,cax=cax,label='log image counts')
```

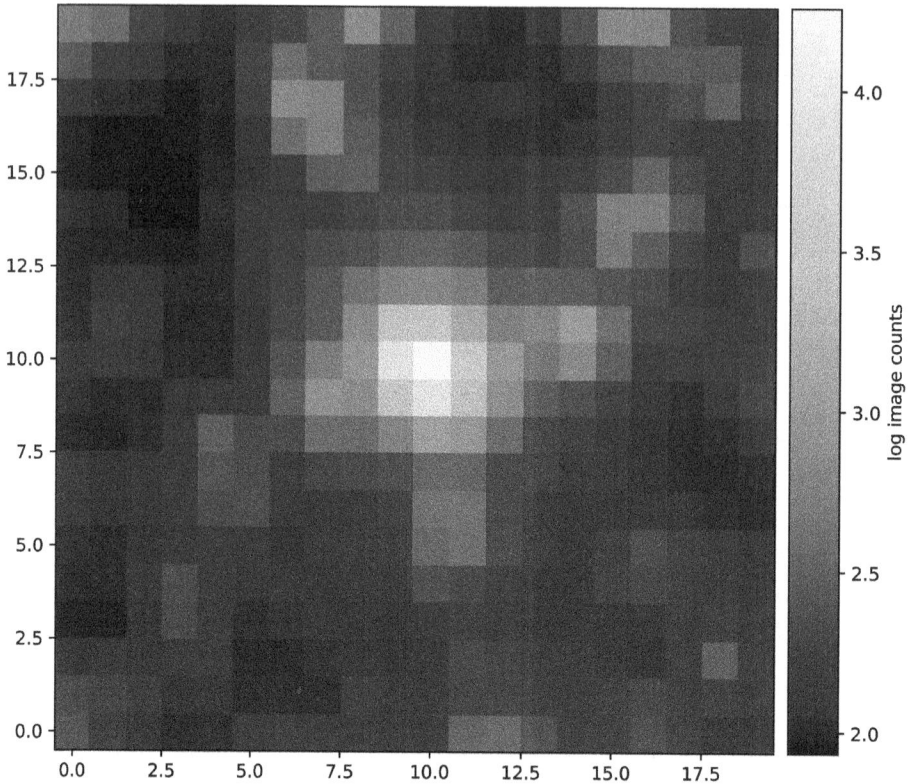

Figure 5.25. Image of a star, plotted via the `imshow()` command, with a colorbar indicating the numerical values represented by colors in the image.

The key step here was the combination of the `make_axes_locatable()` function and the `ax_divider.append_axes()` method. The first is an abstract object that knows the boundaries of the plot axis (`ax`), and its append method can create a new axis to the top, right, left, or bottom of itself, with a size in percentage of the plot's width and a padding between the two (also as a percentage of the plot's width.)

Once we have this new axis, we call the `plt.colorbar()` command and input the mappable (the output of `ax.imshow()`) and the colorbar axis we just created. Here I also pass a label to reflect what quantity the colorbar is describing.

Given that to the left, right, top, or bottom of a figure is where we almost always want to place a colorbar, this method of creating its axis is very convenient. As a note, this method of appending axes can be used for more than colorbars—nothing about the actual axis creation specified it had to be used for this purpose.

5.14 Summary

In this chapter, we learned how to create visualizations of data in Python using the Matplotlib library. While many other libraries exist for this purpose (many optimized for, e.g., web browsers and interactive figures), Matplotlib remains the de-facto standard of all figures published in astronomical journals, and provide near-infinite flexibility for customizing two-dimensional, static figures. We covered the basic figure and axes objects that underpin our plots, as well as how to modify the defaults to suit the needs of our visualizations. Once you have a handle on these basics, best practices include reading up on visualization theory, as well as examining published papers to find figures you find to be well-made both aesthetically and for conveying a given point. Learning to implement any such features into your own figures is a great way to build out your own personal style.

References

Hunter, J. D. 2007, CSE, 9, 90

Rougier, N. P. 2021, Scientific Visualization: Python and Matplotlib https://www.labri.fr/perso/nrougier/scientific-visualization.html

Astronomical Python
An introduction to modern scientific programming
Imad Pasha

Chapter 6

Numpy

6.1 Introduction

Nearly every Python script written by any astronomer (and nigh on anyone who writes Python code for science or industry) begins with the line `import numpy as np`. The Numerical Python library (Harris et al. 2020) is an integral, foundational set of functions used throughout almost any code. The basic element or building block of Numpy is the `array`, which is so important to the burgeoning scientist that you should consider it a core data type.

Once you have an array of values (or several), the following sections describe some of the common ways we use Numpy to carry out operations on our arrays. I want to stress again that this is only a sampling of some of the most basic use cases. But the nature of research coding is to learn by doing. Once you are familiar with these basics, and the loop of

1. having a coding task to accomplish,
2. identifying the Numpy (or other library) function that will accomplish it, and
3. successfully implementing it,

you'll begin exploring the broad, and deep, set of code that Numpy provides on a case by case basis.

The Numpy library offers two primary advantages: a litany of useful functions for the creation and manipulation of numerical data, and the ability to carry out operations on that data at much higher speed (computationally) than pure-Python, looping-based methods. The second advantage is known as *vectorization*, and it will be discussed in much more detail in Part III. For now we will focus on the functions that make our life easier.

6.2 The Array

Before we can carry out operations, we first need our container to hold data. For Numpy, this is the `array`.

We can define an array using Numpy:

```
import numpy as np

array1 = np.array([])
array1
```

```
array([], dtype=float64)
```

In this case, we have initialized an empty array. Notice something interesting about the representation of the array (i.e., what is shown when we print it). Beyond indicating it is an array, and showing us the array values (in this case, []), we also have an indication of a `dtype=float64`. This is Numpy telling us that by default, it initializes an array to contain values which are stored as 64-bit floats. That is the highest fidelity we typically use (followed by 32-bit floats).

The reason Numpy is telling us this is because Numpy arrays are *typed*. For many Python datatypes (e.g., lists), types are defined dynamically, on the fly, and can thus be mixed freely. This freedom when coding is traded off with execution speed. Arrays are useful not simply as containers of numbers, but because operations between arrays are fast to compute. Numpy accomplishes this by utilizing compiled code written in C, but this requires types to be known ahead of time. Thus, Numpy forces us to "chose" a data type for our array. *Most of the time*, we don't have to worry about this directly: we plug numbers into arrays, do our math, and everything works (and quickly)! Later in this book, when we talk about speeding up code, we'll return to this topic.

Arrays, when defined, have several useful *attributes* that we might query. Let's define some new arrays that actually contain values:

```
array1 = np.array([1,2,3])
array2 = np.array([[2,3,4],
                   [5,6,7],
                   [3,0,4]])
```

Here we have created two arrays, one of which is one-dimensional and one of which is two-dimensional. To create an array with additional dimensions (known as *axes* in code parlance), we insert a "list of lists." That is, for each set of enclosing brackets, we establish a new axis to extend into—notice the three sub-lists above have an extra set of enclosing brackets. Printing an array will show it, if it fits in screen:

```
array1
```

```
array([1, 2, 3])
```

```
array2
```

```
array([[2, 3, 4],
       [5, 6, 7],
       [3, 0, 4]])
```

We can confirm the dimensionality and shapes of these arrays:

```
print(array1.shape)
print(array2.shape)
```

```
(3,)
(3, 3)
```

The first has a shape of (3,) because it is three elements. The second array has a shape of (3,3). We can also now ask for the **dtype** of our arrays.

```
array1.dtype
```

```
dtype('int64')
```

Because I inserted integer numbers when defining the arrays, the dtype is a 64 bit integer. What if we need floats? We can accomplish this in two ways. Adding decimal points to the numbers on definition, like this:

```
array1 = np.array([1.,2.,3.])
array1.dtype
```

```
dtype('float64')
```

Or, we can *cast* the values to be floats using a method that each array has

```
array1 = np.array([1,2,3]).astype(float)
array1.dtype
```

```
dtype('float64')
```

We can extend this logic to other reasonable types as well:

```
array1.astype(str)
```

```
array(['1.0', '2.0', '3.0'], dtype='<U32')
```

Now we've converted our values to strings. Note that for a given axis of a Numpy array, *all* values must conform to the same type. If you need a container to have a mixed set of types (like a string, a float, and a dict) then use a list or dictionary.[1]

Above, we queried the shape of our array, which is one of the most useful attributes to check when setting up or debugging research code. Sometimes we are also interested in the size (the number of total elements in the array). If we check the size of our array2, we see:

```
array2.size
```

```
9
```

Sometimes we don't need to know the exact shape (the lengths of each axis) but rather need the number of dimensions:

```
array2.ndim
```

```
2
```

Often, we find that an array we have read or created is the transpose of the array we actually need. One can compute a transpose, but Numpy arrays have an attribute that stores their transpose:

```
print(array2)
print(array2.T)
```

```
[[2 3 4]
 [5 6 7]
 [3 0 4]]
[[2 5 3]
 [3 6 0]
 [4 7 4]]
```

In this example, we can see the columns have become the rows and vice versa.

---

[1] You may also run across Numpy record arrays, which have named columns and can have different types for each column. You can read more about this in the section on tabular data, but in general, we normally use more fully featured structures for that type of task.

Beyond useful attributes, Numpy arrays have numerous methods (attached functions) that are useful in many contexts. We've already seen one above: the .astype() method, which casts the values of an array to a different type.

When we defined array2, we created a 3 × 3 (ndim 2) array. Sometimes, it is valuable to *flatten* this array to a single dimension. We can do that as follows:

```
array2.flatten()
```

```
array([2, 3, 4, 5, 6, 7, 3, 0, 4])
```

Why might this be useful? One reason is that several methods of arrays return flattened values—for example, there are methods to give us the indices of the minimum or maximum values in the array:

```
print(array2.argmin())
print(array2.argmax())
```

```
7
5
```

Here, the methods are telling us the 8th element (index 7) is the minimum value, and element 6 (index 5) has the maximum value. But if we try to use our indexing rules to actually get, say, the minimum value, we'll get an error:

```
array2[7]
```

```
IndexError
Traceback (most recent call last)
In[16], line 1
----> 1 array2[7]

IndexError: index 7 is out of bounds for axis 0 with size 3
```

We can parse this IndexError—we are asking for element 7, in the 0th axis of an array that only has 3 elements. On the other hand, knowing these are flattened indices, we can simply flatten the array and index so that:

```
array2.flatten()[array2.argmin()]
```

```
0
```

(this is not the fastest way to retrieve this value: running np.min(array2) or array2.max() would provide it. But those do not give you the index of the max or min.)

Can we go the other way, and turn a flattened array into one of a certain shape? Yes!

```
array3 = np.array([5,6,7,8,9,10,11,12,13,14,15,16]).astype(float).
  ↪reshape(3,4)
array3
```

```
array([[ 5.,  6.,  7.,  8.],
       [ 9., 10., 11., 12.],
       [13., 14., 15., 16.]])
```

As we can see, I can define a 1D array, cast it to float, and *reshape* it to a new shape. This only works because 3×4 is an appropriate shape for an initial array with 12 elements. Other combinations are also possible:

```
array3.reshape(4,3)
```

```
array([[ 5.,  6.,  7.],
       [ 8.,  9., 10.],
       [11., 12., 13.],
       [14., 15., 16.]])
```

```
array3.reshape(2,6)
```

```
array([[ 5.,  6.,  7.,  8.,  9., 10.],
       [11., 12., 13., 14., 15., 16.]])
```

```
array3.reshape(3,3)
```

```
- - - - - - - - - - - - - - - - - - - - - - - - - - - - - - - - - - - - - - - - -
ValueError
Traceback (most recent call last)
In[21], line 1
----> 1 array3.reshape(3,3)

ValueError: cannot reshape array of size 12 into shape (3,3)
```

As expected, attempting to reshape to 3×3 fails since 12 elements are not divisible this way. There are also plenty of linear-algebra related methods that allow us to treat arrays like matrixes—for example, we can grab the diagonal elements of a square array:

```
array3.diagonal()
```

```
array([ 5., 10., 15.])
```

There are many more methods to explore for Numpy arrays, but they are better learned in the context of some real data. So for now, we will continue with our introduction to Numpy.

6.3 Precision

It's worth stopping briefly to introduce the concept of precision. When we query the dtype of an integer or float, we'll see a number after it, e.g., float32 or float64. This number refers to the memory usage of the value, being either 32 bit or 64 bit. 64 bit variables use twice as much memory as 32 bit ones. These memory allocations are related to the bits of precision: 24 bits of precision for a 32 bit float (23 plus a sign), and 53 for a 64 bit float (52 plus a sign).

Precision in code informs questions of how long, or precise, of a number can the computer store without rounding errors and other issues leading to incorrect calculations. Take the following example:

```
import numpy as np

print(int(np.float32(123456789)))
```

```
123456792
```

We can see that in this case, a 32-bit float was not long enough to accurately represent this value, and when we converted it to an integer, we got a different value (due to rounding errors). This is obviously bad (imagine if this integer was a galaxy ID; if we read in a file with the IDs and they were read in as float32 by default, when we went to retrieve them our IDs would be nonsense.

If we switch to float64, we have ample memory to handle this case.

```
print(int(np.float64(123456789)))
```

```
123456789
```

How can we know when we're at risk of this issue? In general, the number of digits of precision for a given bit is given by

$$n \sim 2^b - 1 \tag{6.1}$$

where b is the bits of precision. We know that 32-bit floats have 23 bits of precision (plus one for the sign), meaning that in general, you can trust 32-bit values up to integers of $\sim$8388607, which is seven digits (so 1.123 456 789 would also be a problem). Our ID above had 9 digits, hence the last few digits being affected in the calculation. On the other hand, 64 bit floats, with 53 bits of precision, can handle values up to 9×10^{15}. This means we have precision out to ~15 decimals.

Here we can see the same precision difference illustrated for values after the decimal:

```
np.float64(0.0024757564245)
```

```
0.0024757564245
```

```
np.float32(0.0024757564245)
```

```
0.0024757565
```

Again, if we needed the full precision of that number, a 32-bit float would not suffice. Finally, we can see how much deeper we can go with a 64-bit float before we hit precision issues:

```
np.float64(0.00247575642458884888)
```

```
0.002475756424588849
```

In general, a 64-bit float will get you ~15–17 decimals of precision.

The takeaway here, to some extent, is to simply use `float64` and `int64` (which are usually the default) for everything, unless you have a problem so specific that the memory allocation difference actually matters.

6.4 Key Library Functions

There are hundreds (if not more) Numpy functions for all sorts of tasks. We cannot possibly cover them all—the best way to learn the depth of the library is to seek out the relevant functions when your research calls for them. But several categories of function are ubiquitous and useful enough to list. This list contains a few representative examples of the different categories of functions that can be found within Numpy. Their unifying principle is that they concern or operate on `arrays`.

- `np.arange()` and `np.linspace()` both facilitate the automatic creation of arrays of a given length, start value, end value, and either step size or number of steps. These are incredibly useful for creating 1D grids on which to evaluate models, and you'll see them used throughout this text.
- `np.meshgrid()` and `np.mgrid[]` both facilitate the evaluation of functions on multidimensional grids by creating coordinate arrays based on the grid's size and shape (other similar grid functions exist).
- `np.random` is a module of random generators to quickly create random numbers from uniform, normal, or other distributions.

- `np.zeros()`, `np.zeros_like()`, `np.ones()`, `np.ones_like()`, and `np.empty()` allow one to create arrays of certain shapes and sizes (e.g., the size of another extant array) containing zeros, ones, or nothing.
- `np.vstack()`, `np.hstack()`, `np.column_stack()`, `np.dstack()`, and `np.row_stack()` are commonly used instances of a set of many more "stacking" functions that allow multiple arrays to be composed into more complex shapes. Similar functions exist to split arrays using similar rules.
- `np.isfinite()`, `np.isinf()`, and `np.isnan()` all help condition on the values of the data we are working with, allowing us to prune (or find) generally unwanted values.
- `np.masked_array()` and its associated functions facilitate performing calculation on an array for which some of its values are masked.
- `np.sin()`, `np.tan()`, `np.floor()`, `np.ceil()`, `np.sum()`, `np.nansum()`, `np.median()`, and `np.exp()` all handle mathematical operations—including rounding, and computing statistical measures while ignoring NaNs. These mathematical operators can act on whole arrays, and almost always allow one to select which axis of an array to operate on.
- `np.polyfit()`, `np.polyval()`, `np.polynomial()`, and their associated functions facilitate the creation, manipulation, and fitting of polynomial functions of different orders.
- `np.unravel_index()`, `np.sort()`, `np.argsort()`, `np.argmin()`, and `np.argmax()` are examples of functions dedicated to storing, modifying, and evaluating properties of array values or array indices.
- `np.percentile()`, `np.nanpercentile()`, `np.std()`, `np.histogram()` and `np.histogram2d()` all concern the calculation of statistical measures.

We will use many, but not all, of these functions throughout this text. You can learn much more about each beyond our brief introductions here by reading the Numpy documentation for a given function online. The documentation usually also includes usage examples.

For now, however, we will turn the scenario around and take on a research problem—and that problem will drive our use and manipulation of Numpy arrays.

6.5 Research Example: An Exoplanet Transit

In the examples above, we defined arrays by manually typing in every value. That is not very realistic to how we gather data—one could, in theory, say, jot down into an array the voltages measured on a lab set up, one by one, for different settings, similar to manual data entry into a spreadsheet. But far more often, we receive data from an instrument, or produce it programmatically.

For example: you may have seen this classic schematic of an exoplanet transit (Figure 6.1). The transit method, as it is called, is a means for the detection and characterization of a planet orbiting another star via the "dip" in light that occurs when the planet passes (transits) between the observer at Earth and the star.

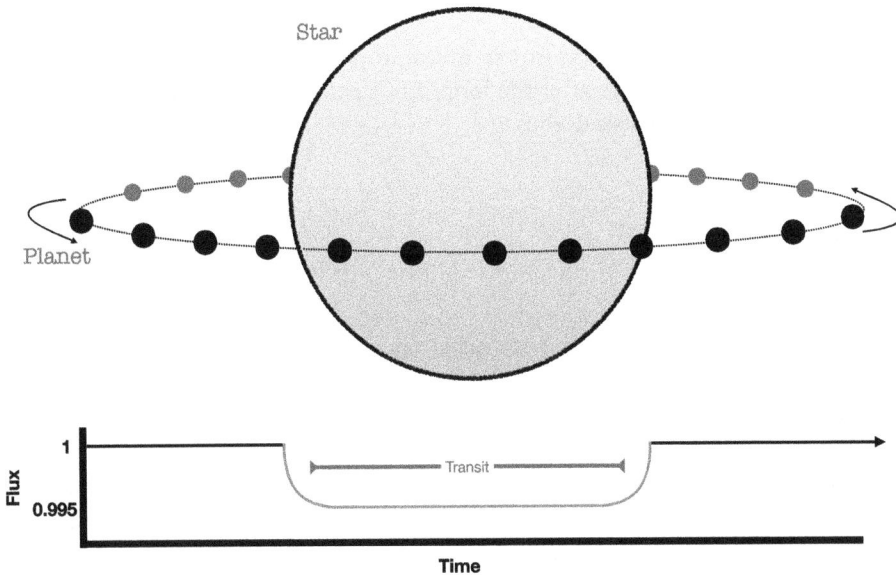

Figure 6.1. Cartoon demonstrating an exoplanet transit. When we look toward extra-solar planetary systems that are edge on, the planets in the system spend some time in between us and the star, blocking some of the star's light. We can use these dips in brightness to both find, and characterize, exoplanets.

Despite blocking only a tiny fraction of the star's light in most cases, sensitive telescopes (especially those in space, such as the Kepler and TESS satellites) can detect such changes.

Amazingly, given some cutout images of an exoplanet-hosting star (in the form of 2D Numpy arrays) over time, we can actually extract a transit using only some simple array operations in Numpy. Let's do that now. We'll be using data from the TESS satellite, in particular, of the star HAT-P-11, for which there is a measured exoplanet transit. Because we need many images of the star over time (in this case, roughly 1000), I am going to use a *loop* to read them all in.

```python
import numpy as np
from astropy.io import fits

imlist = []
jd = []
for i in range(1237):
    fn = f'../../BookDatasets/tess/HAT-P-11/{i}.fits'
    imlist.append(fits.getdata(fn))
    jd.append(fits.getheader(fn)['btjd'])
imlist = np.array(imlist)
jd = np.array(jd)
```

For ease, I've read each image (and a timestamp from the header) into lists, which I then turn into Numpy arrays after the loop. Let's examine what we've loaded, using some of the methods discussed above:

```
print(jd)
print(jd.shape)
print(imlist.shape)
```

```
[1683.36712646 1683.38800049 1683.40881348 ... 1710.17987061]
(1237,)
(1237, 20, 20)
```

As we can see, we have our timestamp array jd with 1237 elements, and then our imlist, which also has 1237 elements, but each of those elements is a 20×20 image, making a final shape of (1237,20,20). You can imagine this like a loaf of sliced bread —the full array is the loaf, each image is a slice, and we can choose whether to extract a slice, multiple slices, or "drill down" through all slices for a subset of pixels. A schematic of this is shown in Figure 6.2, with our array shape and some indices of the three dimensional array indicated.

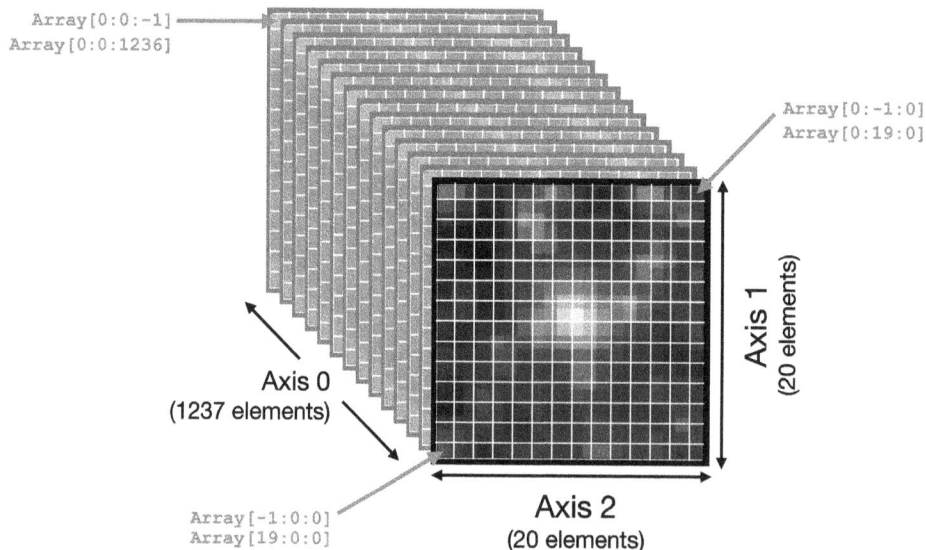

Figure 6.2. Overview of the array shape we are working with in this example. The two axes of length 20 are the image dimensions, and the long, or stack axis is the collection of all the images. For several of the indices, I show what the appropriate indexing would be to retrieve the value of that array element. Note that this array structure was set by *us*; Numpy does not care which order we decide to arrange our axes. But for our ease of conceptualizing our indexing and slicing, it often makes sense to group axes that correspond to a single entity (such as grouping the dimensions of each image here).

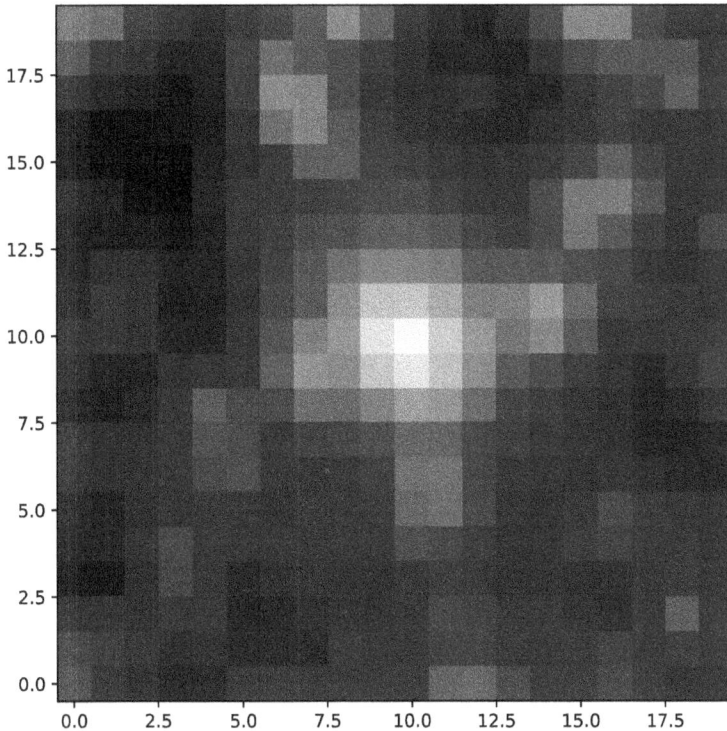

Figure 6.3. TESS cutout image of HAT-P-11, accessed using the `lightkurve` package Collaboration et al. (2018).

We should examine one of these images more closely to see what it looks like:

```
import matplotlib.pyplot as plt

fig, ax = plt.subplots(figsize=(7,7))
ax.imshow(np.log10(imlist[0]),origin='lower');
```

Since the shape is N (images) by 20 by 20, I indexed the 0th element to retrieve the first 20 × 20 image (indexing `imlist[0,:,:]` would be equivalent). You can read that as "index the first of 1237 elements along axis 0, then all elements along the other two axes."

In Figure 6.3, we can see the individual cutout from the TESS satellite for one single image in our stack. We can clearly see a star here in the center, with some other (fainter) stars and darker patches that are probably empty sky.

Exercise 6.1:
It is a highly valuable exercise to be sure you can extract arbitrary slices of an array.
1. Create a subarray of the 3rd through 10th image, but only the first 3 rows/columns of those images.

2. Create one of the central 5 × 5 pixels, but only for images with even numbered indices.

Such exercises are not just abstract: let's see how we need this indexing to continue extracting our exoplanet transit signal. Based on the schematic in Figure 6.1, you might imagine we only need to measure the brightness of the central star in each image (assuming the images were taken over a period of time during which a transit occurred).

Unfortunately, real data is never so clean as the schematic. We can try this anyway to see what happens (trying the simplest solution is always a reasonable starting point in science). There are many ways to decide which pixels we are going to associate with the star—this process is known as choosing an *aperture*. We could choose only the central, brightest pixel, the 9 central pixels, or even the whole image. We could lay down a circle (under the reasonable assumption that stars are circular intrinsically) and compute the sum within it, accounting for partially included pixels. We could use statistical measures of the star, such as its full width at half maximum (FWHM), to estimate the radius of such an aperture. We could even fit a full 2D model of the light (as a Gaussian or Moffat or other profile) and extract a flux or radius from that.

Here, we'll do something simple, but I wanted to highlight that the choice (and right course of action) is nontrivial. For our purposes, let's select the central five rows and columns of the image to be our "cropped" version. How would you index this? The result of such a slice selection is shown in Figure 6.4.

The red box in Figure 6.4 shows the pixels we have selected as being "part of our star." It is worth the reminder: Python and Numpy indexing *includes* the lefthand index and stops *before* the righthand index. Indexing 8:13 will give elements 8 through 12. Thus, our "cropped array" of only star pixels is created as

```
cropped_array = imlist[:,8:13,8:13]
```

Note the initial colon and comma—we want *all* 1237 images, but only the restricted range of *each* image. So now, we just need to know the sum of the pixels in our `cropped_array` for each image.

This gives us an opportunity to learn about axis selection in Numpy functions. The function we need is, helpfully, called `np.sum()`. This function allows you to compute the sum of all elements in an array or list:

```
l = [1,2,3,4]
np.sum(l)
```

10

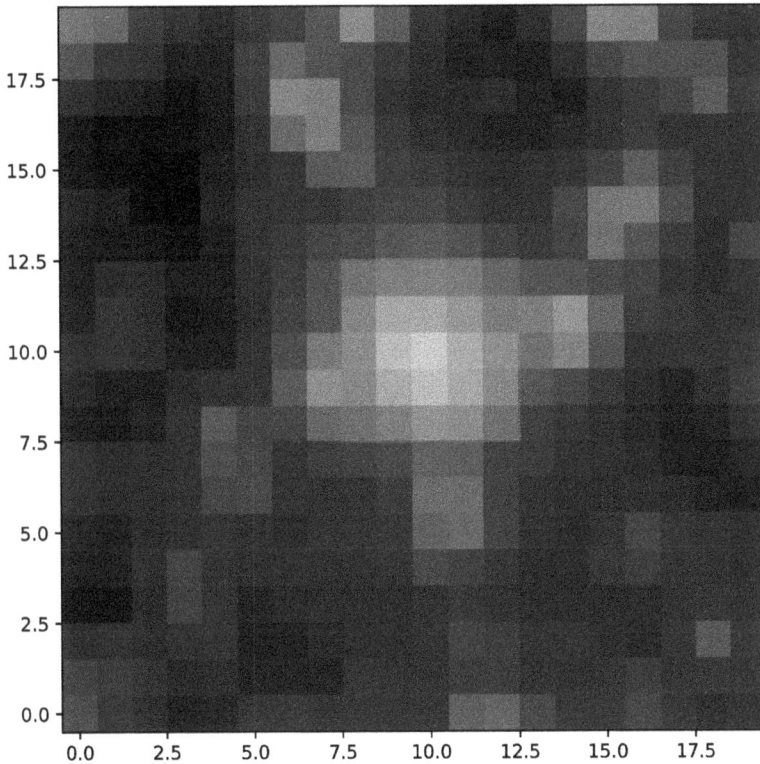

Figure 6.4. Same image as Figure 6.3, but now with a red mask covering the region we will be cropping out and considering as our aperture for the star.

When we have more than one axis in our array (such as our `imlist` or `cropped_array`), the sum function will collapse those axes:

```
np.sum(imlist)
```

```
203091860.0
```

This large sum is the sum of *every pixel* in *every image* in the full `imlist` stack. Sometimes this is what we want, but often we actually only need to compute the sum along one or more of the axes. In this case, we want the sum of each of the 1237 elements in `cropped_array` (each of those being a 5x5 array). To accomplish this, we need to specify the axis we want to sum along. The 0th axis is the "depth" axis of our image stack, so that's not what we want. We can see if we try it:

```
np.sum(cropped_array,axis=0).shape
```

```
(5, 5)
```

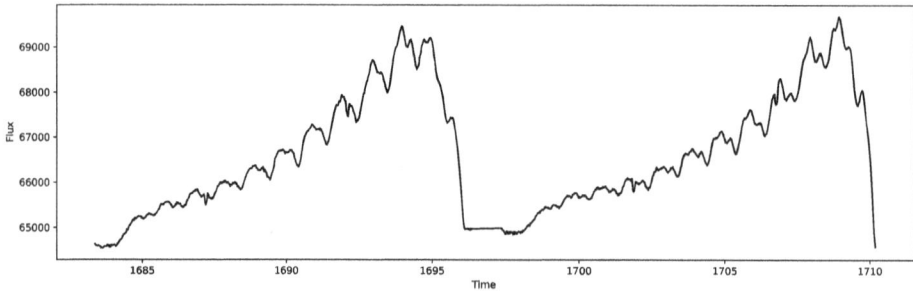

Figure 6.5. "Raw" extracted light curve of HAT-P-11. The signal of the exoplanet blocking the star is buried in the much larger variations in flux that are primarily instrumental in nature.

This gives us *one* (5 × 5) array. We want, instead, an array of shape (1237,), in which each 5 × 5 image has been summed to one number. That means we need to sum both along the first and second axes of the image:

```
flux = np.sum(cropped_array,axis=(1,2))
flux.shape
```

```
(1237,)
```

We now have an array of the right shape.[2] We can thus plot the measured fluxes (brightness) against the time of each observation (which I extracted into the jd array earlier) (Figure 6.5).

```
fig, ax = plt.subplots(figsize=(13,3))
ax.plot(jd,flux,'k,')
ax.set_xlabel('Time')
ax.set_ylabel('Flux');
```

This does not look much like the schematic in Figure 6.1.

Much of the structure in this plot represents variations in the measurements not due to actual variation in the star, but rather instrumental effects. Luckily, we can remove much of this variation (Figure 6.6). Let's return to the imlist array, and instead of extracting the central 5 rows and columns, let's extract a patch that looks to be mostly empty sky:

```
bg = imlist[:,0:4,12:17]
fig, ax = plt.subplots(figsize=(7,7))
ax.imshow(np.log10(imlist[0]),origin='lower')
ax.fill_between([-0.5,3.5],11.5,16.5,color='r',alpha=0.3);
```

---

[2] Array shapes being correct, much like dimensional analysis, is often a helpful indicator that an operation has been carried out properly.

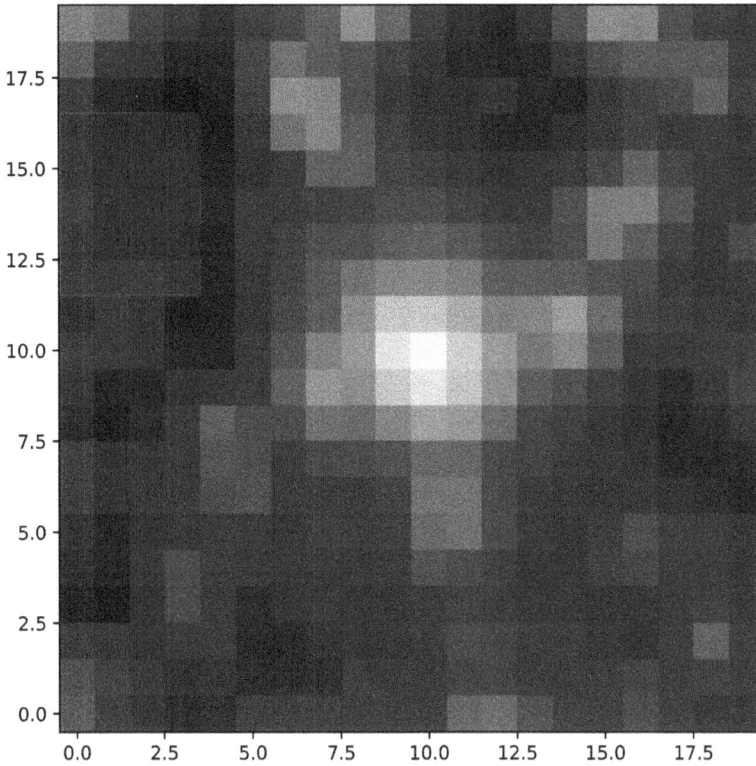

Figure 6.6. Same image as Figure 6.3, but now with the mask covering the region we will use for estimating the background variation.

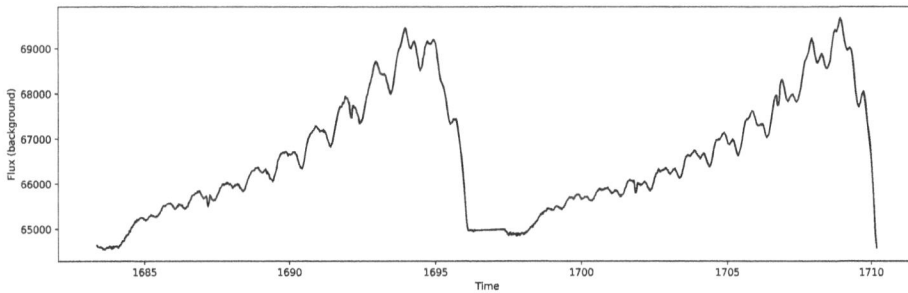

Figure 6.7. Flux as measured in the background pixels over time. We can see that this looks nearly identical to Figure 6.5, demonstrating that those variations were not originating in the star itself.

The red patch now covers pixels which (hopefully) don't contain signals from another star, and should just represent the overall variation from frame to frame (Figure 6.7). We can play the same summation trick and view the results (Figure 6.7).

```
flux_bg = np.sum(bg,axis=(1,2))
fig, ax = plt.subplots(figsize=(13,3))
ax.plot(jd,flux,'k,')
ax.set_xlabel('Time')
ax.set_ylabel('Flux (background)')
```

As we can see, the background pixels track this same large scale structure. So by subtracting it from our stellar flux measurement, we will have a much cleaner view of what is going on.

In order to carry out the subtraction, I will repeat our axis-collapse from above, but use the `np.mean()` method instead—this provides the average of the pixels rather than their sum. This value will represent the mean background *per pixel*. This is important as our background aperture contains a different number of pixels as our primary aperture. The value we need to subtract from each computed sum is the per-pixel background times the number of pixels in the primary aperture (that is, 25).

```
mean_bg = np.mean(bg,axis=(1,2))
flux_minus_bg = flux - mean_bg*cropped_array[0].size
```

Notice we don't hard-code the 25, but use the `size` attribute of the 0th element of cropped array as a way to query the number of pixels in our aperture. We can now view our "corrected" light curve (Figure 6.8).

```
fig, ax = plt.subplots(figsize=(13,3))
ax.plot(jd,flux_minus_bg,',',color='k')
ax.set_xlabel('Time')
ax.set_ylabel('Flux (background subtracted)')
```

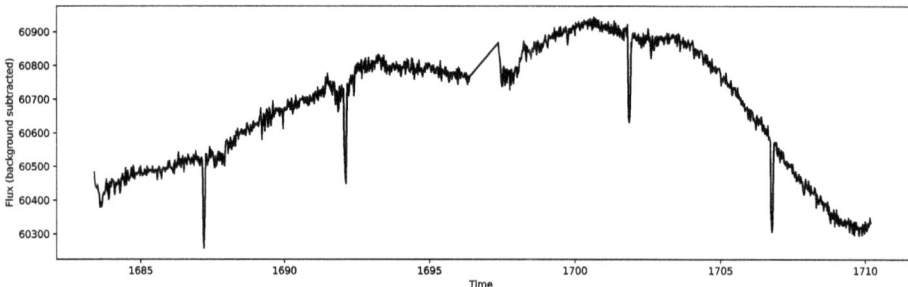

Figure 6.8. Background subtracted light curve for HAT-P-11. In the absence of the larger variations, we now have the dynamic range needed to see the exoplanet transits. The sharp linear feature in the figure center is `Matplotlib` connecting two points which in reality were on either side of an observing gap.

All of a sudden, the variation in the light curve doesn't seem so strong! We can now also see the end goal: the dips (there are four) in this figure are our exoplanet! We can see this more clearly by *phase folding* the light curve—this means dividing out the orbital period of the planet so these dips line up.[3] I'll use the known period to do this, along with the Python *modulo* operator, which provides the remainder of a division.

```
period_days = 4.887802443
folded_time = (jd % period_days)
fig, ax = plt.subplots(figsize=(13,3))
ax.plot(folded_time,flux_minus_bg,'.',color='k',ms=1)
ax.set_xlabel('Fraction of Phase')
ax.set_ylabel('Flux (background subtracted)')
```

After phase folding, we can see all the dips line up, clearly showing the transit. Notice that we're plotting the same flux array—we have simply replaced our *x*-axis with a cycling set of values. If we look closely at Figure 6.10, we can see that the overall variation in Figure 6.9 has translated into a lack of normalization between these different segments, e.g., one has a normalization of roughly 60 500, while the highest flux segment has a normalization of roughly 60 900. If our light curve in Figure 6.9 had been flat (other than the exoplanet dips), Figure 6.10 wouldn't have this issue. As it is, we want to correct this, and renormalize each of these five segments to a common normalization. Unfortunately, while we can see these five lines by eye, they are still one single flux array.

Thus, we are going to need to chop up our full (background subtracted) flux array in to the chunks we see on our plot. To know where to make these delineations, we can use our folded time array itself (Figure 6.10, top).

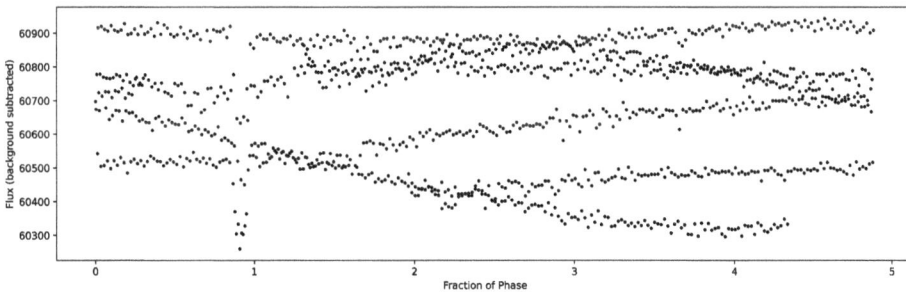

Figure 6.9. Background-subtracted fluxes for the star, now plotted against the "phase-folded" time. This lines up the four transits spread across the light curve in Figure 6.9 at the same location.

---

[3] Aligning the transits also helps characterize its shape, as each individual dip may only have a few measurements at different moments of the transit.

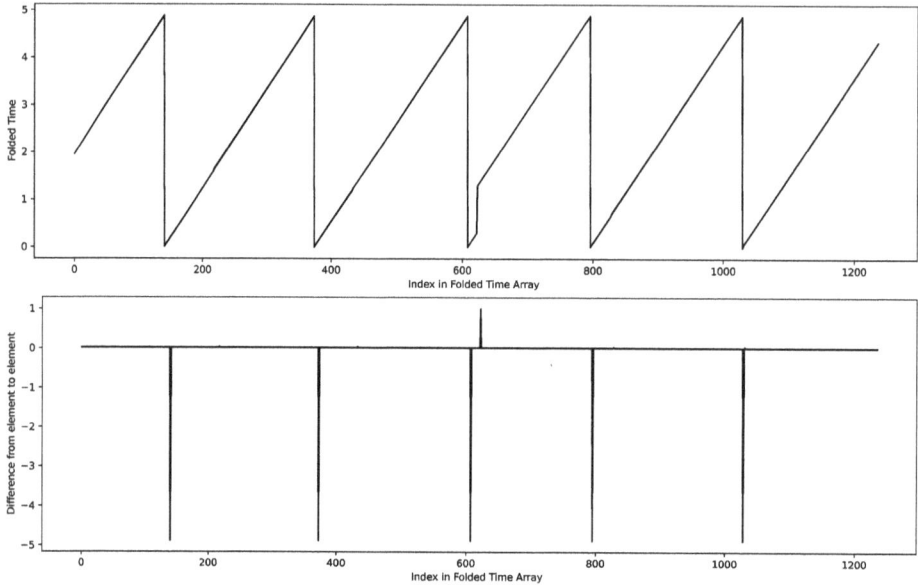

Figure 6.10. Top: Index (element number) of the folded time array, versus the folded time. The drops occur when the folded time cycles back around to its starting point. Bottom: Visualization of the difference array, showing that everywhere the difference is small and positive except where the cycles occur.

```
fig, ax = plt.subplots(figsize=(13,3))
ax.plot(folded_time,color='k',ms=1)
ax.set_xlabel('Index in Folded Time Array')
ax.set_ylabel('Folded Time')
```

As we can see, the phase folding means that as we increase in index number, there are a few key locations where the value of the folded array drops back down to nearly zero and starts over. We need to identify the indices of where this happens. Let's look at the difference between each element and the element before it, by subtracting every element in the array by the value preceding it (except the 0th index):

```
difference_array = folded_time[1:] - folded_time[:-1]
fig, ax = plt.subplots(figsize=(13,3))
ax.plot(difference_array,color='k',ms=1)
ax.set_xlabel('Index in Folded Time Array')
ax.set_ylabel('Difference from element to element')
```

The result of this is seen in the bottom of Figure 6.10. Similar to taking a derivative to find changes in slope, this "finite" difference has nicely highlighted all the locations where our array reset. Put another way, every element is increasing (hence the subtraction to the element before is a small positive number) *except* where it drops down, and the difference becomes negative for one single index. We can now use the np.where() function to identify the location of these negative values:

```
ind, = np.where(difference_array<0)
ind
```

```
array([ 140,  373,  607,  795, 1028])
```

The where() function returns the indices where some condition is True. This is similar to the masking we've covered, but instead of returning a boolean array of length 1237 with true and false values for each element (indicating the true-ness of the condition at those locations), we can now retrieve the indices we need to slice (or chop) the array at those values.

```
print(folded_time[140])
print(folded_time[141])
```

```
4.879834119718751
0.012844665000001143
```

After some quick examination, we can see that indeed, the drops happen between the identified indices and their righthand neighbors. Since there are only five "chunks" to make, we can index manually for convenience:

```
composite_array = np.array([folded_time,flux_minus_bg])
chunk1 = composite_array[:,:141]
chunk2 = composite_array[:,141:374]
chunk3 = composite_array[:,374:608]
chunk4 = composite_array[:,608:796]
chunk5 = composite_array[:,796:1029]
chunk6 = composite_array[:,1029:]
```

To make things easier (i.e., to chunk both the times and the fluxes at once), I created a composite array of the two together (ask yourself: what is the shape of that array?) (Figure 6.11). The 0th axis has length 2 for the two inserted arrays, so we want both of those, and the other axis is the full length of the flux (or time) array,

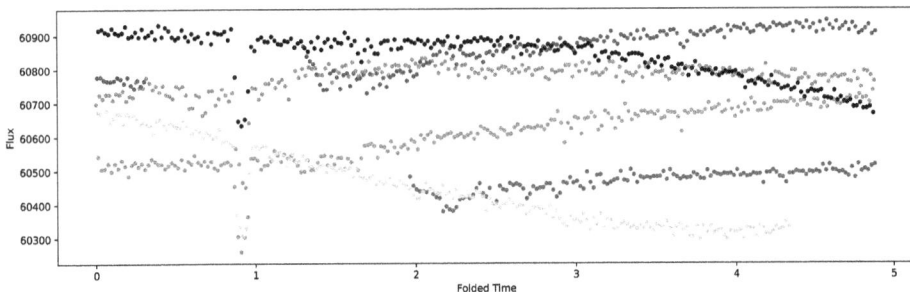

Figure 6.11. Same as Figure 6.9, but now plotting the five separated segments.

which is where we want to chunk on those indices we found.[4] We can now plot each chunk separately to see if we were successful:

```
fig, ax = plt.subplots(figsize=(13,3))
ax.plot(chunk1[0],chunk1[1],'.',color='blue')
ax.plot(chunk2[0],chunk2[1],'.',color='red')
ax.plot(chunk3[0],chunk3[1],'.',color='green')
ax.plot(chunk4[0],chunk4[1],'.',color='purple')
ax.plot(chunk5[0],chunk5[1],'.',color='black')
ax.plot(chunk5[0],chunk6[1],'.',color='orange')

ax.set_xlabel('Folded Time')
ax.set_ylabel('Flux')
```

I plotted each segment using a different color, and we can see we have correctly separated each different chunk in our full array. We can also see now that two of the sections, once folded, did not have any data during the timespan when the transit occurred: the blue curve and the purple curve (chunks 1 and 4). Let's plot again, removing those (Figure 6.12).

```
fig, ax = plt.subplots(figsize=(13,3))
ax.plot(chunk2[0],chunk2[1],'.',color='red')
ax.plot(chunk3[0],chunk3[1],'.',color='green')
ax.plot(chunk5[0],chunk5[1],'.',color='black')
ax.set_xlabel('Folded Time')
ax.set_ylabel('Flux')
```

Now that we have separated chunks, it is going to be much easier to normalize them to each other. At the absolute simplest, we could divide each by their first

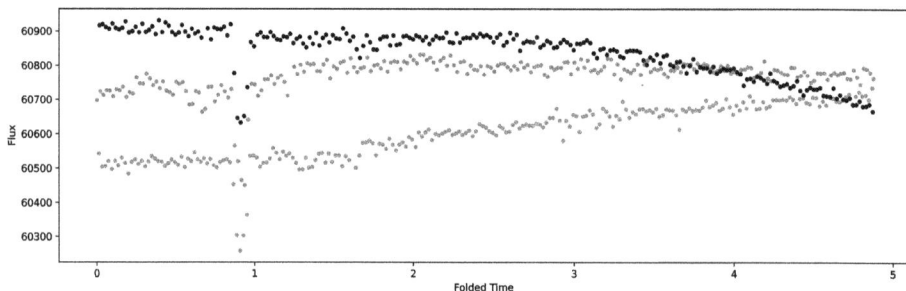

Figure 6.12. Same as Figure 6.11, but with only those segments that contain some measurements during the transit and are relatively flat in that region.

---

[4] In research code, we'd want to not hard code these numbers.

element (or the mean of their first N elements, where the curves are relatively flat). Let's do that now. At this stage, I am also going to remove the orange curve— because it has a definite slope, normalizing by a constant will not be enough to bring it in line with the other segments. Later, we'll learn how to correct this by fitting a polynomial to the continuum and dividing that out.

```python
chunk2_norm = chunk2[1] / np.mean(chunk2[1,:20])
chunk3_norm = chunk3[1] / np.mean(chunk3[1,:20])
chunk5_norm = chunk5[1] / np.mean(chunk5[1,:20])

fig, ax = plt.subplots(figsize=(5,5))
ax.plot(chunk2[0],chunk2_norm,'.',color='black')
ax.plot(chunk3[0],chunk3_norm,'.',color='black')
ax.plot(chunk5[0],chunk5_norm,'.',color='black')
ax.set_xlabel('Folded Time')
ax.set_ylabel('Flux')
ax.set_xlim(0.6,1.2)
```

We now have something that looks a lot like our schematic (Figure 6.13)! You can see that the dip in light (in percentage terms) is very small; during the transit, we received 0.996 (99.6%) of the light as when the planet is not in front of the star. Put another way, TESS detected a $\sim$0.004 (0.4%) dip in the light!

As a final step, let's combine the (good) chunks back into one array that we can save as our final, measured light curve. With lists, the addition operator performs concatenations, but we know that for arrays it performs element-wise addition. So for this, we'll need np.concatenate, which lets us provide sub-arrays to be combined. Because we already separated our chunk fluxes from the times, we can work separately with the times and fluxes (one could combine them if desired).

```python
final_transit_fluxes = np.concatenate([chunk2_norm,
chunk3_norm,
chunk5_norm])
final_transit_times = np.concatenate([chunk2[0],chunk3[0],chunk5[0]])
final_transit = np.array([final_transit_times,final_transit_fluxes]).T
```

When we plot this, we will get a pretty plot like above. But as you may have noticed, the times in our final array are not in order. You can see the confirmation of this in the left panel of Figure 6.14; the folded time array here is cycling.

```python
fig,ax=plt.subplots()
ax.plot(final_transit_times)
ax.set_xlabel('index in array')
ax.set_ylabel('folded time')
```

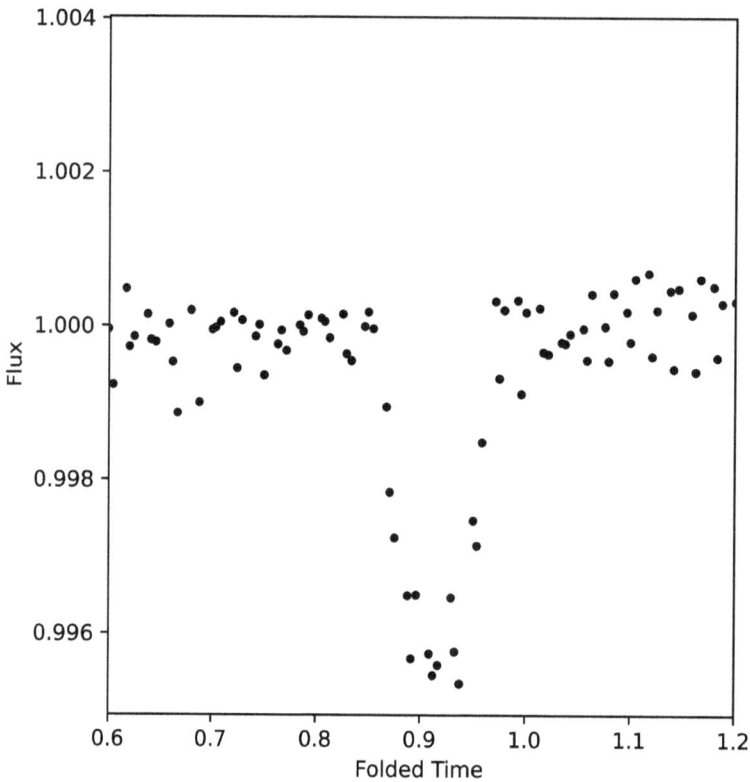

Figure 6.13. Normalized, aligned, and zoomed-in view of the transit. This now closely resembles the schematic in Figure 6.1, save some point to point scatter (from a mix of sources, but in the simplest case, Poisson noise).

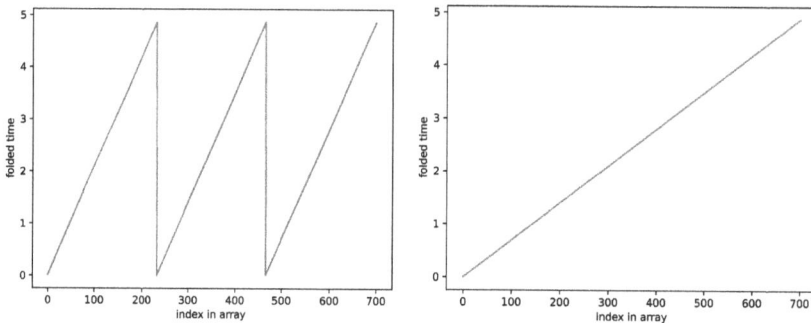

Figure 6.14. Left: Now that we have concatenated several separate chunks together, we see the "sawtooth" structure of cycling values. Now, however, we wish to *sort* these all into a single chronological array. Right: Result of sorting operation on these times—now, they increase monotonically.

In this case, we don't need to play our differencing game because we aren't trying to chunk the array—we don't care where these jumps happen, we just want to sort the entire array. It is straightforward to sort arrays (e.g., via `np.sort()`). Here, we want to *arg* sort (that is, sort an array such that one column is increasing). You may have noticed I transposed the `final_transit` array above; this was to make the times and fluxes be columns (i.e., a final shape of (N,2)). We can sort the whole array by the 0th column by indexing the full array using the indices returned by the `.argsort()` method run only on the 0th column:

```
final_transit_sorted = final_transit[final_transit[:,0].argsort()]
fig,ax=plt.subplots()
ax.plot(final_transit_sorted[:,0])
ax.set_xlabel('index in array')
ax.set_ylabel('folded time')
```

We can check whether the sorting of the times also successfully sorted the fluxes simply by plotting a line chart of the transit—If adjacent points in time are connected, then we know we have sorted our array successfuly (Figure 6.15).

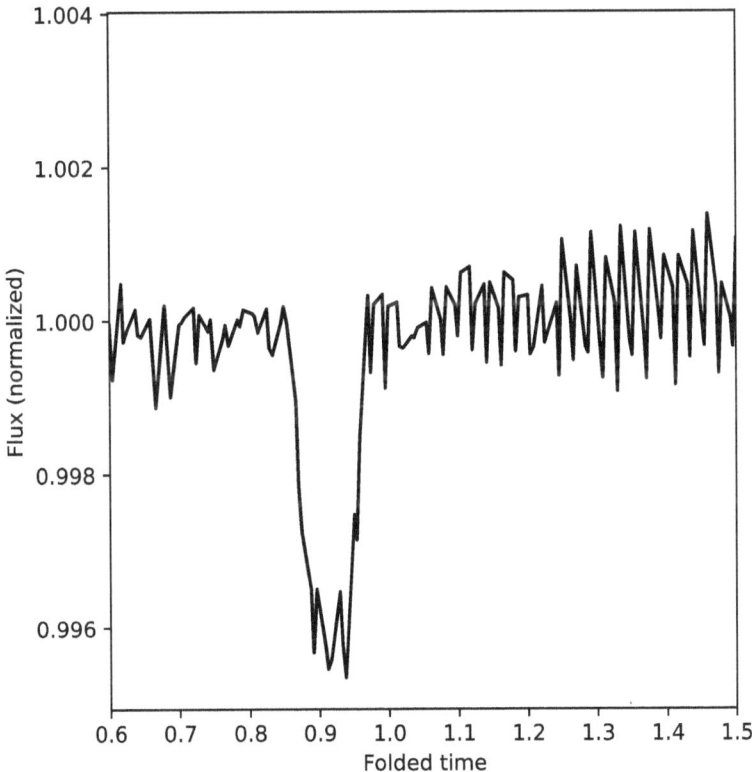

Figure 6.15. Generally, we would plot this transit with points, but here we take advantage of the fact that `Matplotlib` connects adjacent array points together by default—if our array fluxes were out of order but the times in order, we would get a jumble of lines here.

```
fig, ax = plt.subplots(figsize=(5,5))
ax.plot(final_transit_sorted[:,0],final_transit_sorted[:,1],color='k')
ax.set_xlim(0.6,1.5)
```

They were! By plotting lines (which connect all points in the plot) instead of dots, we confirm that the ordering of the data in our array is correct. If we wished, we could now use, e.g., the np.savetxt() function to save our final times and fluxes into a text file on our computer for safekeeping. Re-ordering the data to be chronological also has the benefit that if we now wanted to fit some kind of model to this light curve, we could do so.

How sure are we that this transit is real? By eye, it seems highly unlikely to be spurious (that is, a random few outlier points that are actually noise). It *could* theoretically be sunspot activity, which can also temporarily darken a star. The latter is a question of systematic uncertainty—that is, is the measurement real but the interpretation wrong—and the former is a question of statistical uncertainty. Using a few more Numpy functions, we can address the first.

To get an idea of how big of a deviation from statistical noise our detection represents, we need to know two things: the depth of the transit (by eye, roughly 0.005), and the magnitude of the noise. If we assume that the point to point scatter in the flat, "no-transit" part of the lightcurve is random and Gaussian (more on that later), then the standard deviation of those points is a reasonable estimate of the noise. Let's plot the elements in our final array up to (but not including) the transit (Figure 6.16).

```
beginning_elements = final_transit_sorted[:123,:]
fig, ax = plt.subplots(figsize=(5,5))
ax.plot(beginning_elements[:,0],beginning_elements[:,1],'.',color='k')
```

I've chosen to use this chunk of the light curve because we normalized things to roughly the same section. If we were carrying out this analysis properly, we would need to be careful that our normalization here has not artificially increased or decreased the scatter. But assuming it is reasonable, we can obtain the statistical measure of the standard deviation using np.std():

```
std = np.std(beginning_elements[:,1])
print(f'Noise: {std}')
```

Noise: 0.00028693376279735226

According to our measure, points within 0.0002 of 1 would be considered consistent with being noise (i.e., we could not trust a signal with that kind of magnitude). How many standard deviations away from the mean is our transit depth?

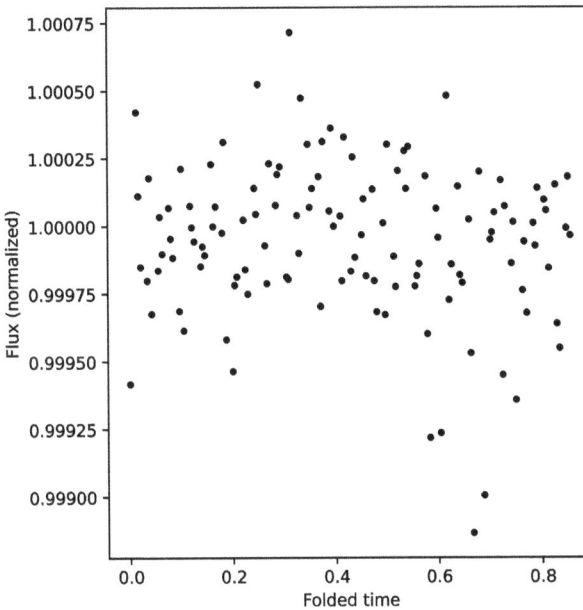

Figure 6.16. Light curve with only the first 122 elements plotted (the next point begins the drop into the transit). These points we assume should intrinsically be 1 in normalized flux, and hence their spread (scatter) is an estimate of the observational uncertainty.

```
0.005/std
```

```
17.425624476026766
```

That's a 17σ detection!

As a note, our standard deviation measure also gives us a way to characterize the uncertainty in the transit depth; if we trust it, the measurement of the depth would be roughly 0.005 ± 0.0002.

6.6 Summary

Throughout this short exercise, we used numerous Numpy functions, paired with arrays and their methods/attributes, to *do* science. There are *hundreds* more functions within the Numpy library that this example did not require us to use, from inserting and appending into arrays, generating arrays of values, and more. This text is not a glossary of those functions; numerous cheat sheets exist to serve this purpose, and in general, any operation you could conceive to carry out on an array of numerical values exists somewhere in the Numpy (or SciPy) library. What is more critical to take away is:

1. The array is a core data-storage unit in Python when working with scientific data.

2. Arrays are typed, and can be multiplied, divided, added, subtracted, exponentiated, logged, and more with each other element by element with high computational speed.
3. Arrays have shapes and can have many dimensions. Many operations, from indexing to aggregations (like means or sums) can be carried out on an entire multidimensional array or only on certain axes, as specified by the user.

The rest of this textbook will leverage Numpy continuously, utilizing and re-utilizing these core principles, and introducing new useful functions as they become relevant.

References

Harris, C. R., Jarrod Millman, K., van der Walt, S. J., et al. 2020, Natur, 585, 357
Lightkurve Collaboration 2018, Lightkurve: Kepler and TESS Time Series Analysis in Python Astrophysics Source Code Library, ascl:1812.013

Astronomical Python
An introduction to modern scientific programming
Imad Pasha

Chapter 7

SciPy

7.1 Introduction

SciPy (scientific Python) is a powerful open-source library for scientific computing and data analysis in Python. It offers a comprehensive suite of functions and modules that enable efficient and reliable numerical routines for a wide range of scientific applications. Where Numpy is primarily an array creation and manipulation library, SciPy has functionality that extends to more complex algorithmic tasks. In short, we handle our data with Numpy, and perform analyses with various tools, SciPy included.

SciPy has grown extensively over the years, and has far too many features and tools, covering a wide domain of scientific applications, to cover here. For the average astrophysicist, SciPy's most useful functions include those for

- numerical integration, differentiation, and solving ordinary differential equations,
- optimization algorithms for finding the minimum or maximum of a function,
- linear algebra operations (e.g., matrix operations, solving linear systems of equations, eigenvalue problems, singular value decomposition),
- signal (image) processing (e.g., filtering, Fourier transforms, interpolation and smoothing), and
- statistical analysis tools (e.g., sampling from single and multivariate Gaussians, assessing statistical measures, fitting distributions etc.).

In this section, we will cover some of the more immediately applicable tools under the SciPy umbrella. Overall, I would argue that most astronomers' data processing tasks rely primarily on Numpy and Astropy, with a few SciPy functions sprinkled throughout. The key difference between Numpy and SciPy is that Numpy focuses on arrays and their manipulation and relatively simple mathematical operations with high efficiency, while SciPy implements more complex algorithms for

doi:10.1088/2514-3433/acfa9ach7   7-1

accomplishing tasks we often need to carry out in research. We won't have space to cover much of what exists in this library, but we will be able to see a few examples that highlight a few of its core modules.

7.2 Numerical Integration

In calculus, we learn how to compute analytic integrals of mathematical functions. This is a common task in astronomy as well. However, we are often faced with functions for which the analytic integral is not known. Additionally, we sometimes want the area under a curve for which we do not know the generating function at all. For both of these cases, we turn to the task of *numerical integration* (i.e., estimating the integral), and the SciPy package has functions for this task which serve most straightforward cases.

Let's use a concrete example to highlight these different tools (and which will also be useful in the next section): a spectrum of an emission line.

A common task, given a spectrum of a galaxy, is to measure the flux (energy per time per area) received in an *emission line*.[1] Recall (or learn now) that a spectrum has units of flux density: energy per time per area *per wavelength* (or equivalently per frequency). An emission line, which is emitted by gas moving at different velocities within a system, has a certain width. To get a flux from the flux densities, we'll need to *integrate* over the width of the line, to rid ourselves of the per-wavelength unit (Figure 7.1).

Below, I'm going to quickly make some fake data for us to use, which roughly approximates what a real galaxy spectrum might look like:

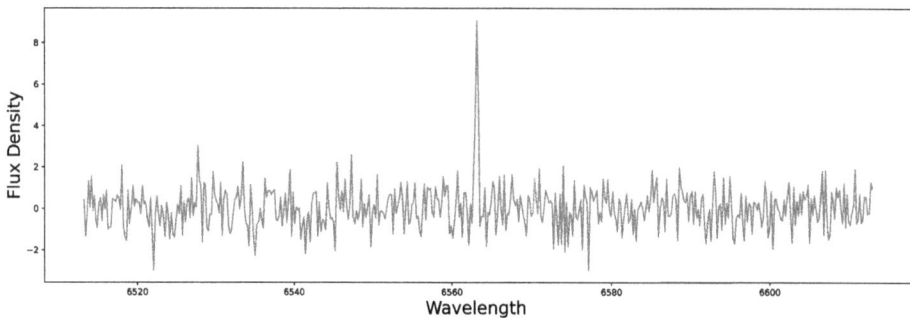

Figure 7.1. Mock spectrum of an *Hα* emission line.

---

[1] Emission lines such as *Hα*, [NII], and [OIII] encode information about the physics going on within the galaxy, e.g., its rate of star formation.

```
import numpy as np
import matplotlib.pyplot as plt
wl = np.linspace(6500,6600,500)+13
flux = np.zeros_like(wl)
flux[250]=10
flux[249]=4.95;flux[251]=5.01
flux[248]=1.2;flux[252]=1.1
flux[247]=0.11;flux[253]=0.10
noise=np.random.normal(0,1,size=flux.size)
flux+=noise

fig, ax = plt.subplots(figsize=(15,5))
ax.plot(wl,flux)
ax.set_xlabel('Wavelength')
ax.set_ylabel('Flux Density');
```

We can see clearly the emission line of interest here, a mock of $H\alpha$ at 6563 Angstroms. The line spans several wavelength elements on our detector. To get the flux of the line, we want the sum of the area under the curve for the section of the spectrum covered by the emission line. Let's zoom in to get a slightly better view of the region we should integrate over:

```
fig, ax = plt.subplots(figsize=(15,5))
ax.plot(wl,flux)


ax.set_xlabel('Wavelength')
ax.set_ylabel('Flux Density')
ax.axvline(wl[246])
ax.axvline(wl[254])
ax.set_xlim(6550,6570);
```

We can see from Figure 7.2 that the vertical bounding lines represent some (reasonable) bounds within which to compute the integral.[2]

The simplest, and most data-driven estimate we can make of this integral would be to to simply sum the values we've associated with the line (being careful to multiply by the width of one pixel, in our wavelength units). This has the value of being a direct measurement, but may also reflect details of our observational setup not intrinsic to the line when it was emitted from the distant galaxy.

One step beyond a simple sum (which, in integral terms, would be the midpoint rule) would be to make some sort of interpolative estimate about the shape of the underlying data at resolution finer than our grid, here defined by the pixels in our detector.

---

[2] Choosing these bounds is nontrivial and shouldn't, in general, be done by eye. Common choices for integration width are the full-width at half maximum of the emission line, or to fit a Gaussian (see below).

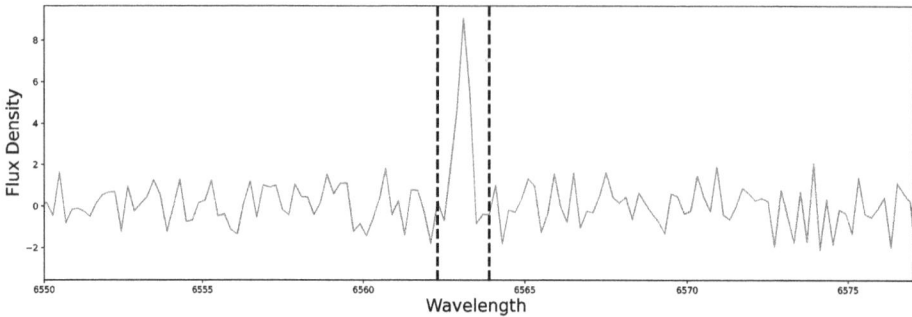

Figure 7.2. Zoomed in region of the *Hα* line, with the region we will be integrating over bounded by vertical dashed lines.

The `SciPy` package has several different algorithms we can use to compute the integral given this input (known as sample-based input). Here I'll cover two: trapezoid rule, and Simpson's rule. Other methods are used similarly.

Let's do the integral. I'm going to index to create sub-arrays which only have the emission line (i.e., within our vertical lines above), and then feed them into the integrate functions:

```
from scipy.integrate import trapezoid,simps
sub_wl = wl[246:254] # range chosen by eye
sub_fl = flux[246:254]

integral_trapz = trapezoid(sub_fl,sub_wl)
integral_simps = simps(sub_fl,sub_wl)

print(f'Trapezoid Rule: {integral_trapz:.4f}')
print(f'Simpsons Rule: {integral_simps:.4f}')
```

```
Trapezoid Rule: 5.1628
Simpsons Rule: 5.2523
```

The two methods have different ways of estimating the area, but produce similar results.

All of these methods are direct estimates of the flux given the data — i.e., we have not assumed a model, or intrinsic shape of the emission line. This has benefits, but also the drawback that our flux measurement may be impacted by, e.g., the gridding of our detector. The asymmetric, sharp, somewhat triangular shape of this line is not what the galaxy emitted. In a moment, we will instead fit a function (a Gaussian) to the line to estimate its flux (as well as its position and width). There is some motivation for asserting the line has an intrinsic shape that is close to Gaussian (or similarly shaped curves); we will not cover this explicitly here. But before we get to that we can prepare by learning about how `SciPy` can integrate Python functions directly.

To fit a Gaussian to our data, we'll need a function that returns the value of a Gaussian with certain parameters for any input x. Recall the formula for a Gaussian form:

$$g = Ae^{-(x-\mu)^2/2\sigma^2}$$

where μ is the location (center) and σ is the standard deviation (width). I've also added an arbitrary scaling, A, which controls the amplitude. The following code *defines* a Python function—we will discuss this much more fully in Chapter 9, so feel free to glance ahead if you want more context to function creation. For now, given this function, we will use it as we use any other function we have imported thus far.

```
def gauss(x,amp,loc,sigma):
    return amp * np.exp(-np.power(x-loc,2.0)/(2*np.power(sigma,2.0)))
```

We can test out our function by attempting to create a Gaussian with some parameters and plotting it. I will use parameters similar to our emission line's true values for convenience (Figure 7.3).

```
xx = np.linspace(6560,6565,1000)
gauss_test = gauss(xx,10,6563,0.25)
fig, ax = plt.subplots(figsize=(15,5))
ax.plot(xx,gauss_test)
ax.set_xlabel('Wavelength')
ax.set_ylabel('Flux Density')
ax.set_xlim(6550,6570);
```

What is the area under *this* curve? Well, we could use the trapezoid or Simpson's rule above on this inherently gridded sampling we've plotted, but there turns out to be a more accurate estimate. We'll now take advantage of SciPy's *functional* integrators, which numerically integrate cases in which we have an explicit function (like our Gaussian) but may not know its analytic integral. Note that we can of course only numerically compute definite integrals.

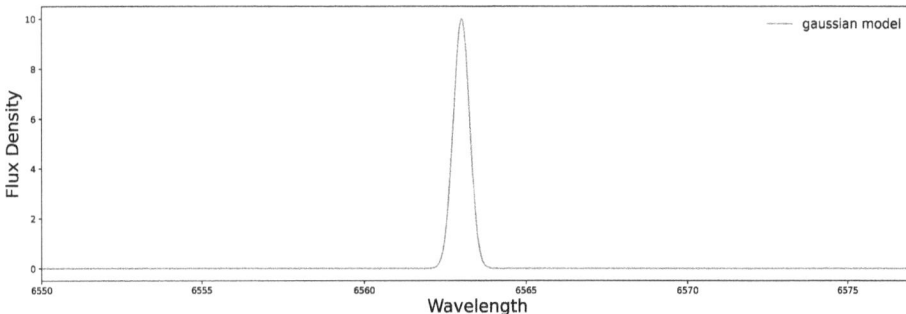

Figure 7.3. Example of the output of our gauss function, for which I have chosen parameters somewhat similar to what our "real" line has.

The most general SciPy integrator for this is quad, which uses a standard Gaussian-quadrature method. This should work for you most of the time until you start getting into more advanced territory. The quad function assumes you have some known, defined function, for which the first argument is x, the input value (or vector) and the rest of the arguments are the parameters, if any. In our example, x will be wavelengths, and the amplitude, location, and sigma are the arguments.

```
from scipy.integrate import quad

res = quad(func=gauss,a=6560,b=6565,args=(10,6563,0.25))
print(f'Numerical Integral with Quad: {res[0]:.4f}')
```

```
Numerical Integral with Quad: 6.2666
```

How close is this to the true answer? Luckily, for a Gaussian, we have an analytic expression for its full integral:

$$\int_{-\infty}^{\infty} Ae^{-(x-\mu)^2/2\sigma^2}dx = A\sqrt{(2\pi\sigma^2)}$$

which we can use to check.

```
true_integral = 10 * np.sqrt(2*np.pi*0.25**2)
true_integral
```

```
6.2665706865775
```

Looks like our numerical integral did a great job estimating the area under the Gaussian!

So, to summarize: if you have a Python function (in 1D) which can compute an output for any input value, you can integrate it over some bounds with quad. Meanwhile, if you have actual data arrays (samples) and don't know the functional equation that generated those samples (i.e., you gathered a spectrum), you can estimate the area under a section of that data using simps or trapezoid (or any of the other options in SciPy).

7.3 Optimization

As we discussed above, numerical integration gives us the ability to estimate the flux of the emission line based on the data we have gathered. But if we believe the actual underlying shape of the line can be defined by some *model* (such as a Gaussian), we may wish to *fit* a model to our data; that is, to find the parameters (A,μ,σ) which produce a curve most consistent with our data. We will talk much more about fitting in a later chapter, but for now, we'll use this example to explore SciPy's optimization module.

Optimization is simply the task of finding a set of parameters which maximizes the value of a function. For example, the optimized value of x for the function $f(x) = x$ is infinity, while for $f(x) = -x^2$, the optimized value is $x = 0$.

Optimization is closely related to fitting, because when fitting, we generally create something called a *loss function* which represents how close some model is to our data. We want to minimize this quantity, so by adding a minus sign, we can turn a fitting problem into an optimization problem. The metric for determining the closeness of our data to some model is up to us, but a χ^2 is a common choice.

Let's use SciPy to find the parameters of the Gaussian that best describes our emission line. I'll import the scipy.optimize.curve_fit() function to do this:

```
from scipy.optimize import curve_fit

res = curve_fit(f=gauss,xdata=wl,ydata=flux,p0=[10,6563,0.25])
res
```

```
(array([8.94793200e+00, 6.56312604e+03, 2.29044917e-01]),
 array([[ 6.63492106e-01,  8.65860692e-06, -1.13135863e-02],
        [ 8.65860692e-06,  5.79807463e-04, -3.04261178e-07],
        [-1.13135863e-02, -3.04261178e-07,  5.79189560e-04]]))
```

We feed the function the name of our Python function to fit, the data (wavelength and flux), and 0, an initial guess at the parameters. This guess is often needed because given the full range of possible values, the optimizer sometimes will not converge. The first output of res is the best fit values, the second is a covariance matrix which contains some (usually not that useful) information about the uncertainty in the fit. Let's plot a Gaussian with the best fit parameters over our data:

```
fig, ax = plt.subplots(figsize=(15,5))
ax.plot(wl,flux,label='data',color='k')

params = res[0]
ax.plot(xx,gauss(xx,*params),label='model',color='r')
ax.set_xlabel('Wavelength')
ax.set_ylabel('Flux Density')
ax.legend()
ax.set_xlim(6550,6570);
```

Figure 7.4 demonstrates our optimizer seems to have done a reasonable job converging on parameter values that describe our data well. If you were satisfied with the fit, you would then compute the area of *this* Gaussian, and quote that as your emission line flux. This Gaussian fit is also a better estimate of other parameters of the line (like its position, which can be used to determine a redshift) than is possible with the raw data.[3]

There are a few drawbacks to using the curve_fit() method. One is that it can be a bit finicky, and sensitive to the input guess, particularly with noisy data. This

---

[3] Note that here, we assume the only transformation of the intrinsic Gaussian is the gridding of our pixels — more advanced fitting may attempt to *forward model* effects from the atmosphere, telescope optics, and detector to actually predict the shape one would see in gathered data.

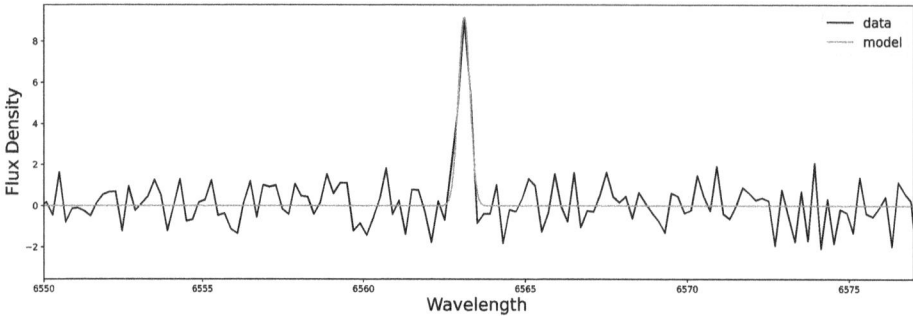

Figure 7.4. Final view of our mock data with the best-fit model Gaussian from `scipy.optimize.curve_fit()` overlaid in red. There may be a bit of confusing code above: using an asterisk on the `*params` list when feeding it into the `gauss()` function. This is a process known as unpacking, and is discussed fully in Chapter 9.

makes it challenging to integrate into a more automated pipeline (say you need to measure the flux of hundreds of emission lines across many spectra, and cannot inspect each along the way to tweak the input guess). Another is that it has no robust way to truly estimate the uncertainty on the parameters. In your analysis, you need to be able to quote the flux as $F = X \pm Y$, (or indeed, have different uncertainties in the upward and downward direction). The covariance matrix returned by SciPy might be a good starting point, but has certain assumptions baked in, and at *best* can only tell you about the formal (statistical) uncertainties between the model and single set of data. Often, the largest uncertainty comes on the values in the data set, which may have both statistical and systematic uncertainties, Thus, the covariance matrix is usually not the full picture. There are many ways to estimate uncertainties, e.g., one can use a method of perturbation and re-measuring, or move to a Bayesian framework in which you adopt priors on the parameters and obtain not just the best fit, but the set of models (and thus parameters) consistent with the data. We will cover this much more extensively in the chapter on inference (Figures 7.5, 7.6 and 7.7).

Exercise 7.1. An SDSS Spectrum

In this exercise, you will try to recreate the steps above, this time to measure an emission line's location in real data. Let's use the **astroquery** package to retrieve the Sloan Digital Sky Survey (SDSS) spectrum for a star-forming galaxy. This will preview several functions and tools we will learn in upcoming chapters; feel free to paste these lines yourself, or refer to the **Astropy** chapter for more. We start by creating a We start by creating a coordinate from the galaxy name, then query the region of that coordinate searching for spectra in the database. We then take the ID of the first

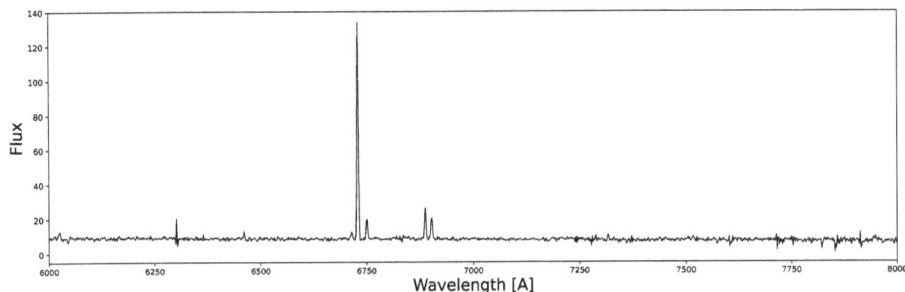

Figure 7.5. A spectrum of a star forming galaxy from the SDSS. In Exercise 7.1, you you will fit an emission line in this galaxy.

match, use the `get_spectra` method to retrieve the list of HDU objects, and pull the spectrum itself from the first extension.

```python
from astropy.coordinates import SkyCoord
from astroquery.sdss import SDSS
coord = SkyCoord.from_name('J140404.9+005953.3')
xid = SDSS.query_region(coord, spectro=True)
spectra = SDSS.get_spectra(matches=xid)[0]
spec = spectra[1].data
```

Let's plot our spectrum to see what we have (Figure 7.5). The variable `spec` is a record-array, which can be indexed by column name. Wavelengths are stored (by default) in log angstroms, so I will exponentiate to get angstroms.

```python
fig, ax = plt.subplots(figsize=(16,5))
ax.plot(10**spec['loglam'],spec['flux'],lw=0.5,color='k')
ax.set_xlabel('Wavelength [A]')
ax.set_ylabel('Flux')
```

If you've learned anything about galaxies, you may recognize several features in this spectrum; namely, as a star-forming galaxy, this system has a collection of emission lines driven by photoionization from star formation. The brightest is $H\alpha$, which we can see near $\sim$ 6563 Angstroms, sandwiched between two [NII] lines and next to two [SII] lines near 7000 Angstroms. There are also some emission lines in the blue—$H\beta$ and [OIII] are readily visible around 5000 Angstrom. The feature at $\sim$5500 Angstroms is an artifact.

Let's zoom in on the region around $H\alpha$, to see what kind of model we might need to create to fit a velocity (wavelength shift) to this data. Fill in the line of code for `ind` using `np.where()` to select the region of the wavelength array between 6600 Å and 7000 Å, and then plot it to get something similar to what I show in Figure 7.6.

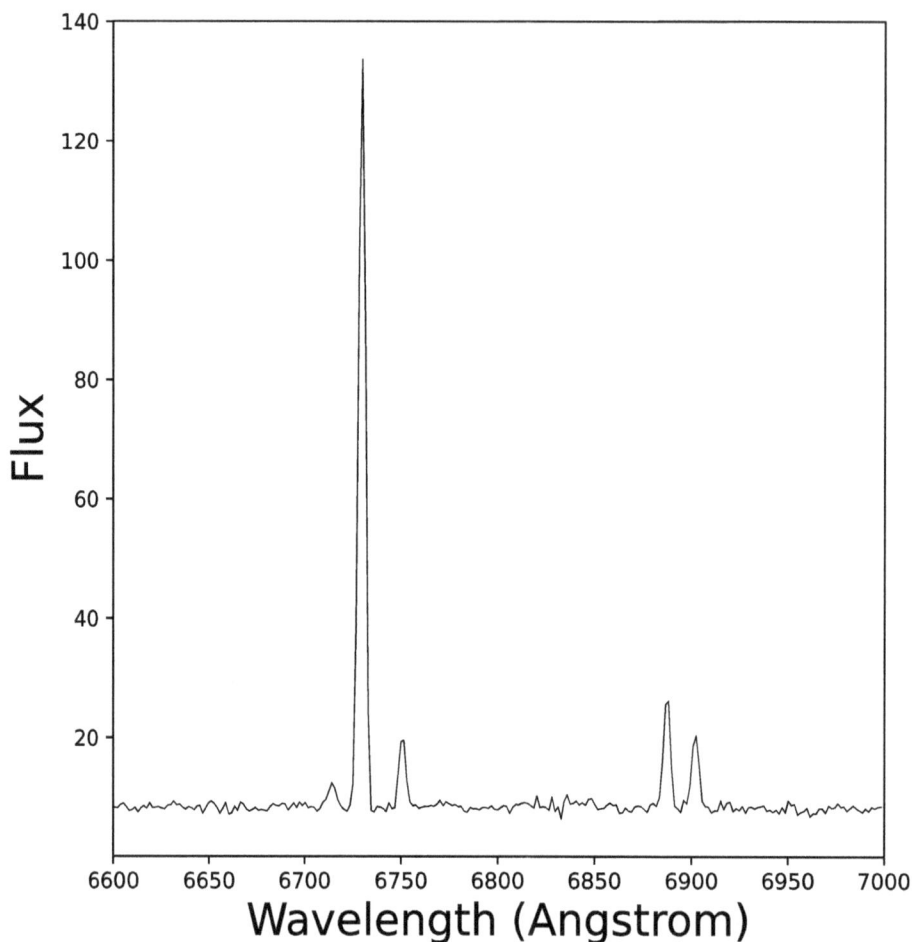

Figure 7.6. Zoom in on the region of the SDSS spectrum surrounding Hα, [NII], and [SII].

```
import numpy as np
ind, = # FILL IN HERE
wl_use = 10**spec['loglam'][ind]
fl_use = spec['flux'][ind]
#Plot Here
```

What kind of model should we fit to this spectrum? In general, one might consider, e.g., fitting all five of these emission lines at once, constraining their $\Delta\lambda$ between one another and fitting for a single shift (allowing each line to bit fit by a Gaussian with some amplitude and width). For this example, however, we are going to simply use the Hα line. You can use the **gauss()** function we created above.

Our goal is to infer the best-fit values for location, scale, amplitude, and offset to match the data—though the scale, amplitude, and offset can be thought of as *nuisance parameters*—we ultimately don't care about their values, and are only interested in the precise location of the line, but will need to fit the other parameters as well to determine it.

We will need our **Gauss** function, as well as a starting guess for the parameters, to initialize the fit. Based on your zoomed in version of the plot, try to choose reasonable starting values. Remember to use the shortened arrays, not the full spectrum!

```
from scipy.optimize import curve_fit
guess = # FILL IN
fit =curve_fit( #FILL IN
```

Make a model using the best-fit parameters and see how it compares to the emission line. You should get something similar to what I show in Figure 7.7.

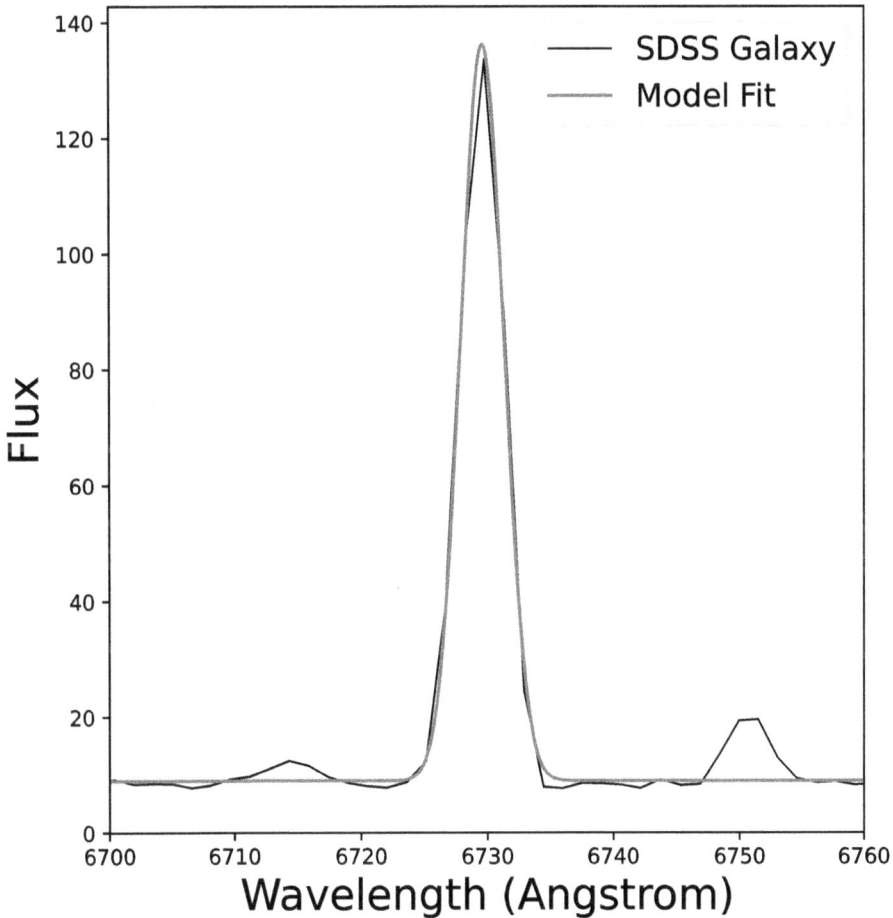

Figure 7.7. Spectrum from SDSS with the best-fit model from SciPy plotted over it. We can see it closely matches the line.

We can see that overplotting the model over the data, we have successfully returned model parameters which fit the emission line well. Note that the region near the peak where the true data seems "angular" is a reflection of the fact that the resolution of the spectrograph is too low to capture the rapid changing in flux for the line. This is why fitting a curve (in this case a Gaussian) to *all* the points in the emission line provides a better estimate of the line center than, say, choosing the pixel with the highest flux, which we can see is slightly redward of our actual best-fit center.

Now free to ignore our nuisance parameters, we can simply take fit[0][0] as our location, and convert this to a redshift. You can use the following function to convert your fit value:

```
def z_from_lam(lam_emit,lam_measured):
    lam_shift = lam_measured/lam_emit
    return lam_shift -1

z_from_lam(6564,fit[0][0])
```

0.025228667428977447

You should find a redshift for this galaxy of $z \sim 0.0252$. The actual redshift denoted on SDSS is 0.0252 ± 0.0001. Is your measurement within the uncertainty quoted by SDSS?

One thing you may notice is that while we have found the single best fit value for the redshift (using this line), we fit the line with a Gaussian-shaped curve, so we have assumed that this shape is at least a reasonable approximation to the intrinsic shape of the emission line. In some contexts for astronomical line fitting, this is *not* a good assumption. Additionally, we have little to no characterization of the *uncertainty* in our redshift measurement.

While the *formal* uncertainties returned by the curve fit will be very small, curve fit doesn't actually *know* about what might be the dominant sources of uncertainty in our result—in cases like this, for example, the wavelength calibration may have an uncertainty orders of magnitude larger than those in our fit. A wavelength calibration (determining the mapping between detector pixels and wavelength) is itself a fit, so uncertainties there will *propagate* to our own fit. While we may fit the emission line nearly perfectly, if the actual wavelength values in our array have large uncertainties, then our redshift determination does as well.

This is why in science, we spend (usually) significantly more time and effort attempting to understand and characterize the many sources of uncertainty propagating through our pipelines than we do actually making the measurement itself.

7.4 Statistics

In some ways, observational astronomy is just statistics wearing a trench coat. An understanding of statistical distributions and properties is essential for interpreting astronomical data. It is thus no surprise that both the `scipy.stats` module and the even more specialized `astropy.stats` modules are very useful to our analyses.

There are two primary avenues by which we use these modules: statistical *measures*, and statistical *modeling*. A statistical measure might be asking for the standard deviation of an array—or for a more complex statistic, such as a robust weighted mean, or a biweight scatter. Statistical *modeling*, on the other hand, involves generating statistical distributions for use in some form. This task is often paired with some sort of statistical framework, such as using the Bayesian formalism to fit models to data.

The `scipy.stats` module provides us the ability to generate a random sample from an N-dimensional distribution,[4] or to compute its probability density function or cumulative density function. It also gives the ability to compute binned statistics of any form on, say, a two dimensional grid of data.

In this section, we'll explore a small subset of what is possible with the `scipy.stats` module—but in practice, your specific research task will determine if and how you leverage this library.

7.4.1 Distributions

We've already seen one example of a `scipy.stats` distribution object: in Section 5.10, when we created a multivariate normal to sample models from the predicted covariance matrix of a polynomial fit.

While the `Numpy` library has a `random` module for generating pseudo-random numbers from some common distributions, `scipy.stats` gives us far more control over which distributions we use and how we use them.

Let's return to our multivariate Gaussian from before and explore it in more detail.

As a reminder, we created our object via:

```
from scipy.stats import multivariate_normal
import numpy as np

cov = np.array([[1,0],[0,1]])
norm = multivariate_normal(mean=[0,0],cov=cov)
```

Here we have created a covariance matrix with no covariance (i.e., no off-diagonal terms), set the location (mean) of the normal to the origin, and the standard deviation in each direction to be 1. We can visualize this, as before, using `np.mgrid` and a `contourf`:

---

[4] For example, a Gaussian—but the module has dozens of different distributions to choose from.

```
import matplotlib.pyplot as plt

x,y = np.mgrid[-6:6:0.1,-6:6:0.1]
pos = np.dstack((x,y))
fig, ax = plt.subplots(constrained_layout=True,)
ax.contourf(x,y,norm.pdf(pos))
ax.set_box_aspect(1)
```

As we can see in Figure 7.8, the probability density function of this multivariate Gaussian peaks at (0,0), has no covariance, and 68% of the probability is contained within the circle that has a radius of 1.0 (at least, it seems so by eye). If we modify the standard deviation in either axis, we'll get elongation either up and down or side to side, and if we add off-diagonal covariance terms, we will get a diagonal distribution.

The PDF() method of a distribution object can tell you, for some position in the parameter space of the PDF, the likelihood of that position.

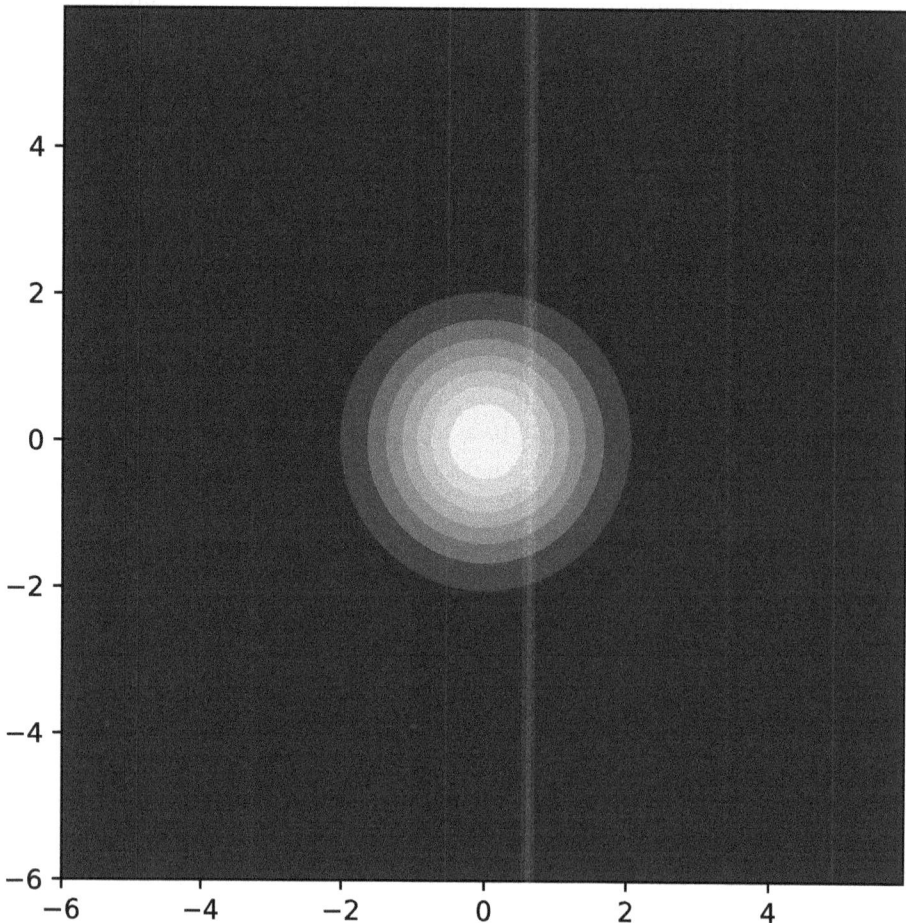

Figure 7.8. Result of creating a multivariate Gaussian with a location of (0,0) with a standard deviation of 1 with no covariance.

The inverse task—randomly sampling some value from the PDF—is accomplished with the rvs() method (random variable sample).

```
norm.rvs()
```

```
array([ 2.47123146, -0.66851784])
```

Because we have a two dimensional distribution, a sample from it has two values. Here, we seem to have drawn an *x*-value that (generally) has quite low probability (it is outside of 1σ). The other coordinate, however, is well within. In theory, if we sampled enough times from something which had this distribution, we would "recreate" the PDF. Let's try a less intense version of that, and see where 100 randomly drawn samples would fall in this parameter space:

```
fig, ax = plt.subplots(constrained_layout=True,)
samples = norm.rvs(size=100)
for i in samples:
    ax.plot(i[0],i[1],'x',color='w',ms=10)

ax.contourf(x,y,norm.pdf(pos))
ax.set_box_aspect(1)
fig.savefig('scipy-stats-2.pdf')
```

A few of our random samples fall far away from the center of the distribution (Figure 7.9), but as expected, most fall somewhere in the 2σ region of the distribution.

Thus far, we have evaluated the PDF for some value, and obtained random samples from the distribution. We can also ask the question "what is the probability that a random sample takes a value less than X". Such questions are often used when assessing the statistical likelihood that some experimental result can reject a null hypothesis, or whether two distributions are statistically different from one another (e.g., a Kolmogorov–Smirnov, or KS, test). In order to answer it, one needs to know the *cumulative distribution function* (CDF). This can be computed by integrating the PDF, and every SciPy distribution has a method that can return this for you. For ease of interpretation, we'll switch to a 1D normal distribution:

```
from scipy.stats import norm
norm_1d = norm(loc=0,scale=1)

print(f'Prob x < 5: {norm.cdf(x=5)}')
print(f'Prob x < 2: {norm.cdf(x=2)}')
print(f'Prob x < 1: {norm.cdf(x=1)}')
```

```
Prob x < 5: 0.9999997133484281
Prob x < 2: 0.9772498680518208
Prob x < 1: 0.8413447460685429
```

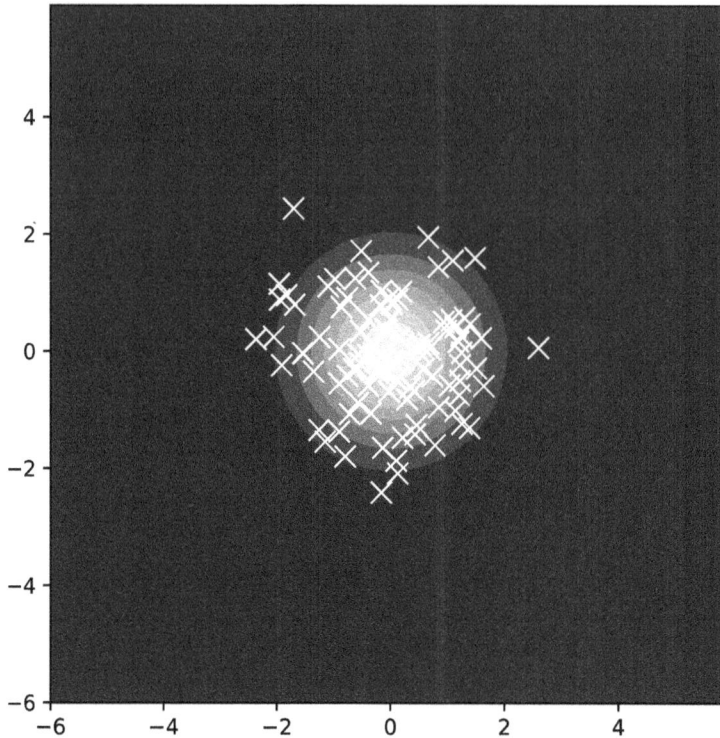

Figure 7.9. One hundred random draws from the distribution, plotted over the PDF.

As we can see, the probability that a random sample from this distribution takes a value less than 5 is essentially 1, while there is only a 84% chance a random sample has a value less than 1. Recall: we set σ for our distribution to 1, such that 68% of the probability is contained between -1 and 1. The remaining 32 percent is split between the values above 1, and the values below -1. The values below -1 contribute the CDF (they are less than our chosen X), while the portion >1 does not. You should be able to convince yourself that 84% of the total probability sits below a value of $+1$. This math may help:

```
(1-.68)/2 + 0.68
```

```
0.8400000000000001
```

There are several other useful functions related to the CDF, such as the survival function (SF) which is defined as 1-CDF, and the percentile point function (PPF), which is the inverse of the CDF. All of these can be directly computed with methods of the `stats` object. One can also compute point statistics such as the distribution's mean, median, and variance, moments of any order, or even *fit* a distribution to data.

For a 1D Gaussian with mean 0 and $\sigma = 1$, these calculations are not so impressive (we know them off the top of our heads, or know how to calculate them).

But when in real research we are faced with a multidimensional covariant distribution that may not even be Gaussian, having the ability to compute these quantities, generate samples, and evaluate PDFs is incredibly useful.

7.5 Summary

In this chapter, we introduced the SciPy library. Where NumPy is our go-to library for storing numerical data and performing mathematical operations on them, SciPy provides some higher level, more complex algorithms that allow us to carry out tasks including numerical integration, statistical distribution analysis, and curve fitting.

There are other useful SciPy functions we did not cover in this chapter; in particular, there are useful methods for carrying out image tasks like smoothing and convolving, which are very useful to astronomers. In general, while all scripts we write will invoke NumPy (and usually Matplotlib), only some will need to invoke SciPy, and usually at the level of individual functions or sets of functions within the broader library.

Astronomical Python
An introduction to modern scientific programming
Imad Pasha

Chapter 8

Astropy and Associated Packages

8.1 Introduction

The `Astropy` library (Astropy Collaboration 2013, 2018, 2022) is an extensive and well-maintained repository of useful functions and classes for astronomers. `Astropy`'s greatest strength is in its most basic components, which greatly ease working with astronomical data of all kinds. As described on the `Astropy` documentation page, it is "a community effort to develop a common core package for Astronomy in Python and foster an ecosystem of interoperable astronomy packages."

In this chapter, we will cover some of the useful core elements of the `Astropy` package, as well as highlight that interoperability with several `Astropy`-affiliated and `Astropy`-coordinated packages (e.g., `photutils`, `astroquery`). The distinction between these designations regards whether the codebase in question is formally controlled by the Astropy Collaboration. But all coordinated and affiliated packages are designed to work seamlessly within the ecosystem of `Astropy`'s core functionality.

8.2 Units and Constants

Units, in many ways, are the bane of an astronomer's experience. We have a litany of units, some of which are quite complicated (like erg s^{-1} cm^{-2} $arcsec^{-2}$ $Å^{-1}$), some of which are logarithmic (like magnitudes), and some of which do not have "straightforward" conversions (such as F_ν to F_λ). On top of that, many physical equations utilized in astronomy include constants (which themselves have sometimes-complicated units).

It is thus quite useful that `Astropy` supplies libraries of units and constants. Attaching units to something makes it a `Quantity`, and `Quantity` objects can be used in equations together (added, multiplied, exponentiated, etc.) and `Astropy` will track the units under the hood, producing an output `Quantity` that "knows" the units it is in.

That output `Quantity` has a handy `.to()` method which allows you to convert to other units—and helpfully, `Astropy` will raise an exception if the units you have

doi:10.1088/2514-3433/acfa9ach8

specified are inconsistent with the units of the input quantity. This makes it much easier to catch when a mistake in an equation or calculation has occurred.[1]

There are several ways to attach units to a value, but the easiest is as follows. Let's say that we were told that the luminosity of a star was 3 $L_\odot$, and the star was 1.3 kpc away in the Galaxy. What is the flux of that star we would detect, in $erg\ s^{-1}\ cm^{-2}$?

This is a problem that might appear on an introductory undergraduate astronomy problemset, and would typically involve some hand-calculator unit conversions, namely, we'd need to convert solar luminosities to $erg\ s^{-1}$, and we would need to convert *kpc* to *cm*. With `units`, we don't have to worry about such trivialities:

```
L = 3*u.Lsun
d = 1.3*u.kpc
F = L / (4*np.pi*d**2)
F.to(u.erg/u.s/u.cm**2)
```

$5.6793093 \times 10^{-11}\ \frac{erg}{s\ cm^2}$

There are other ways to define a `Quantity` (e.g., with `x = u.Quantity(value,'unit')`, but simply multiplying by the unit as above works just fine most of the time.

One more introductory question: can we plot a blackbody curve for a source with a temperature of 5600 K?

We know Planck's Law:

$$B_\nu(T, \nu) = \frac{2h\nu^3}{c^2} \frac{1}{e^{h\nu/k_B T} - 1}$$

By adding in some of the `astropy.constants`, which are unit-having quantities, we can easily do this:

```
import matplotlib.pyplot as plt
import astropy.constants as ac
import astropy.units as u

def Bnu(T,nu):
    return (2*ac.h*nu**3) / (ac.c**2 * (np.exp(ac.h*nu/(ac.k_B*T))-1))

T = 5600 * u.K
wl = np.linspace(0.1,3,1000) * u.micron
nu = ac.c / wl # convert wl to frequency

B_plot = Bnu(T,nu)

fig, ax = plt.subplots(figsize=(7,7))
ax.plot(wl,B_plot)
ax.set_xlabel(r'Wavelength ($\mu$m)',fontsize=20)
```

---

[1] Just be sure when using units this way that every quantity going in that is not dimensionless *does* have units attached.

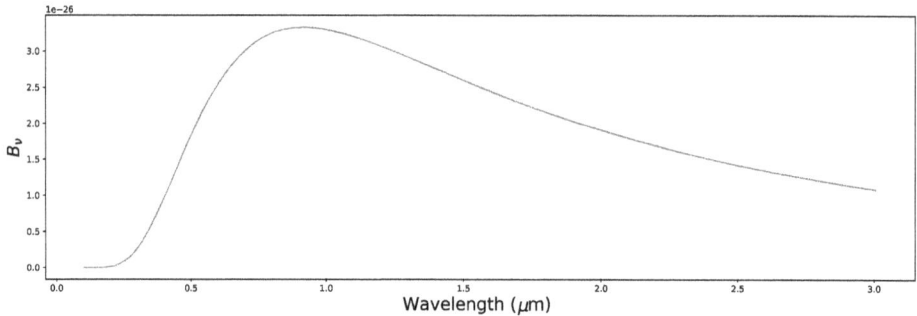

Figure 8.1. Planck curve derived using `Astropy` units and constants.

What units are we plotting on the Y-axis of Figure 8.1? We can easily look at `B_plot` to see what units it has:

```
B_plot
```

$[2.7607048 \times 10^{-33}, 5.2300762 \times 10^{-33}, ...] \frac{J\ m}{\mu m^3}$

If we wish, we can convert these units to something else:

```
B_plot.to(u.erg/u.s/u.cm**2/u.Hz)
```

$[2.7607048 \times 10^{-12}, 5.2300762 \times 10^{-12}, ...] \frac{erg}{Hz\ s\ cm^2}$

If we had done our math wrong, and our units weren't right (let's say our output was supposed to be a luminosity), we would have an error:

```
B_plot.to(u.erg/u.s)
```

```
...

UnitConversionError: 'J m / micron3' (spectral flux density) and 'erg /
    s' (power/radiant flux) are not convertible
```

For many functions, Astropy quantities can be passed through them without issue. But for some, this is not the case. For example, if we wanted to take the log of our temperature variable:

```
np.log10(T)
```

```
...

UnitTypeError: Can only apply 'log10' function to dimensionless⌴
    ⌴quantities
```

We get a warning that we can only apply `log10` functions to dimensionless quantities. At any time, we can "strip" the units from a `Quantity` by querying the `.value` attribute:

```
print(T)
print(T.value)
```

```
5600.0 K
5600.0
```

So, if you are working within a pipeline or framework for which unit quantities are causing problems, it is possible to perform all the necessary unit conversions and checks, then strip the units before feeding into the external tool. Assuming the tool documents what units will come out at the *end* of that process, you can then manually add units back in.

A final piece to note about units: there are some conversions in astronomy that do not make sense when looking at the units directly, but because of the way those units are defined, are possible. The classic example of this is parallax:

$$d = \frac{1}{p}$$

This formula works only when p is in arcseconds and d is in parsec, (it is, indeed, linked to how a parsec is defined). But if we try to use this in code, we'll have an issue:

```
p = 2*u.arcsec
d = (1/p).to(u.kpc)
```

```
...
UnitConversionError: '1 / arcsec' and 'kpc' (length) are not convertibl
```

For a set of these common, definition based conversions, **Astropy** has an *equivalencies* module which facilitates these conversions:

```
d = p.to(u.kpc,equivalencies=u.parallax())
d
```

0.0005 *kpc*

Note that we can directly convert the parallax angle into a distance. Another classic example is spectral units, which often need to switch between wavelength and frequency units. In our example above, we explicitly used the speed of light to compute the conversion. But we could also do it directly:

```
nu = wl.to(u.Hz,equivalencies=u.spectral())
```

The use cases can be more complex as well. Once you hit that level, the documentation for **Astropy** will be the best place to find more.

8.3 Cosmological Calculations

When computing astrophysical quantities measured beyond the local universe, calculations must take cosmology into account due to the expansion history of the universe. Many of these formulas will be taught in your first cosmology course, but thankfully, you do not have to implement them in Python yourself.

In order to carry out calculations that rely on cosmology, we must first *choose* a cosmology. Different instruments and surveys, e.g., the 7 or 9 year WMAP measurements, or the Planck 2018 measurements, have slightly different measured values for the relevant cosmological values (H_0, Ω_m, Ω_Λ, T_{cmb}, etc). For many calculations, the exact cosmology we choose doesn't strongly matter—though if you are working with simulation data, you should obviously use whichever cosmology the simulation was run with.

For these examples, I'll use the WMAP-9 values, to be consistent with the `Astropy` documentation.

```
import numpy as np
from astropy.cosmology import WMAP9 as cosmo
```

Once we have a cosmology imported (or created), we can carry out standard calculations, including comoving distance:

```
cosmo.comoving_distance(z=0.55)
```

2080.4774 Mpc

These calculations become important when we attempt to measure distance-dependent quantities, e.g., the $H\alpha$ emission line luminosity for a galaxy at non-zero redshift. To convert the flux of that line into an intrinsic luminosity, our formula

$$L = 4\pi D_L^2 F$$

now uses a luminosity distance instead of a proper distance. Put another way, the luminosity distance captures the actual flux we would measure from a source of intrinsic luminosity L in the presence of an expanding universe. This quantity is related to several other cosmological distance and size measurements, including the angular diameter distance.

We can calculate the luminosity distance to a source at redshift z via:

```
cosmo.luminosity_distance(z=2.1)
```

16855.095 Mpc

Exercise 8.1. Galaxy Flux

What would be the flux received from a source at redshift 2.5 (in erg s^{-1} cm^{-2}) if its intrinsic luminosity was 10^{10} $L_\odot$?

Another common measurement task involves estimating the physical size (in kpc or pc) of a galaxy or feature in a galaxy at intermediate or high redshifts. For this, we need to know the angular diameter distance — a quantity which relates the on-sky angle subtended by an object and its true physical size, in the presence of a given cosmology. In a Euclidean scenario (e.g., in the very local universe or on Earth) this relation would simply be

$$l_{feature} = D \tan(\theta) \approx D\theta,$$

where l is the physical length of the feature and D is the distance to the object.

A fun fact about angular sizes in our universe is that objects that are farther away appear smaller (as in our daily experience)... to a certain redshift. That turnover is around redshift $z \sim 1.6$, beyond which objects actually begin to appear slightly *larger* again! This fact is of what allows us to still make meaningful morphological measurements of galaxies at extreme redshifts.

We have a similarly useful `Astropy` function to compute the angular diameter distance:

```
d_A = cosmo.angular_diameter_distance(z=0.8)
d_A
```

1574.6296 Mpc

The relation between angular diameter distance and physical scale is

$$d_A = \frac{x}{\theta}$$

so we can compute the physical size of a feature on a galaxy at this redshift. Let's assume our feature is measured by HST to be 0.1 arcseconds across. The physical scale would be:

```
((0.1*u.arcsec.to(u.rad))*d_A).to(u.pc)
```

763.40196 pc

Our feature is thus ~0.763 kpc across.

Exercise 8.2. Getting Bigger Again

At what redshift beyond $z = 0.8$ would a feature of 0.1 arcseconds in angle once again correspond to ~0.763 kpc in physical units?

It may be helpful to define an array of redshifts (from 0 to ~10 and the corresponding sizes for this problem. Plotting size as a function of redshift will help you visualize how to obtain the solution.

8.4 Coordinates

For anyone working with astronomical data, coordinates are essential. Depending on what you are studying, you may use one of several different coordinate systems to uniquely identify different locations on the sky. Astropy provides an extremely useful module to handle coordinates, allowing you to store, plot, and carry out conversions relying on coordinates. The default selected coordinate system is the equatorial system, which is centered on either the Earth or Sun, using the celestial equator and poles as fundamental zeropoints, and units of right ascension (or hourangle) and declination. The second most common way to describe the positions of astronomical objects is the galactic coordinate system, which is centered on the Sun and uses the galactic plane as its fundamental plane and the line connecting the Sun to the galactic center as its longitude axis.

Let's create a SkyCoord object. One helpful static method of this object class[2] is the ability to create a coordinate based solely on the name of an astronomical object, assuming the name is recognized by the large, online astronomical database Simbad.

```
from astropy.coordinates import SkyCoord
coordinate = SkyCoord.from_name('M81')
coordinate
```

```
<SkyCoord (ICRS): (ra, dec) in deg
    (148.8882194, 69.06529514)>
```

Let's see what happens if we try a "bad" name:

```
coordinate = SkyCoord.from_name('not-an-object')
```

```
...
NameResolveError: Unable to find coordinates for name 'not-an-object'
    using http://cdsweb.u-strasbg.fr/cgi-bin/nph-sesame/A?not-an-object
```

So why are coordinate objects so useful? It looks on the surface like our coordinates are just two floats: 148.888 219 4 and 69.065 295 14. Our coordinates have these values in the equatorial coordinate system with RA and Dec in degrees, but we may also need to know our coordinates using different units—e.g., RA in hours, minutes, and seconds, and DEC in degrees, minutes, seconds. We can retrieve those values with a built-in method:

```
coordinate.to_string('hmsdms')
```

```
'00h10m26.11464248s +12d28m40.06794061s'
```

We can also change between entirely different coordinate systems. We can convert our coordinates to galactic coordinates easily:

---

[2] We will learn more about static methods when we discuss Python classes.

```
coordinate.galactic
```

```
<SkyCoord (Galactic): (l, b) in deg
    (107.52256361, -49.16327228)>
```

And of course, we can change the *units* of *l* and *b* as desired from degrees to other angular units.

Beyond the convenience of computations and representations of themselves, SkyCoord objects are also useful when you need to compute quantities that depend on *multiple* coordinates. For example, an extremely common task as an observer is the computation of the offset (angle) between two coordinates. This computation comes into play when, e.g., you are attempting to point a telescope at an object too faint to be seen in short exposures (and for which you are preparing to take very deep imaging or spectroscopy). Telescope pointings are not perfect, and so the solution to this problem is to instead slew to a brighter point source (usually a star) that is close to the invisible target. Careful alignment of the slit or imager onto that star can then commence, followed by a *blind offset*, in which the telescope is nodded the exact, known offset between the star and the target.

More than this, we most often need the offset in a specific form, such as arcseconds east, and arcseconds north, meaning the single angle between the two points must be decomposed into components which align with the equatorial coordinate grid.

Luckily, what would otherwise be a somewhat involved manual calculation can be carried out trivially using Astropy:

```python
import astropy.units as u
offset_star = SkyCoord(ra=148.88917,dec=69.06114,unit='deg')

offset = offset_star.spherical_offsets_to(coordinate)
offsets = [i.to(u.arcsec) for i in offset]
offsets
```

```
[<Angle -1.22275076 arcsec>, <Angle 14.95851347 arcsec>]
```

These quantities are exactly what we would need to execute a blind offset at the telescope. There are numerous other convenience functions accessible to coordinate objects that facilitate similar types of computation.

8.5 Astroquery

For many years, the only way to work with astronomical data sets, and to extract values from a catalog, was to download the entire catalog file to your computer and import it into Python. Otherwise, surveys may have made a webform available online in which you could type conditions and a server would return a data file to you.

Now, there are many large scale astronomical surveys which have a code-based API for requesting and receiving subsets of those survey catalogs directly within

Python analyses. So rather than loading a several-gigabyte file of objects and performing a conditional search, you can use a package like `astroquery` to ask for an online server to deliver only the catalog entries that match some conditions.

This package, which is affiliated with `Astropy`, supports an ever-growing list of services it is compatible with which can be queried with a reasonably uniform API.

A few which might be the most useful day-to-day include

- the GAIA mission, which provides precise positions and proper motions for millions of stars,
- Simbad, which is the astronomical object database used to parse names in `SkyCoord`s and which has basic information thousands of objects,
- NED (NASA Extragalactic Database), which has information about extragalactic objects as well as extinction measures for different sightlines,
- MAST, the STScI archive with data from JWST, HST, TESS, Kepler, and (soon) the Nancy Grace Roman telescope, and
- SkyView, which is a cutout service often useful for making finder charts.

Usage for each differs slightly, but the documentation can be followed to retrieve most general queries using those services. In the following research example, we'll be utilizing the GAIA TAP+ database to automatically find blind offset stars and compute the relevant offsets.

8.6 Research Example: Automatic Offsets

In our preparations for a given observing run, we may wish to find a suitable offset star to use for each of our targets. We know that we want to select stars that are both bright enough to use (and see in short guider exposures) but also which have small offsets to our targets. Put another way, we want to select, from a set of all stars proximate to our target, the closest star brighter than a certain magnitude cutoff.

For this example, I will use the `Gaia` lookup table using an angular cone search to find our potential list. We can easily filter this list for only those brighter than our cutoff. After that, we will create a `SkyCoord` object for these stars, and compute the offsets from each to the target object, and choose the star with the smallest overall offset.

Note that here we will be inserting a vector of coordinates into one `SkyCoord` object—`SkyCoord` objects need not contain only a single coordinate, and in fact, if your situation calls for a set of multiple coordinates to be used for some calculations, you *should* add them all to a single `SkyCoord` object. This is beneficial in certain cases, especially when you would otherwise be looping over multiple coordinate objects to carry out a calculation, due to performance increases that happen under the hood when `Astropy` can treat the internal set of values as arrays.

In the code block below, we will import the Gaia table and set which version of the catalog to use. I'll also create a handy custom exception (see 14.4) to use in our function. The key lines will be within the `find_offset_star()` function.[3]

```python
from astroquery.gaia import Gaia
Gaia.MAIN_GAIA_TABLE = "gaiadr3.gaia_source"
import numpy as np
class NoMatchesError(Exception):
    pass

def find_offset_star(target_coordinate:SkyCoord,
                     cone_radius:u.Quantity = 1*u.arcmin,
                     limiting_magnitude:float=15.0):
    j = Gaia.cone_search_async(target_coordinate, radius=cone_radius)
    r = j.get_results()
    r = r[r['phot_g_mean_mag']<limiting_magnitude]
    if len(r)==0:
        raise NoMatchesError('No stars match the search criteria.')
    star_coords = SkyCoord(ra=r['ra'],dec=r['dec'],unit='deg')
    all_separations = target_coordinate.separation(star_coords)
    ind = np.argmin(all_separations)
    best_star_coord = \
SkyCoord(ra=r['ra'][ind],dec=r['dec'][ind],unit='deg')
    final_offset = best_star_coord.\
spherical_offsets_to(target_coordinate)
    offsets_arcsec = [i.to(u.arcsec) for i in final_offset]
    return best_star_coord,offsets_arcsec
```

```python
coord_star,os= find_offset_star(coordinate)
```

```
INFO: Query finished. [astroquery.utils.tap.core]
```

```python
os
```

```
[<Angle 39.23849223 arcsec>, <Angle 4.08555788 arcsec>]
```

The key pieces of the above "algorithm" are the use of the `SkyCoord.separation()` method for the *list* of coordinate stars (to speed things up),[4] then selecting the star with the minimum distance and returning a single `SkyCoord` to the user, along with the offsets. We also define a quick helper exception which tells the user if the reason the function fails is that there are no stars brighter than the chosen limiting magnitude within the chosen search cone.

---

[3] If you are unfamiliar with functions, see Chapter 9. The code within this function can also be copied and pasted into Python directly, so long as you define some variables matching the function's argument names.
[4] The separation method is similar to the offsets method, but provides the single angular difference instead of the components.

8.7 Research Example: Handling Astronomical Images

Working with astronomical images requires the use of many core elements of the `Astropy`, from `FITS` file handling, to `WCS` (World Coordinate System) and coordinates management, to unit tracking, to cropping and cutouts. Additionally, many basic image measurements can be made with the `Astropy`-affiliated package `photutils` (such as aperture photometry). So astronomical images make for a great testbed for introducing these elements of the library, and thus we will use some images from JWST for this research example.

From the first handmade drawings of early astronomers to glass-plates and now CCD detectors, using telescopes to image objects in the night sky has been a core element of astronomical study. For the last few decades, the Hubble Space Telescope (HST), which orbits above Earth's atmosphere and has a respectable 2.4 meter mirror, has been the workhorse for extremely high resolution imaging of the universe in ultraviolet and optical wavelengths. It has produced some of the most beautiful images of space known, from distant galaxies to nearby star-forming regions. On December 25, 2021, the James Webb Space Telescope (JWST) was launched to a distant orbit. This telescope surpasses HST's light gathering power significantly, with an effective 6.5 meter mirror design, and is tuned to longer wavelengths (near-infrared through mid-infrared) than HST was, opening a new electromagnetic window at high resolution in the nearby universe, as well as unlocking access to light from the first galaxies and stars in the universe, whose light has been redshifted by cosmic expansion far into the infrared (Figure 8.2).[5]

Images taken in astronomical contexts are similar, but distinct, from images one may be used to taken by, e.g., a cellular phone or digital camera. The primary difference is that astronomical detectors are *monochromatic*—each pixel on the detector can be thought of as a simple "light bucket," building up a signal generated by photons from astronomical targets, the sky, and instrumental effects (like dark current). By itself, the pixel has no way of knowing the energy or wavelength of a photon (other than being more sensitive to registering some photons than others). Thus, an astronomical detector, by itself, may be sensitive to a large range of wavelengths but will contain no color information—that is, information about how much brighter or fainter an object is at one wavelength versus another.

In contrast, typical digital cameras actually employ detectors in which individual pixels are independently sensitive to different channels (also known as bandpasses, or chunks of the electromagnetic spectrum).[6] One can imagine a tiny, pixel-sized filter over each pixel, which blocks all light outside of the relevant bandpass (Figure 8.3). Any given pixel thus receives information about intensity (the amount of light hitting the pixel) in only one band. In order to create a full color image

---

[5] You can (almost) always distinguish a JWST image from an HST image as HST images have stars/bright sources with four diffraction spikes, while JWST has six.

[6] Usually these are Red, Blue, and Green, with twice as many green-sensitive pixels as the other two.

Figure 8.2. A JWST "color" image of the edge of a nearby, young, star-forming region (NGC 3324) in the Carina Nebula. The image represents several different images at different wavelengths being combined, and given that all the wavelengths are in the infrared, the human eye would not be able to see *any* of this emission. However, if your eyes were infrared eyes with similar sensitivity as the NIRCam instrument, the relative strength of the redder to bluer bands is likely relatively accurate. The actual mapping of a particular (infrared) color to human-visible color, here, is artistic. *Image Credit: NASA, ESA, CSA, STScI.*

"everywhere," an algorithm is used to infer, (i.e., interpolate) the values for, e.g., the red and blue channels in a green pixel based on the values in the surrounding pixels. These are known as color filter arrays (CFA), of which the Bayer filter design is common.[7]

Why, then, do astronomical cameras not employ this method, and take color images directly? There are several motivations. The first is that we are interested in *quantitative* measures of the brightness of astronomical objects at different wavelengths, ideally, everywhere in our image. In Earth-bound (or artistic) photography, interpolating a color value might work reasonably well, but in a scientific context, we want to avoid this. Second, we are interested in more color channels than the three used in most consumer cameras. An instrument like HST or JWST might have more than a dozen filter-types, of different central wavelengths and widths. If we put all of these on one detector, per pixel, the idea of even interpolating would fall apart.

Luckily, unlike on Earth, where things (often) are moving when we try to snap a photo of them, astronomical objects for the most part are stationary, at least over the course of a night.[8] This means we can slot a filter of some bandpass in front of our *entire* detector, take an image, and then slot a different filter in and take an

---

[7] Inferring the other (e.g., GB) values at the position of one (e.g., R) pixel is a task known as *demosaicing*, for which different algorithms exist.

[8] Some solar system objects are exceptions.

Figure 8.3. Example of a typical layout of pixels in a "color" detector, like those in a consumer camera. Each pixel has a "mini-filter" which permits only red, green, or blue light through. Notice there are ~2x as many green pixels—this is because the human eye is most sensitive at green wavelengths, and so these are acting as "luminance" filters, while the red and blue are providing most of the color information as deviations from that "center." Each red pixel will have no information about how much green or blue light is present at that location, but because there are four blue and four green pixels surrounding it, reasonable values can be interpolated.

additional exposure. Given enough time, we can obtain images in as many filters as we have access to. All will have accurate, quantitative count measures in each pixel for each filter bandpass. These images can later be combined to create either true or false-color images, as well as to make quantitative numerical measurements of color.

Scientifically, a "color" image does not have much meaning, but many valuable properties about an astronomical object (star, galaxy, etc) can be inferred from the amount of light it emits in different bandpasses (usually many more than three). We take images, then measure the gathered signals from such objects (a process known as *photometry*) in different bands, which can then be used in various forms of analysis. As a note, "color" has a specific meaning in astronomy, namely, the flux ratio in two bands (or the magnitude difference, e.g., $g - r$).

Astronomical data come in various *forms*, and also various *formats*. We can distinguish the form between an image (2D data), a spectrum (either 1D or 2D and image-like), a spectral cube (3D), higher-dimensional data, or indeed, catalogs (tabular data). However, it turns out that many of these forms can all be stored within the same file format. That format, for several decades, has been FITS.

FITS stands for Flexible Image Transport System. It has been the de-facto workhorse format for transporting most types of astrophysical data for several decades, and is still heavily in use. *Nearly any* data form you could think of in astronomy *could*, feasibly, be stored within a FITS file, though it is becoming increasingly common for certain types of data not to be.

We will be using FITS in this section in the context of imaging. Once a FITS image has been read into Python, it can be treated as a simple 2D Numpy array. Each pixel of the image registers a value, $F(x_i, y_i)$. As a note, because Python/Numpy use row-first indexing, to index a position (x, y), one would actually take the array and index array[y,x]. A multi-band image, as discussed above, will often be stored in a Numpy array with an extra dimension. It is convenient to think of these as "image stacks."

At any time, we can interrogate the shape of our array (via the array attribute .shape) to infer how data has been stored. For example, if we had a g, r, i image from a detector with 2500×2600 pixels, we would expect to return a shape of either (3, 2500, 2600) or (2500, 2600, 3).

In the following sections, we will build up a working familiarity with the ingesting, displaying, and manipulation of imaging data. We will begin by treating images *only* as two dimensional arrays of pixels, and then move on to adding in information about the sky positions and coordinates associated with those pixels using the framework of the World Coordinate System (WCS). By doing so, we will foray out of Numpy alone and into a suite of Astropy and Astropy-affiliated tools directly oriented around the manipulation of imaging data. We will also discuss the displaying (plotting) of astronomical imaging, which is nontrivial.

A JWST Galaxy

For the duration of the section, we'll be using images of a spiral galaxy captured by JWST in different bands.

At this point, it is relevant to discuss the format in which astronomical images are stored. Typically, we use one of several binary or compressed formats to encapsulate astronomical images, which include both the arrays of pixel data, as well as so-called "headers" which can be thought of as dictionaries of meta-data and ancillary properties about the images. For much of the late 20th century and early 21st century, the astronomical standard was FITS. Despite "image" being in the name, FITS is a standard that can also handle storing tables of information. An additional benefit of FITS files is the ability to store data (and headers) into multiple *extensions*, meaning that multiple bands, or multiple tables, or indeed multiple images of different objects can be stored in a single FITS file.

More recently, several other formats have come into vogue, including the HDF5 standard (hierarchical data format version 5) and ASDF (advanced data science format). Ultimately, while the under-the-hood implementations of data storage vary for these standards, the user-interface is roughly similar. All have specific libraries in Python which handle the loading in of their relevant file type into the typical Numpy arrays we are used to dealing with.

The JWST data we will be working with here was made available by the CEERS collaboration (Bagley et al. 2023) and has been distributed in the FITS format, so we will primarily use this system in this chapter.

To load an FITS image into Python, we will use the astropy.io.fits library (I/O stands for Input/Output). Below, I show two ways of loading the same data:

```
import numpy as np
import matplotlib.pyplot as plt
from astropy.io import fits

#open the file
hdu = fits.open('../../BookDatasets/imaging/jswt_f277W_crop.fits')
# retrieve the image in the 0th extension and its header
image = hdu[0].data
header = hdu[0].header
#close the image
```

In the above snippet, we opened the FITS file, using the variable "hdu" which stands for "Header Data Units". The HDU is indexed like an array, with each index corresponding to a different extension. Each extension has several dot-accessible attributes, including the data, and the header. Finally, we close the image.

A recommended, and more elegant way to accomplish the above, is via a Python-structure called a *context manager*. By opening a file using a context manager, we can ensure that the connection to the file is closed when we are done reading in the needed information.

```
with fits.open('../../BookDatasets/imaging/jswt_f277W_crop.fits') as
    hdu:
    image = hdu[0].data
    header = hdu[0].header
```

Notice here we do not need the close statement. When the indented block of the with-clause ends, it automatically closes the file.

Now that we have an image loaded, let's look at it! I'll do so in the most basic form below, and we'll see that the result is rather underwhelming (Figure 8.4).

```
fig, ax = plt.subplots(figsize=(8,8))
ax.imshow(image, origin='lower');
```

We can see that there is *something* in the image—the top right corner has some dark circles (which are stars), and there are some fuzzy things just barely showing up. Why is this?

Astronomical data often has a very unique property in that the dynamic range across *all* pixels (i.e., the lowest and highest values) is much larger than the dynamic range of *interest* (i.e., the range of values for pixels corresponding to light from the galaxies in the image).

The Matplotlib library has some algorithms for determining automatic defaults for the saturation points in its color scaling, and these often don't work well for astronomical imaging.

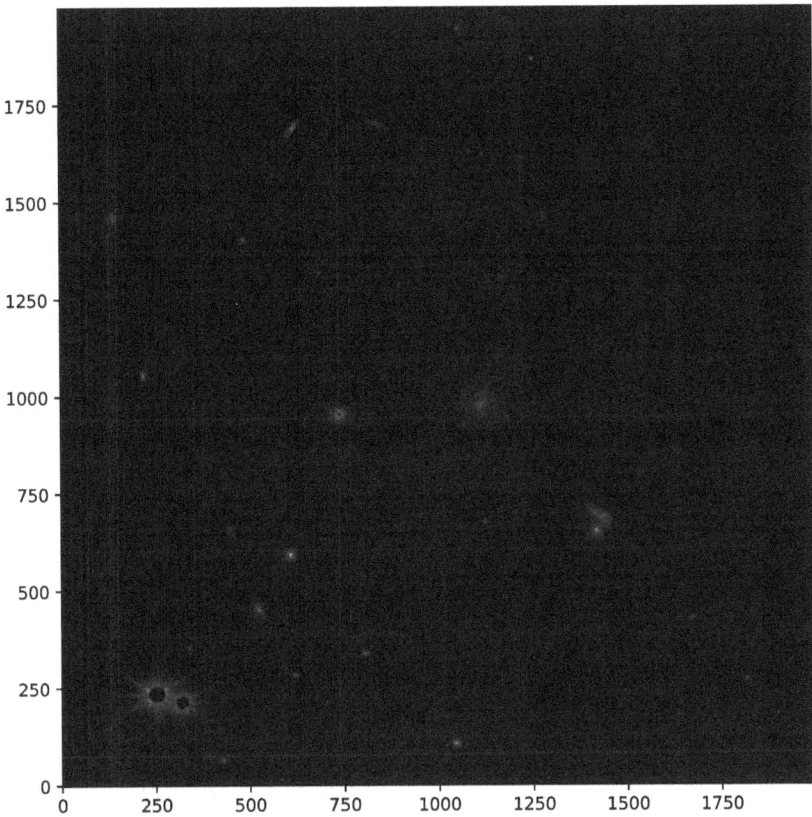

Figure 8.4. JWST image, plotted with no modifications to the color scaling.

There's no "golden" rule for a scaling that will always work, but there are a few tricks that can get us in the right ballpark. The parameters of interest in imshow() are v_{min} and v_{max}, which set the saturation points. Often, a good starting point is to choose values such that

$$v_{min} = \mu_{image} - s \times \sigma_{image} \tag{8.1}$$

and

$$v_{max} = \mu_{image} + s \times \sigma_{image} \tag{8.2}$$

where μ is the mean pixel value in the image, σ is the standard deviation of pixel values in the image, and s is an arbitrary scaling parameter that usually varies between 0.01 and 10.[9] Let's try this out with two scalings for our image:

---

[9] Another popular method is to use percentiles of the image.

```
s = 5
vmin = np.mean(image) - s*np.std(image)
vmax = np.mean(image) + s*np.std(image)

fig, ax = plt.subplots(figsize=(8,8))
ax.imshow(image,vmin=vmin,vmax=vmax,origin='lower');
```

All of a sudden, we can much better see the structure in the image![10] The scaling chosen on the left of Figure 8.5 appears to work well for the structure in the brighter structures (like the spiral arms of the central galaxy), while the scaling on the right better shows faint objects and emission in the outskirts of the larger systems.

Finally, because this is a single band image, it is often best to use a simple color mapping; most traditionally, black to white. This can be done via setting the `cmap` argument of `imshow()` to either "gray" or "gray_r". Below in Figure 8.6, I show a "tweaked" version of the above image, where I have and cropped in on the central galaxy using array indexing and leveraged some slightly more involved `Matplotlib` to scale the color values logarithmically.

One thing to notice in this cropped image is that the process of slicing the array of the larger image *re-indexes* the array. We have a brand new array, which is smaller in dimension than the original, with its own independent set of pixel position indices. When carrying out crops like this, be sure to track whether the indexing you are attempting is based on the new, or old, indices.

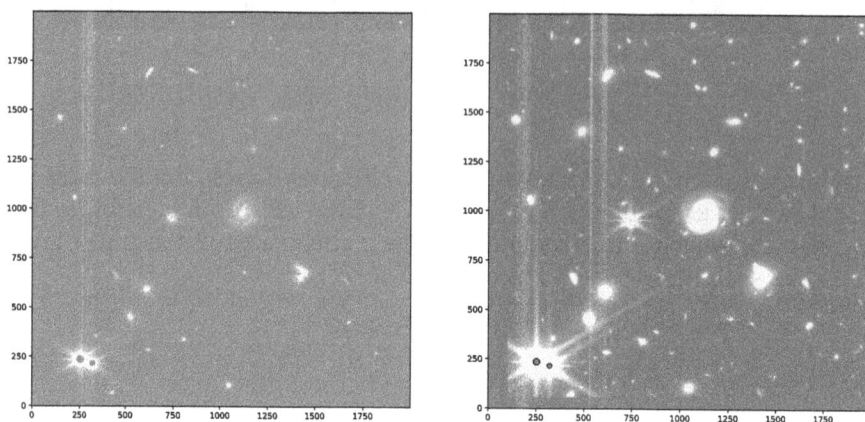

Figure 8.5. Same image as Figure 8.4, but scaled with two different values (with respect to the standard deviation of pixel values), highlighting different structures in the image.

---

[10] You'll notice I've added the `origin='lower'` argument, which flips the image so that 0,0 is in the bottom left-hand corner.

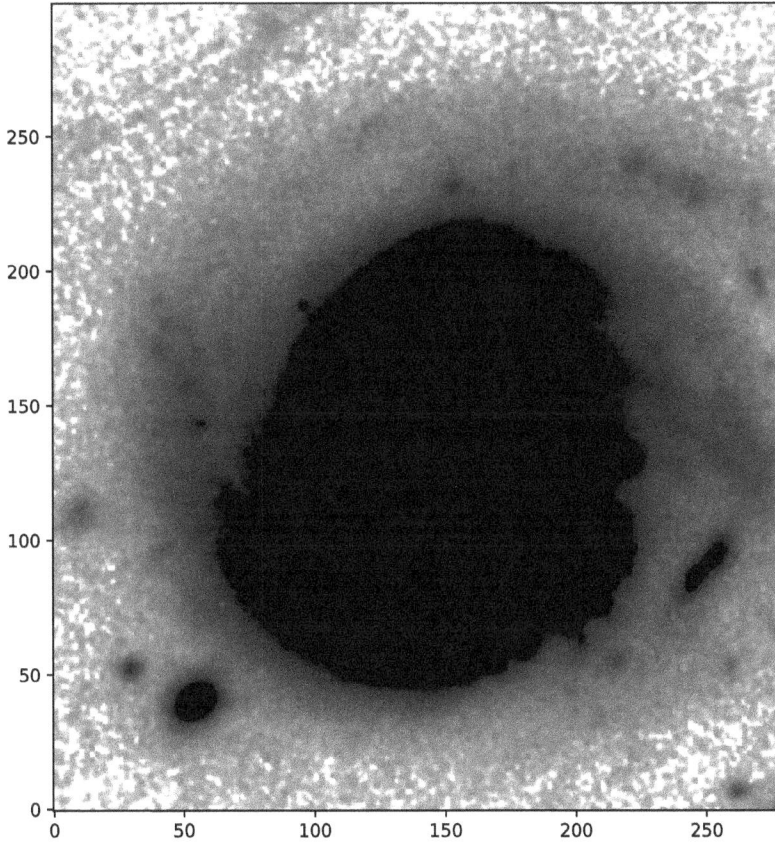

Figure 8.6. Cropped in view of the central galaxy, with colors scaled logarithmically.

Exercise 8.3. Practice with Plotting

Using the same JWST image in the textbook dataset, try to replicate the figure I've made above. Then try out your own scalings, nudging v_{min} and v_{max} until you are satisfied. Try out different color maps as well: `matplotlib` has a ton of fun colormaps—with some experimentation, you can make some truly psychedelic pictures.

8.7.1 The World Coordinate System

Thus far, we have only examined the images based on the 2D `Numpy` arrays comprising the pixel data. Realistically, we need to know, e.g., the coordinates of this galaxy, as well as the pixel scale (arcseconds per pixel) of the image so that we can carry out analyses (e.g., measuring the flux within an aperture of a certain physical radius).

The way we accomplish these tasks is by pairing up our image with an `astropy.wcs.WCS` object. For data that has been *plate solved*, that is, the location and pixel scale fit, the information needed to construct a WCS object is stored in the header of the file.[11]

Helpfully, `Astropy` has a function for constructing a WCS object directly by feeding in a header. We have one for our image, so let's use it.

```python
from astropy.wcs import WCS

im_wcs = WCS(header)
```

You may notice a warning printed if you run this yourself—usually, these can be ignored, so long as the warning indicates the WCS module "fixed" something in the header.

The first thing we can do using our WCS is plot our image using celestial coordinates on the axes, rather than pixel number. To do this, we need to make a *projection* axis, which will take in the WCS and perform the conversion between pixel position and sky position. This is fairly easy to do:

```python
fig, ax = plt.subplots(figsize=(8,8),subplot_kw={'projection': im_wcs})
ax.imshow(image,vmin=vmin,vmax=vmax,cmap='gray_r');
```

Notice in Figure 8.7 that the axes of our plot now show equatorial right ascension (RA) and equatorial declination (DEC). Other projections (like galactic coordinates) are also easily used instead.

When we create figure axes this way, they are actually no longer `Matplotlib` standard axes. Notice, if we print `ax`:

```python
ax
```

```
<WCSAxesSubplot:xlabel='pos.eq.ra', ylabel='pos.eq.dec'>
```

It is listed as a `WCSAxesSubplot`. This object is *similar* to a standard AxesSubplot, but is in fact an `Astropy` wrapper which provides extra tools for manipulating and working with the axes in an astronomical context. We won't need many of these features yet, but I mention this point because a few tasks (such as changing the tick label format, or number of ticks, etc.) is slightly different than normal. Instructions for making these changes are readily available in the `Astropy` documentation.

What else can we do with our WCS object? As mentioned, the WCS is the "translator" between pixel coordinates and world/sky coordinates. As an example, let's assume our central spiral galaxy is not in any survey (unlikely), and we want to estimate its position. There are automated tools for this (`source extractor`, `sep`) but for now let's work manually. From our image, we can easily choose a *pixel position* we think is the center.

---

[11] Raw images straight from a telescope will not have this, or at least, not an accurate version. Tools such as `astrometry.net` can plate solve almost any astronomical image to provide this information.

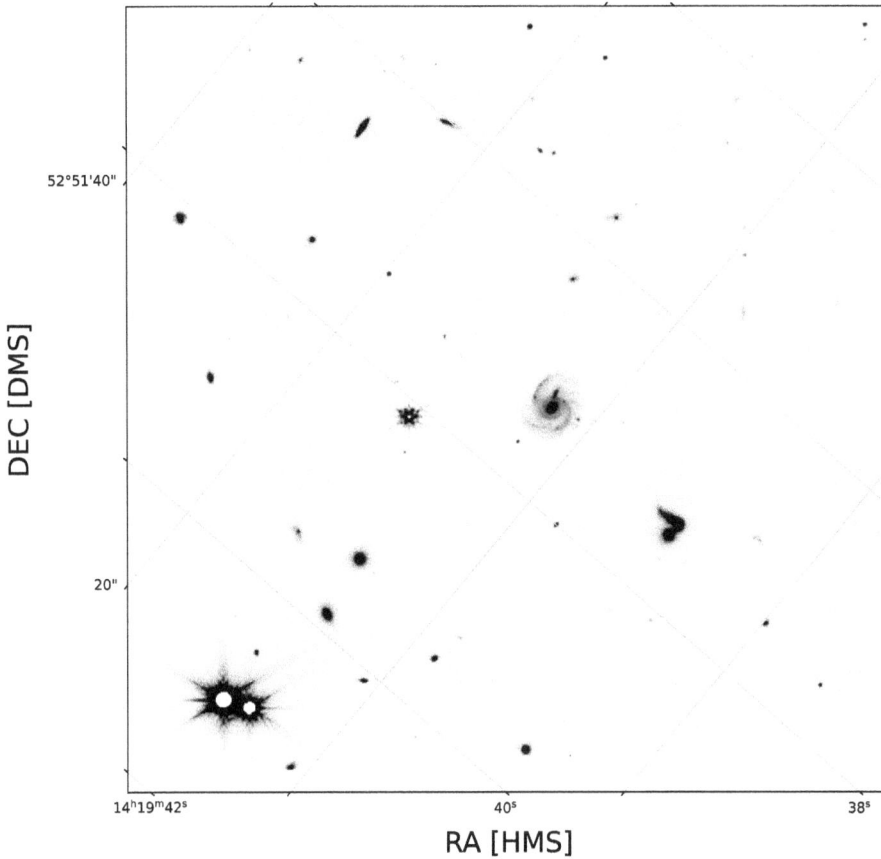

Figure 8.7. Same JWST field as, e.g., Figure 8.4, but now with WCSAxes created to show astronomical coordinates.

To help visualize this, let's create a CircularAperture using the photutils package, which facilitates the creation and use of circles, ellipses, and annuli on astronomical images (Figure 8.8).

```
from photutils.aperture import CircularAperture

# define pixel aperture
aperture = CircularAperture([(1110,980)],r=100)

fig, ax = plt.subplots(figsize=(8,8))
ax.imshow(image,vmin=vmin,vmax=vmax,cmap='gray_r',origin='lower')
aperture.plot(color='r',lw=3)
```

What I've done is created an abstract CircularAperture object, which has a coordinate (in pixels) and a radius (in pixels). Plotting our original image in pixel

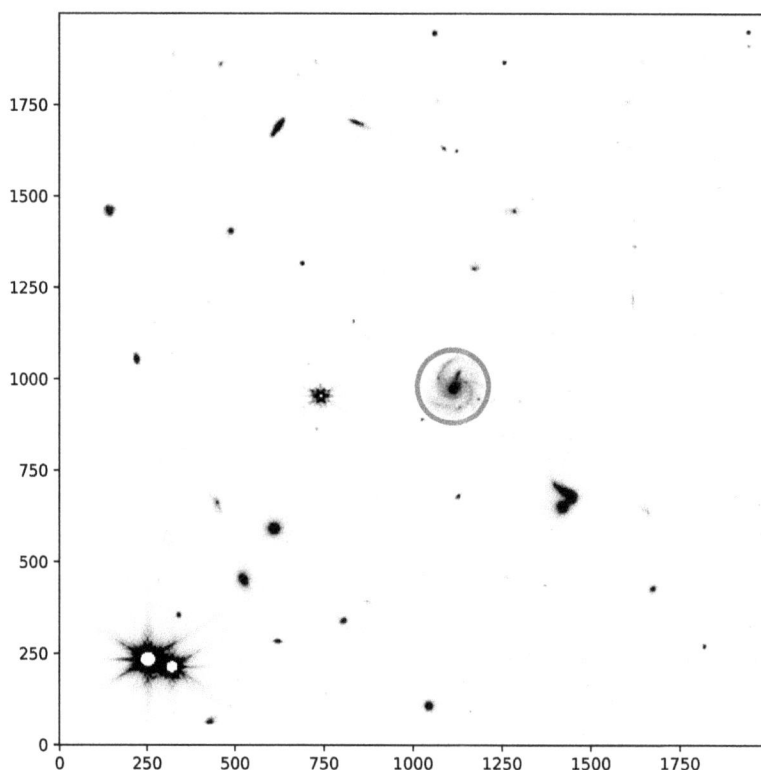

Figure 8.8. JWST image with our circular aperture overplotted.

space, we can use the .plot() method of the aperture to visualize where it is for our image. It seems to nicely (ish) center the Galaxy.

Next, we want to know the circle's position and radius in coordinates on the sky, i.e., it's RA and DEC, as well as its radius in arcseconds (or arcminutes, depending on the scale of your image). Every CircularAperture has a method called to_sky(WCS), which, taking in a WCS object, can convert the quantities of the pixel based circle into astronomical coordinates:

```
aperture.to_sky(im_wcs)
```

```
<SkyCircularAperture(<SkyCoord (ICRS): (ra, dec) in deg
    [(214.9052996, 52.85095396)]>, r=2.999998303527192 arcsec)>
```

As we can see, the output of this conversion is now a SkyCircularAperture, defined by an actual sky position (RA/DEC in degrees) and a radius (in this case roughly 3 arcseconds). We might need to do more conversions (or nicer printing) of our mysterious spiral galaxy's coordinates, so we can easily create a separate SkyCoord (sky coordinate) object.

```
coord = aperture.to_sky(im_wcs).positions
coord.to_string('hmsdms')
```

```
['14h19m37.27190414s +52d51m03.43427078s']
```

In this case, we've used the `.positions` attribute of the SkyCircularAperture to retrieve an `Astropy.coordinates.SkyCoord` object. The positions is plural because both SkyCircularApertures and SkyCoords can nominally contain more than a single coordinate (though the SkyCircularAperture can have only 1 radius). I've then used the `to_string()` method of the SkyCoord object to print the coordinates in the hour-minute-second, decimal-minute-second format.

8.7.2 Image Cutouts

Often, we want to zoom in on a certain patch of a larger image, as I did by manually cropping the array above to get a closer look at the spiral galaxy. Unfortunately, such a simple slicing method doesn't work when we are also trying to keep track of the coordinates of pixels in the image. This is because the WCS object stores information about the image shape, the "origin" coordinate in that image, and the change in coordinate per pixel. When "cropping in" to an image, these values change.

Fortunately, `Astropy` has a sub-library which allows us to create cutouts of images while appropriately converting the WCS objects to be compatible with the newly created crop. This library is `astropy.nddata`, which has the `Cutout2D` class.

Let's start by making a cutout around our galaxy—we know its coordinates, thanks to the operations we carried out above. We also need a box size, for which I'll chose 12 arcseconds (based on the circle we drew having a 3 arcsecond radius). To create a cutout, we need the image (array), sky coordinate, size, and WCS for translation. For size, I'll explicitly here import the `astropy.units` module, which we will learn much more about, but for now, will allow us to define quantities in astronomical (or other) units.

```
from astropy.nddata import Cutout2D
import astropy.units as u

cutout = Cutout2D(data = image,
                  position = coord,
                  size = 12*u.arcsec,
                  wcs = im_wcs)
```

We now have a cutout, so how do we plot it? Our single `cutout` object now contains attributes we can access, including the image of the cutout, and the new WCS. We can access and plot them as follows:

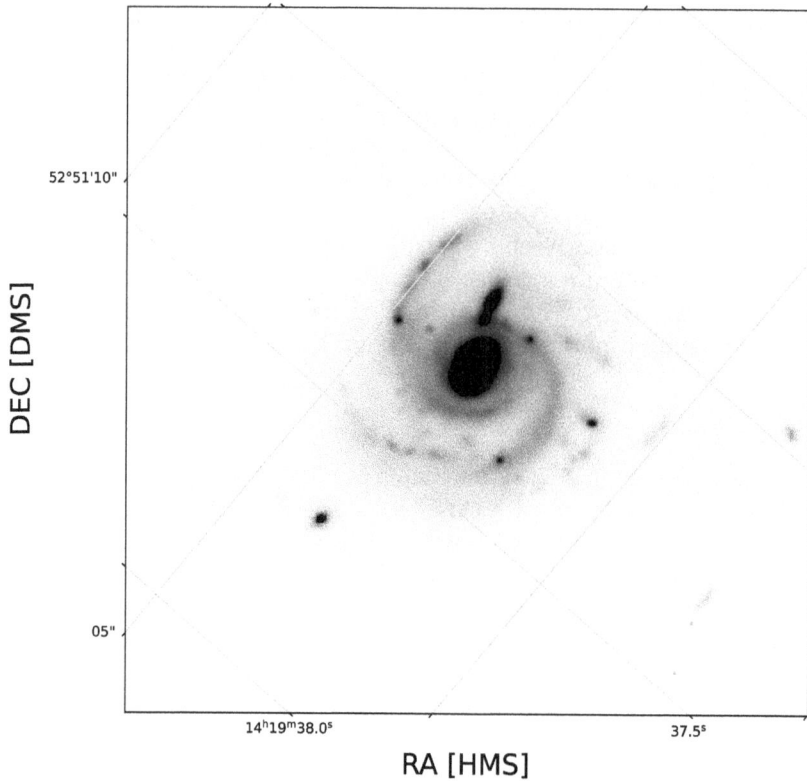

Figure 8.9. Crop in on the Galaxy, but instead of cropping via array slicing, we have created a `Cutout2D`, allowing us to retain information about the WCS in our cropped version.

```
fig, ax = plt.subplots(figsize=(8,8),subplot_kw={'projection': cutout.
   wcs})
ax.imshow(cutout.data,vmin=vmin,vmax=vmax,cmap='gray_r')
ax.set_xlabel('RA [HMS]',fontsize=20)
ax.set_ylabel('DEC [DMS]',fontsize=20)
ax.grid()
```

In Figure 8.9, I've replaced the original image WCS with the new `cutout.wcs` attribute, and am using `cutout.data` to get at the image pixels.

Cutouts are very useful, and in general, we use them like we would our original data, simply remembering that the data and WCS are now attributes of a single object.

8.7.3 Aperture Photometry

We can do more with our aperture than simply estimate the position of our galaxy. We can also measure the amount of flux coming from the system. Back in Chapter 6, when we introduced Numpy arrays via the example of an exoplanet transit light curve, we used a square aperture to sum the flux from a star. This meant we could simply use array indexing to select whole pixels to include in our aperture.

Now we'll graduate to the next step up in photometric measurement. As we've seen, the photutils package has provided convenient methods for creating non-square apertures. Indeed, we aren't limited to circles, but rectangles, annuli, and ellipses are all available.

The photutils package also has a function aperture_photometry(). Given an image and an aperture, it can compute the flux within the object. That is the sum of the pixels within the aperture. This is more involved than the square case, because a circular (or other curved aperture), or an aperture defined in sky coordinates, may *overlap* with some pixels, covering only some fraction of that pixel. That might not seem like a big deal, but for many cases where the brightness slope is very steep, assuming a pixel is entirely inside or outside the aperture is sometimes not accurate enough. The photutils package handles this in several ways: it can use the center of the pixel to simply consider a pixel in or out, but also has a 'code{subpixel} option which will subdivide each pixel into smaller pieces; these are then treated as in or out of the mask with higher accuracy than doing so for the whole pixel. You can select how many subpixels to create. Finally, the default method for aperture_photometry() is the exact method, in which the exact intersection of the aperture with each pixel is computed. This method is slower, but the most precise.

Now that we have some motivation for *why* apertures defined with photutils might be useful, let's actually measure a flux with it. To do so, we'll return to our original CircularAperture object, but add in a CircularAnnulus within which to calculate the sky background, which we will subtract from our flux (Figure 8.10).[12]

```
from photutils.aperture import CircularAnnulus

aperture = CircularAperture([(1110,980)],r=100)
bg_aperture = CircularAnnulus([(1110,980)],r_in=120,r_out=200)

fig, ax = plt.subplots(figsize=(8,8))
ax.imshow(image,vmin=vmin,vmax=vmax,cmap='gray_r',origin='lower')
aperture.plot(color='r',lw=3)
bg_aperture.plot(color='C0',lw=3)
```

Once we have our aperture objects, carrying out the photometry is relatively straightforward:

---

[12] These JWST data have been sky-subtracted, so this step is not *formally* necessary here, but I show it for completeness.

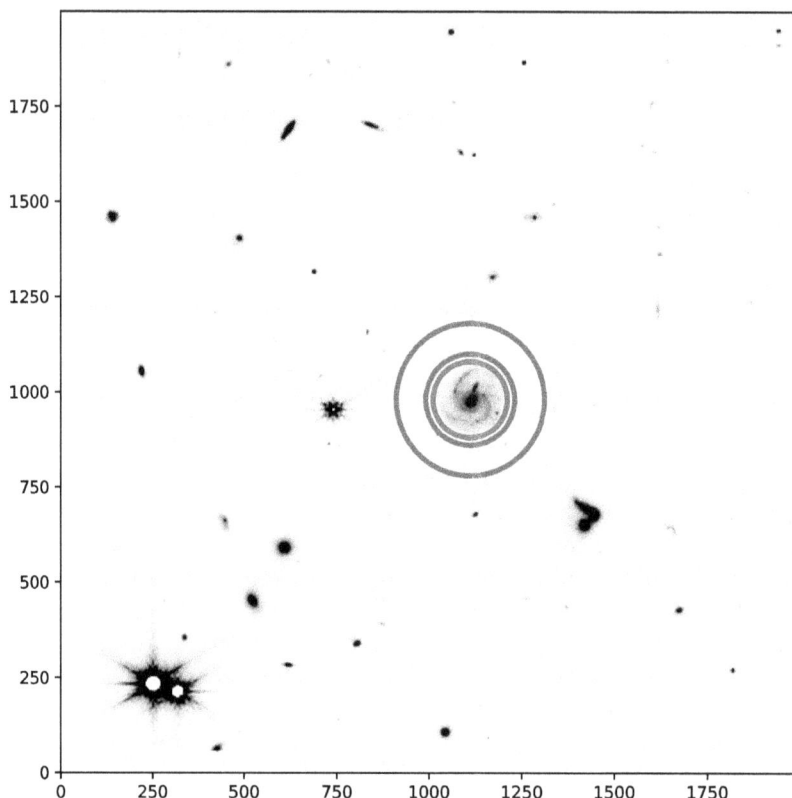

Figure 8.10. Our original aperture (in red), now accompanied by an annulus (blue). We will compute the mean pixel value in the annulus as a local estimate of the sky background, while measuring the sum within the red aperture as our flux measurement for the Galaxy.

```python
from photutils.aperture import ApertureStats, aperture_photometry

phot_table = aperture_photometry(image, aperture)
flux = phot_table['aperture_sum'][0]
bg_stats = ApertureStats(image, bg_aperture)
bkg_mean = bg_stats.mean[0]
print(flux)
print(bkg_mean)
```

8312.259999858908
0.0074764478284599645

The `aperture_photometry()` function returns a table object which has the flux for each aperture input—we only had one, so I indexed the column with the sum we are after and grabbed the 0th element. For the annulus, we don't want the *sum* in the annulus, we want an estimate of the local background value—here, I chose to use the mean for that (median might also be reasonable). The `ApertureStats` class can be used to easily compute those (and other statistical) values. Again, we only have one annulus, so I index the 0th value of the output.

We can see that the background level is very small compared to the flux from our galaxy. But there is one step left before we can subtract it from our flux measurement. The mean background value is the value *per pixel*. Much like the case with the exoplanet transit, we need to multiply by the area of our primary aperture. Thankfully, our aperture knows its own area, so this is easy to compute:

```
print(f'Aperture Area: {aperture.area:.4f}')

final_flux = flux - bkg_mean*aperture.area
print(f'Final flux measurement: {final_flux:.3f}')
```

```
Aperture Area: 31415.9265
Final flux measurement: 8077.380
```

One can proceed to use ever more complex techniques for aperture photometry. The photutils package itself has several methods for more advanced methods, and once one is dealing with, e.g., crowded fields in Hubble Space Telescope imaging, this package ceases to be usable at all, and more involved tools are needed.

Exercise 8.4. Spectral Energy Distributions

The Book Datasets repository contains images the JWST field we have been working within three HST bands and four JWST bands. Write code to measure the aperture flux as we did above for F277W in all seven bands (you can use the same aperture radius in each band). The units of the image files are MegaJansky per steradian (MJy/sr). These units are energy/time/surface area at Earth/subtended area on sky/frequency (e.g., erg s^{-1} cm^{-2} arcsec^{-2} Hz^{-1}. When we take an aperture sum, we are integrating out the per on-sky area. We then have a flux density (technically, F_ν). Make a figure which shows your measured photometry vs the central wavelength of each of the image bands.

You've now made a ubiquitous figure in astronomy: the spectral energy distribution (SED).[13] The varying strength of brightness across the different wavelengths can be fit to extract information about the Galaxy—for a detailed introduction, see Conroy (2013).

8.7.4 Combining Images

Thus far, we've been using only a single band of data (F277W, centered at ~2.77 micron). You may have noticed that the data provided has several available bands, spread across HST and JWST. How do we construct a color image?[14]

---

[13] Technically, a true SED has units of energy, so you could multiply each flux density by the frequency of the band to get νF_ν.

[14] It should be noted here that color is an abstract concept. All JWST wavelengths are redder than the human eye can see. We must map visible colors to these bands, and interpret the output images appropriately for that mapping.

First, we need to load in our several bands. We'll use three for this example, to create a "standard" RGB image. For clarity, I will re-load the image we've been working with to use a consistant naming scheme, and create cutouts for each image around our galaxy immediately. Since this is a repetitive task, I'll create a quick function to do this.

```python
def load_and_cutout(band_name,coord,size):
    with fits.open(f'../../BookDatasets/imaging/jwst_{band_name}_crop.
 fits') as hdu:
        image = hdu[0].data
        wcs = WCS(hdu[0].header)
    cutout = Cutout2D(data = image,
                 position = coord,
                 size = size,
                 wcs = wcs)
    return cutout

f150w = load_and_cutout('f150W',coord=coord,size=12*u.arcsec)
f277w = load_and_cutout('f277W',coord=coord,size=12*u.arcsec)
f444w = load_and_cutout('f444W',coord=coord,size=12*u.arcsec)
```

Feel free to examine the structure of this function. There's some brittleness — I assume a file-name convention, which I know here to be true, for picking out different files — but... beyond this, I rely on inputs for the coordinates and sizes of cutouts to make.

To construct a three-color image, the most straightforward tool is the make_lupton_rgb function in Astropy.

This function takes in images (arrays) to be mapped to the r, g, and b channels, as well as some extra tuning arguments to determine how best to combine and stretch the dynamic ranges to make a "good looking" image. This is hardly scientific, and such images are mostly heuristics.

```python
from astropy.visualization import make_lupton_rgb

rgb_image = make_lupton_rgb(f444w.data,f277w.data,f150w.data)

fig, ax = plt.subplots(figsize=(8,8))
ax.imshow(rgb_image,origin='lower')
fig.savefig('imaging-jwst-color-1.pdf')
```

The resulting image is shown in the left-hand side of Figure 8.11. It appears rather green to the eye, mostly because there was more flux received in what we are treating as the central band.

We can create a better image by tuning the additional parameters, and (arbitrarily) scaling the strength of some of our bands—the results of this are shown on the right in Figure 8.11.

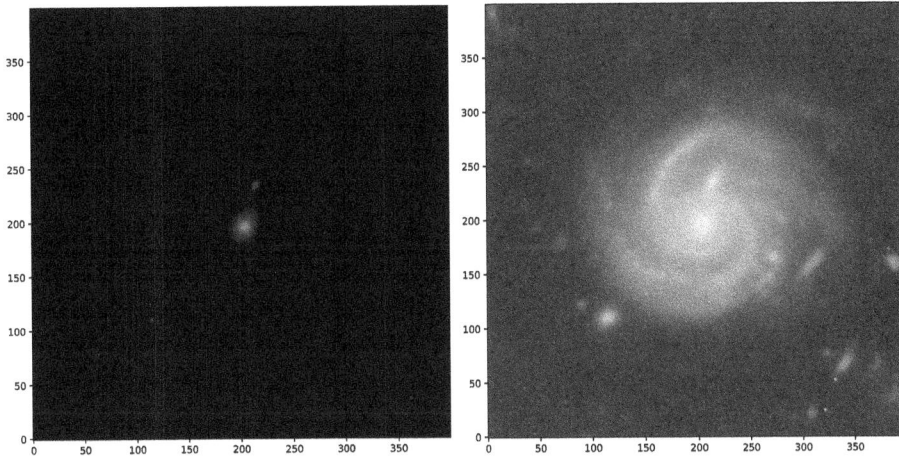

Figure 8.11. Left: JWST color image with the default lupton function output. Right: Same image, but with custom scalings of the image values as well as tweaks to the `make_lupton_rgb()` parameters.

```
rgb_image = make_lupton_rgb(2.0*f444w.data,f277w.data,f150w.
    data,minimum=np.mean(f277w.data)-0.5*np.std(f277w.data),stretch=0.
    1,Q=9)
fig, ax = plt.subplots(figsize=(8,8))
ax.imshow(rgb_image,origin='lower')
fig.savefig('imaging-jwst-color-2.pdf')
```

This looks closer to what we might "expect" for such a galaxy. The entire galaxy is somewhat red (we strengthened the R-mapped band), but this is somewhat reasonable as the whole galaxy is "red" to the human eye (so red we can't see it). Meanwhile, there is some information encoded between the bluer and redder parts of the image. Once again, I'll emphasize that producing color images like this is more an art than a science, and generally, we do not carry out any science directly on such images, though they can be useful to guide our insights into what might be worth further study.

8.8 Summary

In this chapter, we covered the core utilities within the Astropy package, which facilitates our interaction with astronomical data.

The core utilities covered here were

- units and constants, which are useful broadly beyond use astronomical problems, but also contain many astronomy specific terms.
- cosmology utilities for calculating relevant distances and sizes in the universe, and
- imaging utilities, ranging from the fits library for handling astronomical image/data files, and a set of tools for plotting and measuring quantities in images.

There is plenty more to explore in the Astropy library, much of which you will find as the need arises. I encourage any budding astronomer to read through the documentation and tutorial sites to see what is available.

References

Astropy Collaboration, 2013, AAP, 558, A33
Astropy Collaboration, 2018, AJ, 156, 127
Astropy Collaboration, 2022, ApJ, 935, 167
Bagley, M. B., Finkelstein, S. L., Koekemoer, A. M., et al. 2023, ApL, 946, L12
Conroy, C. 2013, ARA&A, 51, 393

Part III

Intermediate Applications and Patterns

Introduction

At this point, we have covered the majority of what aspects of Python are needed to begin working with astronomical data and writing simple analysis routines. As those routines grow more complex, we will need to better organize our code to allow for personal and external readability, testability, flexibility, and modularity.

There are three *primary* paradigms used by the typical Python programmer to organize the logic and execution of code.

1. *Procedural Programming*. This paradigm involves the writing of a list of instructions that tell the computer how to carry out a set of calculations. Procedural programming is what we have been doing thus far in this text; a Python script or notebook containing variable declarations and calculations fits this description. Procedural programming has the advantage of requiring no extra code structure or boilerplate to implement, but has the disadvantage of lacking flexibility, modularity and testability.

2. *Functional Programming*. This paradigm involves decomposing a problem into multiple functions (and critically, *only* functions), and carrying out a task by taking an initial set of inputs and passing them through one function, which returns values to be passed directly into the next. Functional Programming has many advantages over procedural scripts, namely modularity and testability, as we will discuss.[1]

3. *Object-Oriented Programming*. In this paradigm, we create our own Python objects, with methods and attributes and thus internal state, and we carry out operations to modify that internal state. Object Oriented Programming is highly flexible, while remaining modular and testable, though not as strictly as in the functional case.

To give a simple example: the `list.append()` method is object oriented. A method of each list object knows how to append a value to it, and the internal state of the list is modified such that the same list object now contains the new value. In contrast, Numpy arrays do not have an append method, and the `np.append()` function takes the array to be appended and the new value, and returns a *new array* with that value appended, without modifying the original. To mimic the object-oriented approach, one would overwrite a variable via, e.g., `x = np.append(x, 1)`.

Of course, Numpy hasn't been written to be entirely functional—we *know* that arrays have many useful attributes and methods. This highlights the fact that unlike some other languages, Python is *multiparadigm*, and most Python programmers find it useful to mix together elements of all three styles listed above in their code.

---

[1] There are stricter and less-strict interpretations of the functional paradigm.

I begin this part of the text with this description as we are about to dive into both functional programming and object oriented programming, which will be tools to leverage in more realistic research environments. Some things we might attempt (such as vectorization) require a more functional approach, others (like developing software packages) encourage object-oriented approaches. Both will generally mix in a small amount of procedural code around these structures. So without further delay, let's dive in!

AAS | IOP Astronomy

Astronomical Python
An introduction to modern scientific programming
Imad Pasha

Chapter 9

Functions and Functional Programming

9.1 Introduction

The next step forward in the abstraction of how your code will be structured is known as *functional programming*. This refers to the collection of code into specific self-defined functions, with bite-sized tasks that you can verify are working properly (unit testing) and which allow for easy debugging. Formally, the functional paradigm asserts that no internal states should be modified, i.e., every operation should be a function taking an input and returning a new value. We generally don't use the strictest form of functional programming in day-to-day code, but there are aspects of functional programming that are valuable to use. I want to emphasize that while we will discuss a functional approach in this chapter and talk about its benefits, this does not mean my recommendation is to write all code this way!

To employ a functional programming paradigm, we must begin writing our own functions. You already have plenty of experience *using* functions. When we import libraries like Numpy or Matplotlib we are importing *functions* and *classes* other people have written that are stored in those libraries. When I call

```
import numpy as np

x = np.linspace(0,2*np.pi,100)
my_wave = np.sin(x)
```

both np.linspace() and np.sin() are functions that someone on the Numpy team wrote down and placed in the library. We know how to use these functions even though we didn't write them because we read the *documentation*, i.e.,

```
help(np.zeros)
```

```
Help on built-in function zeros in module numpy:

zeros(...)
    zeros(shape, dtype=float, order='C')

    Return a new array of given shape and type, filled with zeros.

    Parameters
    ----------
    shape : int or tuple of ints
        Shape of the new array, e.g., ``(2, 3)`` or ``2``.
    dtype : data-type, optional
        The desired data-type for the array, e.g., `numpy.int8`. ⊔
Default is
        `numpy.float64`.
    order : {'C', 'F'}, optional, default: 'C'
        Whether to store multi-dimensional data in row-major
        (C-style) or column-major (Fortran-style) order in
        memory.

    Returns
    -------
    out : ndarray
        Array of zeros with the given shape, dtype, and order.

    ...
```

The help function above spits out a lot of text, but the most relevant bit is at the top: the documentation shows us (1) what the function does and outputs, (2) what inputs it needs and in which order (and what data type these input should be), and (3) some examples of its use.

In this chapter we will be talking all about functions: how to define them, how to document them, best practices in implementing them, and more. (You have, of course, seen a handful of function definitions sprinkled through the text thus far, e.g., the Gauss function used throughout Chapter 7.)

9.2 Defining Functions

The absolute simplest syntactically legal function we can define in Python would look like this:

```
def my_function():
    pass
```

In the above example, we see that the reserved word def tells Python we are defining a function (see 4.7). We give it a name (my_function in this case) followed by parenthesis containing comma-separated arguments (I have not supplied any).

We end with a colon, the same way we would for a loop or conditional statement. Then, all code associated with our function gets indented (again, just like a loop or conditional statement). For now, I have used the reserved word `pass`, which is a null operator that doesn't do anything but also doesn't throw an error. The `pass` reserved keyword is primarily useful for scenarios in which you want to build the structure or scaffold of your code, defining functions or methods to come back to later.

Let's add some code now. A common task for astronomers is to load a series of `FITS` image files in a directory on our computer and stack up all the images into some storage container (like a multidimensional `Numpy` array), as well as "extra" data from the `FITS` headers like object name, exposure time, etc. We've already seen (several times) how to load a single FITS file using `Astropy`, which has a function for this purpose (`fits.open()`). However, we might want to have our own function which will use `fits.open()` on a whole directory of files at once, and return the data in a format most useful to us.

In short, we want what's called a *wrapper*: a function that combines several other function calls[2] in a useful way (for us as least).

Let's start by defining the base shell of the function:

```python
def load_directory_images(path:str):
    pass
```

I've given the function a recognizable name that tells us what the function does (loads images in a directory), and I've specified the first **argument** to the function. Arguments are, much like in a mathematical function like sine, what gets input into the function in order to facilitate the calculation. In this case, that's a path (string) representing the location of the directory on our hard drive. For sine (e.g., the `np.sin()` function), the argument would be an angle in radians.

The `:str` following the `path` is a *type annotation*: Python does nothing with it (for now), but it informs users of the required input type and some external programs can enforce those types if you have supplied your function definitions with annotations.

As an important note: when we set the name "path", this is the variable name we'll use *inside* our function. It is only internal to our function, and users can supply any variable they want (of any name) to our function. Whatever is in slot number 1 of the argument list will be assigned the temporary variable "path" while it is inside the function.[3]

9.3 Writing Documentation

When we printed out the documentation above for the `Numpy` functions, there was a lot there. When writing functions for our own code, we generally do not need to be that intense—documentation depth should scale with how many people will use a given code (`Numpy` is used by hundreds of thousands of users, so it needs rock solid documentation). However *we should always at least somewhat document our code.*

---

[2] Here I use the word "call" to refer to the act of *using* a function. This is in contrast to the act of *defining* a function. We can define a function that calls (uses) other functions within it.

[3] This is known as local scope. We'll discuss this more momentarily.

Trust me. The most enticing hubris to any programmer is thinking "of course I'll know what these lines of code accomplish (and how) in three months." You won't. Anyone who spends significant time writing code will tell you that documentation is as important, in some sense, as the code itself. We introduced one form of documentation early in this textbook: the comment. Comment lines which annotate single lines of code are incredibly useful when parsing code later. Comments, however, cannot be used to, e.g., generate webpage describing the inputs, outputs, and purpose of every function in a sprawling set of code. For that, we'll need function documentation.

To add documentation to your function, we use triple quotes as follows:

```
def load_directory_images(path):
    '''
    Loads a directory's worth of images into convenient storage units.
    Requires astropy.io.fits.
    '''
    pass
```

Now, we can see that if we run the `help()` command on our function, we get this:

```
help(load_directory_images)
```

```
Help on function load_directory_images in module __main__:
```

```
load_directory_images(path)
    Loads a directory's worth of images into convenient storage units.
    Requires
astropy.io.fits.
```

In short, our own functions can have help pages just like those we import from libraries like Numpy.

9.3.1 Formatting Your Documentation: Best Practices

What I did above technically counts as documentation. But there are a few extra things we really need to make it useful. Let's improve our documentation:

```
def load_directory_images(path):
    '''
    Loads a directory's worth of images into convenient storage units.
    Requires astropy.io.fits.
    Note: All Images in directory must be of same shape.

    Parameters
    ----------
    path: str
        path to the directory you wish to load, as a string.
```

```
Returns
-------
image_stack: array_like
    A stack of all images contained in the directory.
    Array of shape (N,X,Y) where N is the number of images,
    and X,Y are the dimensions of each image.
image_dict: dict
    A dictionary containing headers for each image, the keys
    are the same as the indices of the corresponding
    image in the image_stack
'''

pass
```

Our function is now much better-defined—a user reading our documentation should be able to input the correct argument, and will be aware of some of the function's requirements (though, for user-facing code, *internal* checks that usage is correct is always better than trusting someone has fully read your documentation)!

You may have noticed that I have used a system for formatting the parameters and return values, i.e., name: dtype followed by an indented line with the description. As you move into higher level programming, you may write code that you would like to host on a service such as Github, and which needs an actual documentation website, e.g., on ReadtheDocs.io. You may have used such a site for research code you've downloaded and used. These sites are built using automatic frameworks, e.g., Sphinx, which scrape your entire codebase and build the site for you automatically. These tools require your documentation be formatted a certain way. So, it's never too early to get into the habit of writing documentation recognizable by these tools.[4] Here, I am using the Numpy/Google devised formatting structure, but others exist that are recognizable by these tools.

It may seem like writing all of this documentation takes considerable time and effort, and that is true. How detailed should our documentation be? Sometimes we're just doing some quick exploratory data analysis, and are writing a quick function to extract and plot some data. It seems obvious in this case that stopping to write documentation may not be the best use of our time.

Personally, I have a few thresholds that I use to determine the type of documentation to write. For "quick and dirty" type scripts, I don't bother with documentation or I write a quick one-liner using a comment. For my own research code, which will produce outputs I publish in a paper, I write documentation that is at minimum a detailed description, and if I plan to publish my code along with the paper, then I use full documentation as shown above. If you are writing code to be used by others, then it is *absolutely essential* to write full scale, formatted documentation.[5]

---

[4] There are extensions for editors like VSCode, as well as standalone programs, which can autoformat a docstring for you, or give you a template to fill in.

[5] Another way to formulate these thresholds is that as a quick function you wrote gets used more and more throughout your code, the more valuable it becomes to go back and carefully document it.

9.4 Checking Function Inputs

Often, a useful first step in writing a function is confirming that the inputs adhere to the data types, dimensions, or other rules we've established in our documentation. Why? Because if inputs do not meet these standards, our function may not operate as intended. A common saying in the industry is "garbage in, garbage out". Often, we trust that a user (or us) entering bad data (or wrongly shaped, or typed, data) into our function will cause a *catastrophic* failure which causes the code to stop and throw an error. For example, having two images of different dimensions in our directory will throw an error when we tell Numpy to stack those arrays.

But what is even more insidious, and harder to track down, is when our code inside our function *runs without error*, despite the input being incorrect. If this happens, the function may output incorrect values that get fed into other functions, and tracking down this issue may become challenging.

As usual, input checking is a trade off between time spent writing additional code and further progress on the actual algorithm itself. Once again, I usually implement such steps at the "production code" or "this is going in a paper" stage.

Returning to our image loading function, our only current argument is a string path to a directory location, so there are two things we can check: first, that the input is in fact a string, and second, that the location exists on the computer of reference. Both of these being incorrect would end up raising errors later (when we tried to use Astropy's loading function using path), but we can add these checks ourselves for completeness:

```python
import os
def load_directory_images(path):
    '''
    Loads a directory's worth of images into convenient
    storage units. Requires astropy.io.fits.
    Note: All Images in directory must be of same shape.

    Parameters
    ----------
    path: str
        path to the directory you wish to load, as a string.

    Returns
    -------
    image_stack: array_like
        A stack of all images contained in the directory.
        Array of shape (N,X,Y) where N is the number of images,
        and X,Y are the dimensions of each image.
    image_dict: dict
        A dictionary containing headers for each image, the keys
        are the same as the indices of the corresponding
        image in the image_stack
    '''
    if not isinstance(path, str):
        raise AssertionError('Path must be a string.')
    if os.path.isdir(path) is False:
        raise OSError('Path does not point to a valid location.')
    return
```

Our two checks have now been implemented. First, we check that `path` is an instance of type string—if it isn't, we raise an exception. You'll also notice I've changed the `pass` into a `return`. Returning is what we do at the end of a "finished" function, where we take values calculated in the function and "return" them to the overall code. At the moment, we have nothing to return, so nothing appears after the return statement.

Let's test our input-checking:

```
custom_path = 2
load_directory_images(custom_path)
```

```
        AssertionError                          Traceback (most recent
  call last)
        <ipython-input-27-0d9b0ed7f57c> in <module>
          1 custom_path = 2
    ----> 2 load_directory_images(custom_path)


        <ipython-input-26-f6f134fb9056> in load_directory_images(path)
         20       '''
         21       if not isinstance(path, str):
    ---> 22           raise AssertionError('Path must be a string.')
         23       if os.path.isdir(path) == False:
         24           raise OSError('Path does not point to a valid
  location.')


        AssertionError: Path must be a string.
```

Great! We've shown that our `AssertionError` correctly triggered when `cus-tom_path` was not a string as required.

Now let's create a variable that is indeed a string, but that isn't a location on my computer, and check that `os.path.isdir()` raised our error.

```
custom_path = '~/FolderThatDoesntExist/other_folder/'
load_directory_images(custom_path)
```

```
        OSError                                 Traceback (most recent
  call last)
        <ipython-input-28-9f0c7a13997b> in <module>
          1 custom_path = '~/FolderThatDoesntExist/other_folder/'
```

```
----> 2 load_directory_images(custom_path)


        <ipython-input-26-f6f134fb9056> in load_directory_images(path)
          22            raise AssertionError('Path must be a string.')
          23      if os.path.isdir(path) == False:
   ---> 24            raise OSError('Path does not point to a valid
  ⌐location.')
          25      return


        OSError: Path does not point to a valid location.
```

As a final check, let's put in a string that should work (a real location on my system):

```
real_path = '/Users/'
load_directory_images(real_path)
```

And, as expected, we see that our real path throws no errors.

We're now ready to actually write the function itself!

9.5 Local Scope and Global Scope

Before we go through the actual details of this particular example function, I want to talk about the concept of scope within our Python programs. So far, when working with scripts in which every line is a declaration or calculation or loop or conditional, everything exists within what is known as the *global scope* of the code.[6] That simply means that if we were to run a script in the interpreter, all the variables (at least, their final states, if we overwrite them in the process) would be accessible to us in the interpreter.

One caveat to this is iterators, which are created when you set up, e.g., a `for-loop`. In this case it's even more confusing: the *final* iterator will still be around after the loop, e.g.,

```
for i in range(10):
    continue
```

```
i
```

```
9
```

We see that I wrote a `for-loop` which did nothing but iterate over a list $[0,1,2,3,4,5,6,7,8,9]$. But later, I called the variable "i" and it was still 9, its last value from the loop.

This behavior is part of the reason we use standard iterator variables like `i`, `j`, `k` in our loops... because they're less likely to end up overwriting an important

---

[6] More technically, the variables in one python file when run in an interpreter are in the "module" scope. But when we only have one file we are working with, this is functionally the global, or outermost scope.

variable we want to use later. Of course, if we set up a new loop using i, it will be properly overwritten at the start of the loop.

So, as I've described it, all our variables are swimming together in the big pool that is global scope, and any variable can be accessed anywhere.

While this is convenient (especially when protyping code or carrying out exploratory work), keeping all variables in the global scope is highly prone to bugs—both those which raise exceptions, and those which don't, but produce incorrect results. Imagine the following scenario: a calculation begins with several variables, and then several hundred lines of code proceed to perform operations on those variables. You run your script, which is supposed to produce an image of a nebula, but all you see are NaNs. What happened? Because your script is all one scope, you would need to start interrogating (e.g., by printing) every variable along the way, until you found where one type of expected output was incorrect. Furthermore, if your script has to operate on many similar objects (e.g., in a loop), some of the object in the loop may trigger issues while others don't. All of this is more or less a nightmare to deal with.[7]

If you've ever tried to type an absurdly complicated expression into Mathematica (or Wolfram Alpha), you've seen this effect. When a nonsensical answer emerges, the only way to figure out why is to start breaking down the terms of the expression into small pieces evaluated separately. This is exactly what we want to do with our code, and functions give us the ability to do this.

Functions have what is called **local scope**. This means that any variable defined within a function *stays within that function*. It can't be accessed from outside the function, it can't be modified or overwritten by any code outside the function—it is completely walled off and isolated. Once we take a string and input it to our sample function here as path, for the purposes of the inside of the function, path is totally isolated. This also means that if we have some variable path floating around somewhere in the global scope, it won't accidentally wind up in our function.

There is one huge caveat. Local scope is not *two directional*. Anything accessible in the *global scope* is also accessible in the *local scope* of a function. It is the reverse that isn't true. Observe:

```
a = 3
b = 5
def func(c,d):
    return a+b+c+d

func(1,2)
```

11

Our function only takes 2 arguments, c and d. But inside the function, we *disregard local scope* and utilize a and b as well. As suggested by the name, global scope is truly global, even in functions.[8]

This seems to defy our desire to isolate small units of code (single chunks of calculations) into different, separated functions. So what's the solution?

Solution number 1: simply never call variables inside functions that aren't either inputs to the function or created within the function. For example:

```
a,b = 3,5
def func(a,b,c,d):
    return a+b+c+d

func(1,2)
```

```
        TypeError                         Traceback (most recent
call last)

        <ipython-input-43-00c7caa4d412> in <module>
        3       return a+b+c+d
        4
    ----> 5 func(1,2)


        TypeError: func() missing 2 required positional arguments: 'c'
    and 'd'
```

By specifying that the variables we call "a" and "b" in our function are positional arguments, we have *overwritten* the global scope and told our function that the "a" and "b" it needs to use are ones supplied by the user. Now,

```
func(a,b,1,2)
```

```
11
```

returns the same value, but we knew exactly what was going on along the way.

Another solution, of course, would be to have no variables in the global scope at all, that is, have *everything* isolated into functions. But this is typically impractical, most general use scripts we write will have at least *some* code hanging in the global name space. So solution number 1 is the most solid way to ensure you aren't letting bugs "leak" into your functions. Remember, it is perfectly fine to name a function's input the same name as something being used in the global scope (above, a, for example)—by explicitly defining it as an argument, now only variables passed to the function will be used internally.

---

[8] Most of the time. More on this later.

9.5.1 Debugging with Functions

When you start writing functions, you may run into the issue that debugging becomes more challenging. For example, you write a simple, 15 line function to do some task. You run it, and there's a bug—not an error raised, but the output is weird. But unlike in your script, you can't just look at the intermediate variables in the calculation anymore, because they were in the function!

```
def my_func():
    var = 1
    var2 = 3
    return var + var2


print(var)
```

```
          ⊔
    ------------------------------------------------------------------------------

        NameError                                Traceback (most recent⊔
    ⌐call last)

        <ipython-input-46-2ea387ab95ff> in <module>
    ----> 1 print(var)


        NameError: name 'var' is not defined
```

When we run commands in the iPython interpreter, or Jupyter notebook, etc., we are in the *global* namespace, so we cannot access the variables created inside the function. This often leads to the insertion of a multitude of print statements into our functions to check intermediate steps, but even this isn't ideal; sometimes we need to further examine those variables, interrogate their shape, or length, or other properties.

There are two ways to go with this. When you're starting out, I recommend play testing your code *outside of functions* in the global namespace, tweaking and bugfixing until things work. Then, when you're satisfied, copy that code into a function. Once you get more comfortable with high level programming, there are actually industry solutions, e.g., software that lets you actually "jump into" the namespace of a function and query the variable values. This is highly useful, but not necessary when you have the time and space to just test the code going into functions in a script environment or Jupyter notebook cell first.

Let's get back to our example function. We'll use the glob package to list the FITS files in the directory, set up some holder variables (a list and a dictionary), and then loop over the files, using Astropy to open each file and extract the header and image. These we append into the waiting containers.

```python
from glob import glob
from astropy.io import fits

def load_directory_images(path):
    '''
    Loads a directory's worth of images into convenient storage units.
    Requires astropy.io.fits, glob.
    Note: All Images in directory must be of same shape.

    Parameters
    ----------
    path: str
        path to the directory you wish to load, as a string.

    Returns
    -------
    image_stack: array_like
        A stack of all images contained in the directory.
        Array of shape (N,X,Y) where N is the number of images,
        and X,Y are the dimensions of each image.
    image_dict: dict
        A dictionary containing headers for each image, the keys
        are the same as the indices of the corresponding
        image in the image_stack
    '''
    if not isinstance(path, str):
        raise AssertionError('Path must be a string.')
    if os.path.isdir(path) == False:
        raise OSError('Path does not point to a valid location.')

    files_in_dir = glob(path)
    image_stack = []
    header_stack = {}
    for i,f in enumerate(files_in_dir):
        with fits.open(f) as HDU:
            image_stack.append(HDU[0].data)
            header_stack[i] = HDU[0].header
    image_stack = np.array(image_stack)
    return image_stack, header_stack
```

You may notice some assumptions built into the code, such as that the image and header of the FITS file are stored in the 0th extension of the HDU. For astronomical data from telescopes, this is almost always the case, but this would be an example of personal code in which we knew the format of the FITS images we were trying to load and thus which extension to choose. It's a useful aside, however, to consider that if we were writing general use code for a pipeline that would see many different FITS files of different internal storage systems, we'd need more robust code for dynamically loading them—e.g., we might make the extension an argument of the function.

9.6 Chaining Functions Together

Once you start writing your code into functions, you'll find that the output of function one tends to become the input of function two. For example, I could write a new function:

```python
def median_image(image_stack):
    '''
    Takes a stack of images and returns the median image.
    Parameters
    ----------
    image_stack: array_like
        stack of images, first dimension being image index.
    Returns
    -------
    median_image: array_like
        single image of the median of the input images
    '''
    median_image = np.median(image_stack,axis=0)
    return median_image
```

Now, if I wanted to median the first three images in my full stack, I could feed the following

```python
image_stack = load_directory_images(image_path)
first_3 = median_image(image_stack[0:2])
```

I again want to emphasize that we can call our variables *whatever we want* outside the functions and then feed them in. Often though, the names end up being similar or the same. What matters is the position of the input—i.e., is it the first, second, third, etc, input to the function. *That* is what determines what internal name it gets mapped to while inside the function.[9]

You might be wondering why you would write a function that had a single line of code as its calculation. The short answer is, you wouldn't—my `median_image()` function adds so little beyond your general use of `np.median()` that it isn't worth writing. But usually in astronomy, we don't just want a median. We want, say, a more complicated statistic that requires several lines of code to compute.

As a general guideline, I tend to put something in a function if it performs a single "task" or "unit" of my program which requires more than a few lines, or it is a chunk of code that is re-used extensively in my code. In short, if you find yourself copying and pasting a chunk of code often and changing the variable names for different cases, this is a prime opportunity to construct a function instead.

9.7 The Concept of Main()

So far, we've discussed the way one formats functions, and how to take the output from a function (i.e., listed in the return statement) and save it to a new variable,

---

[9] Until we arrive at keyword arguments, which we will in just a moment.

which can then be passed into other functions. How does this actually flow in a more major script's workflow?

One of the simplest ways is through a `main()` function. Let's say I've written four functions which do the following: load the images from a directory into a stack —clean each image somehow (maybe removing cosmic rays or bad pixels)—align the images (which were, say, dithered)—create "coadds" of the images by stacking them in various ways (mean, clipped mean, median, weighted mean).

Each function assumes general input and has general output. To make it specific, I could write a function, which we often simply call `main()`, like this:

```
def main(image_dir,cleaning_keyword,alignment_keyword,coadd_keyword):
    image_stack, header_stack = load_directory_images(image_dir)
    cleaned_images = clean_images(image_stack,cleaning_keyword)
    aligned_images = align_images(cleaned_images,alignment_keyword)
    coadded_images = coadd_images(aligned_images,coadd_keyword)
    return coadded_images
```

This would usually be the last function defined in our code, and we can see that we here indicate that main takes in all the info needed to run all the functions properly. Assuming that all works, we could then open a terminal, run our Python script, and then simply run something like `final_output = main(inputs)` function to run everything in sequence and get the final output.

We can also specify this function to be run automatically if the script is run. At the bottom of our Python script, below the `main` and other functions, we can add the following:

```
if __name__ == '__main__':
    main(image_dir,cleaning_keyword,alignment_keyword,coadd_keyword)
```

This is a conditional statement that checks whether our current Python file has been run. Essentially, when I open an iPython interpreter and type `run myscript.py`, Python automatically sets a "secret variable" called __name__ to '__main__', because the script is being run. You can put whatever you want inside this block, which is only True if you run the script directly (as opposed to importing functions from it). In this case I've chosen to put a function call to my own `main()` function inside. Now, if I open the interpreter and type

```
run myscript.py
```

It will execute my `main()` call automatically, without me having to type in `main (args)` into the terminal myself.

You may be wondering why you wouldn't simply have a call of your main function at the bottom of your script, without this weird conditional. And you're right: if you did that, the same thing would happen, and running the script would then run your main call, hence running all your functions. But something we haven't talked about yet, but will talk about *in detail* soon, is the idea of `importing` your own functions between Python files. When you begin doing this, it becomes considerably more important to have actual executions tucked away inside these conditionals that only run if our target file is run directly in the interpreter, rather than imported into another script.

9.8 Keyword (Optional) Arguments

Thus far, our discussion of the definition of functions has only included what are known as positional arguments. When I define a simple function like the following:

```
def func(a,b,c,d):
    return (a+b-c)*d
```

We can see clearly that the *position* of the four variables in the argument list matters. Whatever the first number we supply is will be deemed a, the second number I feed will be b, and so on. And this affects the output now, as an order of operations has been established (rather than a simple sum). If we flipped around the order of the numbers we fed in, we'd clearly get a different answer.

There are several other forms of argument, beyond positional. The first is an *optional*, *default*, or *key word* argument (the three terms are used interchangeably). This is extremely useful when we want to obey the golden scope rule above about not using any variables not asked for as arguments, but we *do* know that this value often takes a single value.

To give a concrete example, let's say I want to calculate the sine of some values in my code, and *usually* the angles I'm working with are in radians (which is what np.sin() requires) but *sometimes* they're in degrees. I can write a quick wrapper for my sine function as follows:

```
def my_sin(x,units='radian'):
    if units=='radian':
        return np.sin(x)
    elif units =='deg':
        new_x = x * np.pi / 180.0
        return np.sin(new_x)
```

The way this works is that my function *assumes* units to be "radians" unless otherwise specified:

```
my_sin(np.pi)
```

```
1.2246467991473532e-16
```

We see this returns 0 (to computer precision) as expected. However, if I specify different units:

```
my_sin(90,units='deg')
```

```
1.0
```

The code knew to convert my degrees into radians and then return the np.sin() value.

A cool thing about these types of arguments is that they are *not* positional. If a function has *only* keyword arguments, they can be entered in any order, and if a function has both positional and keyword arguments, then as you might expect, all the positional arguments must come first, then the keyword arguments can be passed in any order. You can think of this like our very first introduction to lists and dictionaries: lists are positional and can only be indexed by element, while dictionaries can be unordered because we index them by key. Python does the same with your function inputs: it uses the index of the positional arguments to assign their variable name within the function, and uses the provided optional keys to bind the relevant arguments. We can see in a quick example that Python will raise an exception if we break this rule, here using a simple `Matplotlib` command:[10]

```
plt.plot(x,alpha=0.9,y,color='red',ms=5,ls='-',label='points')
```

```
File "<ipython-input-58-1fd2328cc66e>", line 1
  plt.plot(x,alpha=0.9,y,color='red',ms=5,ls='-',label='points')
                         ^
SyntaxError: positional argument follows keyword argument
```

In short, if you define a function with three positional arguments and four keyword arguments, it might look like:

```
def func_args(x,y,z,a=1,b=25,c=None,d=False):
    if d:
        return x+y+z
    elif c is not None:
        print('wow!')
    else:
        return x+b
```

The above function is of course nonsensical, but make sure you understand the code flow that occurs. A final note about default arguments: do not use mutable datatypes (e.g., lists)! That is, do not set list_input = [] as the default to a function. The reason is that Python will reuse mutable defaults across function calls. If you want the default to be an empty list, set the argument list_input = None, and then check if the input is None within the function before creating a new empty list.

Exercise 9.1.

What would this function's output be if we ran it
- with only x,y,z entered?
- with only x,y,z,d entered?
- with only x,y,z,c,d entered?

---

[10] I chose this example because `Matplotlib` calls typically have many, many optional arguments that can be set to set a given line, point, axis, etc. exactly as desired.

9.9 Packing and Unpacking Function Arguments

What if a situation arises where we want our function to accept an unlimited number of arguments? To give a simple example: what if I want to write a sum() function that sums up as many numbers as you put into it? Of course, we could write a function that takes a single argument as a list or array, but for the sake of this example, how would we allow the user to enter as many numbers as possible?

The answer is to use Python's *packing* and *unpacking* syntax. For simple iterables in Python (e.g., lists), we can use an asterisk immediately before the iterator to pack or unpack it (much like a generator). When we use the asterisk in a function definition as one of the arguments, it will take any extra arguments passed and will pack them into a list which can be accessed in the function:

```python
def mysum(a,b,*args):
    running_sum = a+b
    for i in args:
        running_sum+=i
    return running_sum
```

```python
mysum(1,2)
```

```
3
```

```python
mysum(1,2,3,4,5,6)
```

```
21
```

In this example, we define all variables passed after b to be packed into a list called args.[11]

We can then access that list; in this example, I iterated through them and added them to the initial sum of the positional arguments. Of course, since they're already a list, a faster method would be:

```python
def mysum(a,b,*args):
    return a+b+np.sum(args)
```

where by faster I mean both fewer lines of code and more computationally efficient.

But what if the extra arguments we want to accept aren't in just any order, and we want to track that somehow? The solution is similar, but instead we'll use the double-asterisk packing format: **kwargs. This tells our function to accept any number of additional *keyword arguments*, like the ones we've been discussing above. For example:

```python
def pretty_print(string,**kwargs):
    print(string)
```

My pretty_print() function now requires a string input... but is happy to accept any other kwargs I throw at it:

---

[11] You need not use the variable name args, which is not special, but this is the standard across many codes and thus improves readability.

```
pretty_print('Hello, world!',subtext='Ive had my morning␣
    ↪coffee',energy_level=5)
```

```
Hello, world!
```

What happened to those extra keyword arguments? Like the example for `*args`, they got packed up, but this time into a dictionary of name kwargs, that can be accessed in the function. Again, the name kwargs isn't special, but is standard to use for this purpose. Let's see how we can utilize a passed keyword argument:

```
def pretty_print(string,**kwargs):
    print(string)
    if 'sep' in kwargs.keys():
        print(kwargs['sep'])
```

```
pretty_print('Hello, World!',sep='----------------')
```

```
Hello, World!
----------------
```

Our function is still agnostic to any extra keywords supplied. However, if one of those keyword arguments happens to be `sep`, our function does something special: it indexes the value of that key from the internal dictionary of kwargs and in this case prints it. Notice that in the packing, the argument *name* becomes a string key in the dictionary, and the argument value is then inserted into the dictionary with that key (without its data type being changed).

The above examples provide a base level of use, but may not seem that exciting. Why not just make `sep` an optional keyword of my `pretty_print()` function? In this example, that would be a reasonable course of action. What if, however, we had a function acting as a wrapper, and one of the *interior* function calls had many optional keyword arguments we might want to access. We could either duplicate *all* of them into our outer function's argument list... or we could add a `**kwargs` which gets passed through to the other function. By doing so, we pack, and then unpack, any number of useful arguments to the inner function, without adding many extra keyword arguments to our outer function. A common example of this is the passing of plotting parameters.

As a final note, packing and unpacking work when we *call* functions as well. If I have some list, `mylist = [a,b,c,d]`, and I *ran* a function with four positional arguments, I could run it via `func(*mylist)`. Python will unpack the list and feed each value into the function as if we ran `func(a,b,c,d)`. The same works for keyword arguments and dictionaries; a dictionary `mydict = {'a':1, 'b':2}` could be unpacked into a function which had two keyword arguments a and b as `func2(**mydict)`. Of course, in this case, the function would need to have no positional arguments.

9.10 Testing Function Outputs: Unit Testing

Earlier, we discussed the testing of inputs to your functions to ensure proper data types or any other restriction your function needs to produce sensible results. What about the output?

When we write functions, the goal is to take a large process (like reducing a set of data from raw images to science spectra) and reduce it into small, repeatable, single-task chunks so we can evaluate that each step is performing properly and independently. During the development of such a code, and such functions, you likely test the functions outputs yourself, manually—i.e., put in some sample data, make sure the output of the function makes sense.

The problem is that code lives and breathes. After inserting your code into a larger framework, you'll find you have to go back and tweak that function, add an extra input or output, modify one part of the calculation. A more advanced, but valuable way to ensure your functions still do what you want them to is by implementing what are known as **unit tests**.

Unit tests are extra pieces of code that throw sample problems with known outcomes at each of your production functions and ensure that the functions are operating as expected. For large scale collaborations with intense pipelines, the amount of code that exists in the unit tests may even exceed, or vastly exceed, the amount of production code actually doing the science! But it is these tests that make the scientists confident in every step of their pipeline, even as it evolves and changes over time.

While that sounds daunting, implementing unit testing is less challenging than it sounds. There are several frameworks that handle the unit testing for you. In this example, we'll be using `pytest`.

`Pytest` is `pip` installable, and simple to use. Simply create a file that starts with `test_` or ends in `_test.py` somewhere that you can access the functions of interest (say, in the same directory as your code—later we'll talk about how to put them in a separate tests directory). Assuming you've done this, inside your Python file for the test, you'll want to import `pytest` as well as your functions. For example, If we had all the functions discussed in this chapter in one Python file called `utility_-functions.py`, then in the first line of my `test_utilities.py` file I'd have

```
import pytest
from utility_functions import *
```

where here I'm simply importing all the functions we would've defined.

Next, we want to define some tests. The basic nature of defining a test is to create a function which runs your production function with some set input and asserts that the output is some known value. For example:

```
def test_load_images_from_directory():
    testing_path = '/
  some_path_I_never_mess_with_to_some_test_fits_files/'
    image_stack, header_dict = load_directory_images(testing_path)
    # I know that there are 7 sample images in that directory,
    # of image dimensions 1200 by 2400
    assert image_stack.shape == (7,1200,2400)
```

In the above example, our function goes over to some testing images I've saved somewhere for this purpose, and tries to load them with my function. I know things about those images—for example their dimensions, and that there are 7 of them. This means that the expected shape of the resulting image stack is (7,1200,2400). You're already used to checking equivalencies using ==, now we assert this equivalency.

Believe it or not, that's it! At least for setting up a simple test. Now, outside in the regular terminal, in this directory, simply type

```
pytest
```

and the software will locate any files that start with `test_` or end in `_test.py` in this directory, run any of the functions within, andreport on successes (places the assert is true) or failures (places where the assertion fails).

Now, any time we make changes to our `load_directory_images()` function, we can simply run `pytest` again to make sure we didn't break anything. (Of course, if we change the number of outputs, we have to adjust our test to reflect that, etc.)

There is a *lot* more to testing—for example, methods to test many inputs all at once, which we'll cover later in the chapter on building packages. But feel free to start setting up some very simple tests for your research code now!

You might be thinking, wouldn't it be great if the testing code just ran automatically any time I changed my research code? Good news, friend! This is the exact purpose of tools which provide **Continuous Integration (CI)**. Essentially, you can set up something similar to `pytest` which actually lives in the cloud and tests your code everytime you push a new commit to Github (or your version control service of choice). While there is no need at the undergraduate level to be trying to both host your personal research code on github AND have it continuously integrated and tested, it's always good to be aware it is an available option once your code gets complex enough to warrant it!

9.11 Type-Hinting

When we define a function in Python, we are not required to explicitly define the data types of the inputs and outputs. This is part of Python's "dynamic typing" framework that makes it convenient to code in (but slower to execute).

Nevertheless, it is often useful to specify what datatypes are expected to be input (and output) from Python structures like functions or classes. In part, it is an additional layer of documentation, allowing users to quickly assess if they are using a function correctly. One can then go deeper—tools such as `mypy` can be used to actually *enforce* type definitions, and we can write unit tests which check that data types are behaving as expected. Note that Python calls this process Type *Annotation* because Python itself does not check types; it simply does not throw syntax errors for doing so.

Let's look at a few examples of type annotations. Let's say we have written a simple function whose job is to fit a polynomial of some order to a set of data, (x, y), evaluate the fit across the domain of x at n equally spaced points, and return $(x_{\text{fit}}, y_{\text{fit}})$:

```
import numpy as np

def fit(x,y,order=1,n_pts=100):
    fit = np.polyfit(x,y,order)
    x_fit = np.linspace(x.min(),x.max(),n_pts)
    y_fit = np.polyval(fit,x_fit)
    return x_fit,y_fit
```

As we have defined it, it can be reasonably inferred that n_pts and order are likely integer inputs—but one might insert a float by accident. For x and y, on the other hand, it is unclear what the input needs to be, for example, is a numpy.array required, or can an equivalent list be supplied?

Within the function, the way we have written it, the input *cannot* be a list, because we employ the x.min() and x.max() methods which are not defined for lists—but are for arrays. We could (and perhaps should, as it is easy) write this function in such a way as to be agnostic to whether the input is a list or array. Either way, this illustrates why a type annotation may be useful. Let's rewrite this function to have both type annotations and doc-string:

```
def fit(x: np.array,
        y: np.array,
        order: int = 1,
        n_pts: int =100)->tuple:
    """wrapper for polyfit that returns fit-line on a linspace over
    domain of input x

    Parameters
    ----------
    x : np.array
        x-values to fit
    y : np.array
        y-values to fit
    order : int, optional
        degree/order of the polynomial fit, by default 1
    n_pts : int, optional
        number of points between min(x) and max(x) to evaluate fit, by
    default 100

    Returns
    -------
    tuple
        x at which fit was evaluated, y, the evaluated fit
    """
    fit = np.polyfit(x,y,order)
    x_fit = np.linspace(x.min(),x.max(),n_pts)
    y_fit = np.polyval(fit,x_fit)
    return x_fit,y_fit
```

To summarize, we can add colons to our function arguments and specify their data type (including non native types, like `np.array`). Slightly better practice would be to `import numpy` and use `numpy.array` as it is more readable, but the use of `np` is so ubiquitous that in this instance, we'll call things appropriately readable. We then add a small arrow after the final parenthesis in the function definition and specify the return type of the function, in this case a tuple.

What if an input to a function is allowed to be multiple types? For example, let's carry out what was suggested above and write this function such that native lists and Numpy arrays are equally valid inputs to the function. How do we now specify the type annotation? To ask a similar question, how do we specify input types that are, say, functions themselves?[12]

For these tasks, we will need to invoke the built-in Python module `Typing`, which will allow us to properly name various type-related requirements.

In the case of our "or" situation (list OR array), we can use the `Union` operator to specify that both are allowable. For the next set of examples, I am going to remove the docstring of our function for space.

```python
from typing import Union

def fit(x: Union[list,np.array],
        y: Union[list,np.array],
        order: int = 1,
        n_pts: int =100)->tuple:
    fit = np.polyfit(x,y,order)
    x_fit = np.linspace(np.min(x),np.max(x),n_pts)
    y_fit = np.polyval(fit,x_fit)
    return x_fit,y_fit
```

The syntax for the `Union` and other type-annotation calls is to use open-close square brackets as shown above. Additionally, it is possible to define new "data types" which are, e.g., the unions of others. If we did that, our code might look like this:

```python
DataLike = Union[list,np.array]

def fit(x: DataLike,
        y: DataLike,
        order: int = 1,
        n_pts: int =100)->tuple:
    fit = np.polyfit(x,y,order)
    x_fit = np.linspace(np.min(x),np.max(x),n_pts)
    y_fit = np.polyval(fit,x_fit)
    return x_fit,y_fit
```

---

[12] For example, the way `scipy.optimize.curve_fit()` has as the input some function which can generate models.

Another useful example of something from the typing library is the `Callable` type, which allows us to indicate that an input is another function (or formally, another object which can be called like a function). One can even go a step further, and specify what data-type needs to be inserted into *that* `Callable` object, using the same syntax as above.

```python
from typing import Callable

DataLike = Union[list,np.array]

def func(in_func: Callable[[DataLike],str]):
    pass
```

In this example, we define that the `in_func` argument of `func` is a `Callable`, and then we also use square brackets to show that the specific `Callable` input needs to take, as an argument, something that is `DataLike`, and return something that is a `str`.

There is a lot more to type annotations, including defining sequences and using user-defined classes, but we will end our discussion here — in general, most of the time, it is relatively easy to add type annotations to our code (and often we don't need many bells and whistles to do it). Later, in the chapter on Software Development, we will discuss more formally how and when one should do this, but suffice it to say that it is generally good practice.[13]

9.12 Summary

If variables are the atoms of code, functions are the molecules—a critical fundamental building block of larger, more complex programs. Learning how to write them, document them, and test them, is a critical step in becoming a better programmer. Here are the takeaways you should have at the end of this chapter:

- Functions isolate chunks of code in a *local* namespace which the rest of your code can't access, making them silos.
- Functions take arguments: You can specify positional, keyword (optional), and even infinite (*args or **kwargs) arguments.
- Inside your functions, you should only use variables made within the function or supplied as arguments—no dipping into the global namespace!
- Inside your functions, you should always add documentation (of some kind) to establish the function's purpose, its inputs, and its outputs
- It's often worth taking a few lines to check that inputs match the requirements of the function and raise errors if they don't.

---

[13] If you find that your function takes in many very different data types for an argument and your `Union` definition is getting unwieldy, this likely points to a poorly designed function.

- Functions return things—if you don't include a return statement, the calculations in the function go away when called. We set new variables equal to the function called with some parameters, and place what we want the function to output in the `return` statement line.
- Just like we check the inputs inside a function, we can check the output of a function using unit testing to make sure it is operating as intended.

Astronomical Python
An introduction to modern scientific programming
Imad Pasha

Chapter 10

Classes and Object Oriented Programming

10.1 Introduction

A common adage in the Python community is that "everything is an object." It is more or less true: most of the elements of Python we have interacted with thus far (e.g., lists, arrays, dictionaries) are what the Python language would refer to as an object.

So what is an object?

In a practical sense, objects are sometimes best thought about in the context of properties and behaviors that define them. A similar adage exists in physics: "a tensor is something that transforms as a tensor." An object in Python is something defined via a "class"— an abstracted scaffold of code which allows us to attach information (called attributes) and functions (called methods) to a particular entity. The most important thing to remember as we work through this chapter is that *you've been working with classes and objects this whole time.* In order to define new ones (in the same way we defined our own functions), we'll need a bit of extra boilerplate code beyond a simple def func() line. The reason is that the behavior of classes (which define objects) is more complex than that of a function.

Let's start with a simple example. If we define a list in Python, that list is an object. It has two key accessible elements—attributes and methods. We've used these before, and they're accessed in a syntax known as ***dot notation***. To add a new value to our list, we might run mylist.append(10). We add a dot to the end of the variable name (an instance of a list), and then use the type of formatting we're used to when calling a function: the function name with closed parentheses containing any arguments. In this case, we won't call append() a function; we'll call it a ***method*** instead. The distinction is purely that a method is a function which *belongs* to a class and which typically operates in some way upon the object itself (in the example above, the list adding a new value to itself).

doi:10.1088/2514-3433/acfa9ach10

Attributes are a means to store information within an object which can easily be queried either by users, outside the class, or from methods within the class itself. We access these via dot notation as well, but don't need the parentheses like we do for a method call. For example, we can access the shape of some `Numpy` array `arr` by calling `arr.shape`. This should also be a familiar operation. So as we can see, one way to say "objects behave like objects" is to say that an object has certain attributes and methods defined for it, which can be used to interact with that object.

There are additional properties that can be defined for objects, beyond methods and attributes. For example, as you may remember, the behavior of adding two lists together, and adding two Numpy arrays together, is markedly different:

```
np.array([1,2,3]) + np.array([4,5,6])
```

```
array([5, 7, 9])
```

```
[1,2,3] + [4,5,6]
```

```
[1, 2, 3, 4, 5, 6]
```

The behavior of two objects when we use the operators (e.g., addition, multiplication) between them is also something defined within the object in its class definition.

Object Oriented Programming, or OOP for short, refers to a style of code writing that relies primarily on the creation and use of objects and their attributes and methods to solve problems. It is important to stress that for some applications (specifically simple research scripts with straightforward, linear structures), OOP is neither necessary nor advantageous to use. Often, a mix of functions and global lines in a script are sufficient. But as your code grows more complex, *or if you want to make it user friendly for others to use*, OOP is a good style to use.[1] This is because classes provide a natural organization to code, and encourage an API (Application Programming Interface) that is reasonable in scope and intuitive to use.

In the following sections, we are going to build up a class from scratch, adding features and complexity as we go. By the time you get to Subclasses and Superclasses, you will have reached the end of what I would deem introductory material on this subject. Classes can still be useful organizational structures if all we do is create some attributes and methods. But they become increasingly more powerful as we begin using some of the more advanced features in the later sections of this chapter.

For our example this chapter, we'll be creating some custom objects to aid in the planning and execution of astronomical observations. Observational astronomers spend a significant amount of time on this task; time at a telescope (e.g., Keck Observatory) is precious, and one needs a lot of information ready to go to at

---

[1] It's also a standard in industry, making it good to be familiar with.

moments' notice while observing. What type of information? A non inclusive list might include

- target lists, with astronomical coordinates and epoch,
- finder charts, indicating what the starfield in the target vicinity looks like,
- blind offset stars (bright stars near a target that would be too faint to see normally), and the offset between those stars and the target
- background "blank sky" fields, usually for sky subtraction
- details about how to set up the instrument (e.g., for longslit spectroscopy, the slit length and width to use, the grating or grism to use to disperse the light, dichroic, etc.)
- details about the observing strategy (exposure times, when to observe each target, priority lists)
- airmass (or altitude) plots, showing the observability of each target throughout the night (and rise and set times)

Depending on the type of observation, there may be even more to keep track of. Clearly, there is some benefit to be found in organizing this information in some kind of container. Since many of the things above require actual computations (e.g., using `spherical_offsets_to` from `astropy`), it might be beneficial to do so in a code environment, rather than a simple document.

As we look at this list of needed information, we can think about how to best organize it. Here, I will suggest we create a class, defining a new Python object, called `Target`. This object will hold all of the information we need to know about each of our targets, such as their coordinates, their offset stars, their finder charts, etc. We can then organize a group of `Target` objects into an `ObservingPlan` of some kind. Let's start with the `Target` class.

10.2 Defining Classes

We define a class much the same way as we do a function, but with one key difference: the initial arguments of a class aren't placed in parenthesis in the definition line, they are placed in a special function known as a *constructor*:

```python
from astropy.coordinates import SkyCoord

class Target:
    def __init__(self,
                 target_name:str,
                 target_coordinates:SkyCoord=None):
        self.target_name = target_name
        self.target_coordinates = target_coordinates
    def some_method(self):
        pass
```

We use the special word `class` to begin our class definition and then provide a name for our class (it is standard practice to capitalize classes to better distinguish them from functions or variables). But *inside* the class (i.e., indented), we begin writing what look like functions. The first of these is the constructor, which is defined by having a dunder (double underscore) `init` dunder.[2] The first argument of this, and all functions inside a class[3] is the word `self`. The word `self` isn't special here; it is the "first position in the method arguments" that makes it special, but self is the standard name to use.

What does `self` do? Recall here that we are defining a new object, and thus need to have a way of assigning, and referencing, attributes and methods. You can think of our class definition here like a scaffold — we are telling Python where to "slot in" information later on, when we actually try to use our class. The `self` variable is how we refer to the class itself. When we set `self.target_name` equal to something, it means later, if we create some object, e.g.,

```
m81 = Target('M81')
```

we can access the target name using dot notation, as we've often done with lists and arrays and every other Python datatype:

```
m81.target_name
```

```
'M81'
```

So as a simple rule: anything you want objects to be able to access as attributes via dot-notation should be defined inside the class by setting `self.ATTR = VALUE`. There is actually one other way to set attributes, which is helpful when we know the attribute name as a string instead:

```
class Target:
    def __init__(self,
                 target_name:str,
                 target_coordinates:SkyCoord=None):
        self.target_name = target_name
        self.target_coordinates = target_coordinates
    def set_coordinates(self,coordinates):
        setattr(self,'target_coordinates',coordinates)
```

We've now modified the dummy method to essentially overwrite the coordinates stored in the class, but rather than setting `self.target_coordinates=coordinates`, we use the special `setattr()` function, feeding `self` as the object to set the attribute of, the string `target_coordinates` as the name of the attribute, and the input `coordinates` as the value. Let's see it in action:

---

[2] Class methods that use this __name__ format are known as dunder methods colloquially.
[3] With several exceptions, e.g., see below.

```
m81_coord = SkyCoord.from_name('M81')

m81 = Target('M81')
m81.set_coordinates(m81_coord)
print(m81.target_coordinates)
```

```
<SkyCoord (ICRS): (ra, dec) in deg
    (148.8882194, 69.06529514)>
```

In this example, where the provided name can be parsed by SkyCoord.from_name(), our class could do the work of finding the coordinates for us. But that won't be the case for every target (e.g., those not in any astronomical catalog). Let's write a short try/except statement to at least *attempt* a name parsing:

```
from astropy.coordinates.name_resolve import NameResolveError
import warnings


class Target:
    def __init__(self,
                 target_name:str):
        self.target_name = target_name
        try:
            self.target_coordinates = SkyCoord.from_name(self.
    ↪target_name)
        except NameResolveError:
            self.target_coordinates = None
            warnings.warn('Coordinates could not be parsed from name;␣
    ↪please set manually.')

    def set_coordinates(self,coordinates):
        setattr(self,'target_coordinates',coordinates)
```

To accomplish this check, as we learned in Chapter 4, we can catch cases when SkyCoord throws a NameResolveError — but instead of stopping the code, as the exception would, we have imported the warnings package and used warnings.warn() to provide the user information without stopping code execution. Let's see this in action.

```
m81 = Target('M81')
print(m81.target_coordinates)
```

```
<SkyCoord (ICRS): (ra, dec) in deg
    (148.8882194, 69.06529514)>
```

The resolve worked for our galaxy with a known name. Let's try with a custom name:

```
custom_galaxy = Target('N15738-J')
print(custom_galaxy.target_coordinates)
```

```
None
  UserWarning: Coordinates could not be parsed from name;
please set, manually.
```

As we can see, the code ran successfully, but warns us that the target has no coordinates set. As a note, `UserWarning` is the default used when we simply call the `warnings` module, but we could also define our own custom warnings if we wished.[4]

Let's add a new method to our class to calculate the offset between a star and our target. Telescopes usually ask for these in arcseconds north, and arcseconds east (so west and south are negative).

```python
import astropy.units as u

class Target:
    def __init__(self,
                    target_name:str):
        self.target_name = target_name
        try:
            self.target_coordinates = SkyCoord.from_name(self.
  target_name)
        except NameResolveError:
            self.target_coordinates = None
            warnings.warn('Coordinates could not be parsed from name;
  please set manually.')

    def set_coordinates(self,coordinates:SkyCoord):
        setattr(self,'target_coordinates',coordinates)

    def add_offset_star(self,coordinates:SkyCoord):
        if not hasattr(self,'target_coordinates'):
            raise AssertionError("Cannot add offset star if target
  coordinates not defined.")
        offsets = coordinates.spherical_offsets_to(self.
  target_coordinates)
        self.offsets = [i.to(u.arcsec) for i in offsets]
```

I've used this opportunity to highlight the `hasattr()` function, which works similarly to `setattr()` but instead checks whether an object has a given attribute and returns a boolean. In this case, because one can initialize a custom object without coordinates, the case might arise when a user attempts to run the `add_offset_star()` method *before* having run the `set_coordinates()` method. This won't do, as we need two sets of coordinates to calculate the offset between them. So here, if `self` does *not* have a target coordinate attribute, we raise an error to specifically indicate to the user that target coordinates need to be set.[5] After that

---

[4] I discuss this briefly in Chapter 14.

[5] Thinking about these types of situations (and minimizing the bugs that can occur as a result) is a critical component of API and software development.

check is handled, we simply use the relevant **Astropy** function to compute the offsets and parse their units to arcseconds.

```python
m81 = Target('M81')
m81.add_offset_star(SkyCoord(ra=148.8282194, dec=69.0529514,unit='deg'))
print(m81.offsets)
```

```
[<Angle 77.17760913 arcsec>, <Angle 44.47520345 arcsec>]
```

It's also worth pausing to highlight an incredibly useful feature of classes. Notice that in our **add_offset_star()** method, we can access **self.target_coordinates**, even though it is not an argument to the method. In essence, classes give us the ability to create a sort of "intermediate" scope: by explicitly setting certain variables as attributes, they become available within *any* of our class methods, without muddying up the global namespace. This feature can also become a bug if we are not careful, so it is important to be selective when deciding what to set as attributes and what to handle via method calls and returns.

10.3 Setters and Getters

Notice that our **add_offset_star()** method does not actually *return* anything (and when we run it, we don't set a variable equal to it). In coding parlance, this method is what is known as a **setter**—when run, it *sets* a particular attribute (in this case, **self.offsets**). We could also write a corresponding method called a **getter**, which is responsible for retrieving the offsets (as opposed to simply using dot notation to access it, as above). That might look like this:

```python
class Target:
    def __init__(self,
                 target_name:str):
        self.target_name = target_name
        try:
            self.target_coordinates = SkyCoord.from_name(self.
 target_name)
        except NameResolveError:
            self.target_coordinates = None
            warnings.warn('Coordinates could not be parsed from name;
 please set manually.')

    def set_coordinates(self,coordinates:SkyCoord):
        setattr(self,'target_coordinates',coordinates)

    def add_offset_star(self,coordinates:SkyCoord):
        if not hasattr(self,'target_coordinates'):
            raise AssertionError("Cannot add offset star if target
 coordinates not defined.")
```

```
        offsets = coordinates.spherical_offsets_to(self.
  target_coordinates)
        self._offsets = [i.to(u.arcsec) for i in offsets]
    def get_offsets(self):
        return self._offsets
```

Here, we modify our attribute to have an underscore, _offsets. Formally, this doesn't do anything, but in general, it is assumed that underscored attributes are "internal" and shouldn't be accessed by users directly. We then add a get_offsets() method which simply returns them.

```
m81 = Target('M81')
m81.add_offset_star(SkyCoord(ra=148.8282194, dec=69.0529514,unit='deg'))
m81.get_offsets()
```

[<Angle 77.17760913 arcsec>, <Angle 44.47520345 arcsec>]

That's not all that interesting—we've added two lines of code to accomplish the same task. But what if we wanted to store the offsets *internally* in one format, but print them to the user in another? We could use our get_offsets() method to make those changes before returning that to the user. For example, let's say we wanted to change the way offsets are returned to be a string of the format 77'' N, 44'' E. We could put this logic into our getter method:

```
class Target:
    def __init__(self,
                  target_name:str):
        self.target_name = target_name
        try:
            self.target_coordinates = SkyCoord.from_name(self.
  target_name)
        except NameResolveError:
            self.target_coordinates = None
            warnings.warn('Coordinates could not be parsed from name;
  please set manually.')

    def set_coordinates(self,coordinates:SkyCoord):
        setattr(self,'target_coordinates',coordinates)

    def add_offset_star(self,coordinates:SkyCoord):
        if not hasattr(self,'target_coordinates'):
            raise AssertionError("Cannot add offset star if target
  coordinates not defined.")
        offsets = coordinates.spherical_offsets_to(self.
  target_coordinates)
        self._offsets = [i.to(u.arcsec) for i in offsets]
    def get_offsets(self):
        return f"""{self._offsets[0].value:.1f}'' N, {self._offsets[1].
  value:.1f}'' E"""
```

Now, the internal self._offsets attribute is still a list of Astropy quantities, but running the get_offsets() method returns a string:

```
m81 = Target('M81')
m81.add_offset_star(SkyCoord(ra=148.8282194, dec=69.0529514,unit='deg'))
print(m81.get_offsets())
```

```
77.2'' N, 44.5'' E
```

As it turns out, there is actually a clever way to allow users to access an attribute *through* a getter method but using the same dot notation as they would for an attribute. That is, from the outside, it appears an attribute is being accessed, but a method is actually being run under the hood. This is accomplished via the @propery decorator. We can convert our above get_offsets() method above to behave this way by adding the @property decorator over it. We'll also change the name of the method to offsets(), because that's the "attribute-like" name we want users to query by:

```
class Target:
    def __init__(self,
                 target_name:str):
        self.target_name = target_name
        try:
            self.target_coordinates = SkyCoord.from_name(self.
    target_name)
        except NameResolveError:
            self.target_coordinates = None
            warnings.warn('Coordinates could not be parsed from name;
    please set manually.')

    def set_coordinates(self,coordinates:SkyCoord):
        setattr(self,'target_coordinates',coordinates)

    def add_offset_star(self,coordinates:SkyCoord):
        if not hasattr(self,'target_coordinates'):
            raise AssertionError("Cannot add offset star if target
    coordinates not defined.")
        offsets = coordinates.spherical_offsets_to(self.
    target_coordinates)
        self._offsets = [i.to(u.arcsec) for i in offsets]
    @property
    def offsets(self):
        return f"""{self._offsets[0].value:.1f}'' N, {self._offsets[1].
    value:.1f}'' E"""
```

Now, when we simply query the "attribute" offsets, we should in fact trigger the getter method and retrieve a string, not the astropy.Quantity list.

```
m81 = Target('M81')
m81.add_offset_star(SkyCoord(ra=148.8282194, dec=69.0529514,unit='deg'))
print(m81.offsets)
```

```
77.2'' N, 44.5'' E
```

As we can see, from the "user side," we access the offsets in the convenient attribute format, but under the hood, our class is performing some modifications to the internally kept attribute before returning it. In a similar vein, we can use this pattern to make any necessary checks or raise warnings and exceptions if internal attributes are not set properly.

10.4 Representation

At the moment, if we print out our m81 object wholesale, we get something rather unhelpful:

```
print(m81)
```

```
<__main__.Target object at 0x7f86183f5de0>
```

This printout tells us we have a Target object and gives us its memory address. But remember, when we print, e.g., a np.array() object, we get a much more useful output:

```
import numpy as np
print(np.arange(3))
```

```
[0 1 2]
```

The difference here is that all Python classes have a __repr__() method that defines what gets printed when we run print() on the object. The default is what we see above. But we can write our own method to make for a more useful view. The writers of Numpy did so for all classes in that library. Let's update our Target class to instead give us a more useful output:

```
class Target:
    def __init__(self,
                 target_name:str):
        self.target_name = target_name
        try:
            self.target_coordinates = SkyCoord.from_name(self.
    target_name)
        except NameResolveError:
            self.target_coordinates = None
            warnings.warn('Coordinates could not be parsed from name;␣
    please set manually.')

    def set_coordinates(self,coordinates:SkyCoord):
        setattr(self,'target_coordinates',coordinates)

    def add_offset_star(self,coordinates:SkyCoord):
        if not hasattr(self,'target_coordinates'):
            raise AssertionError("Cannot add offset star if target␣
    coordinates not defined.")
```

```
        offsets = coordinates.spherical_offsets_to(self.
    target_coordinates)
        self._offsets = [i.to(u.arcsec) for i in offsets]
    @property
    def offsets(self):
        str_out = f"""{self._offsets[0].value:.1f}'' N, {self.
    _offsets[1].value:.1f}'' E"""
        return str_out

    def __repr__(self):
        outstr = f'Target object for {self.target_name} with
    coordinates {self.target_coordinates.to_string()}'
        return outstr
```

Now, when we create an instance of our class and try to print it, we'll get something useful:

```
m81 = Target('M81')
print(m81)
```

```
Target object for M81 with coordinates 148.888 69.0653
```

10.5 Subclasses (and Superclasses)

Thus far, we have created a single class to handle our targets.[6] A useful feature of Python classes, however, is the ability to create subclasses. Subclasses allow us to inherit all of the structure and methods of a class, but add more specific features.

This is best explored with an example. So far, our **Target** class can handle the input of some information that is pretty general to all observations—coordinates, name, and offset star if applicable. Now let us ask the question: what if our observing plan involved taking images, or taking spectra? We could add methods to our class to handle the input of either, e.g., a **add_imaging_parameters()** and **add_spectroscopy_parameters()**. Or, we could create a subclass as follows:

```
class ImagingTarget(Target):
    def __init__(self,target_name:str):
        super().__init__(target_name)
    def add_filter(self,filtname:str):
        self._filtname=filtname

class SpectroscopyTarget(Target):
    def __init__(self,target_name:str):
        super().__init__(target_name)
    def add_slit_params(self,
                        slit_length:float,
                        slit_width:float,
                        slit_PA:float):
        self._slit_length = slit_length
        self._slit_width = slit_width
        self._slit_PA = slit_PA
```

---

[6] Technically, our class is a subclass of the **object** superclass.

In order to make a subclass, we define a new class with a new name, but add parentheses and reference our Target class there.[7] When we create these subclasses, we use the super() function within our constructor to, in essence, run the Target constructor. We then define subclass-specific methods. All of the methods and attributes of Target will be available to both ImagingTarget and SpectroscopyTarget; this is known as *inheritance*.

```
m81_im = ImagingTarget('M81')
m81_im.add_offset_star(SkyCoord(ra=148.8282194, dec=69.
  .0529514,unit='deg'))
print(m81_im.offsets)
```

77.2'' N, 44.5'' E

Here, despite the fact that in the class definition for ImagingTarget, we don't create an add_offset_star() method, we can call it, because it was a method of Target, which we inherit from. But now, since we have made an ImagingTarget object, we *also* have the ability to set a filter for our observation:

```
m81_im.add_filter('G')
```

On the other hand, our imaging object has no method called add_slit_params(), because those are not attributes needed when setting up imaging. We can take this process of subclassing and inheritance as many layers deep as we wish; e.g., we could make a subclass of SpectroscopyTarget:

```
class lrisTarget(SpectroscopyTarget):
    def __init__(self,target_name:str):
        super().__init__(target_name)
        self.dichroic_list = ['D460','D500','D560','D680']
    def set_dichroic(self,dichroic:str):
        if dichroic not in self.dichroic_list:
            raise AssertionError(f'Dichroic not in list. Choose from
  .{self.dichroic_list}')
        else:
            self.dichroic=dichroic
```

Once again, we create a class and indicate the superclass we are inheriting from—this time, from SpectroscopyTarget. Now, we have all methods of Target *and* any specific methods from SpectroscopyTarget at our disposal. We can use this class as we would those; but I have added a method for setting a dichroic (light splitter) that is entirely specific to the Keck/LRIS spectrograph. We could easily make subclasses of SpectroscopyTarget for any number of instruments, giving us the flexibility to hard code certain attributes unique to those instruments.

---

[7] If one wished to be fully complete, our Target class could've been defined with class Target(object); Python does this for us automatically.

10.6 Static Methods

I hinted above that there are special types of method for which self is not the first argument. Static methods are methods of a class that do not require an *instance* of a class object to exist in order to be run. One example of this that we have used throughout this textbook (and in this example) is SkyCoord.from_name(). Notice how we don't have to first instantiate a SkyCoord object and then run the method via dot notation—instead, the class itself, as imported, is dot-notated directly to run the method. In this case, the returned result is a SkyCoord object itself. Put another way, the static methods are bound to the class, rather than to objects of that class (hence the lack of a self reference).

When are static methods most useful? Generally, as in the SkyCoord case, static methods come in handy when we need some sort of utility method that relates to the class. We *could*, in general, just write a separate function to handle that type of utility, but using a static method instead keeps things nicely organized.

Let's see this in an example. As currently written, our Target class has a method that computes the offsets to an offset star. At current we must *first* instantiate a Target object (and make sure it has coordinates set), *only then* can we compute offsets. But sometimes, when sitting at the telescope, we need to compute new offsets, e.g., after adding a last minute target or new offset star. It might be handy to have the *functionality* of our current offset calculator, including parsing the output, but be able to call it as Target.compute_offset(coord1,coord2). Let's implement that feature, and then change our internals to use it.

```python
class Target:
    def __init__(self,
                 target_name:str):
        self.target_name = target_name
        try:
            self.target_coordinates = SkyCoord.from_name(self.
 target_name)
        except NameResolveError:
            self.target_coordinates = None
            warnings.warn('Coordinates could not be parsed from name;␣
 please set manually.')

    @staticmethod
    def compute_offsets(coordinate_1,coordinate_2):
        offsets = coordinate_1.spherical_offsets_to(coordinate_2)
        offsets = [i.to(u.arcsec) for i in offsets]
        return offsets
```

```
    def set_coordinates(self,coordinates:SkyCoord):
        setattr(self,'target_coordinates',coordinates)

    def add_offset_star(self,coordinates:SkyCoord):
        if not hasattr(self,'target_coordinates'):
            raise AssertionError("Cannot add offset star if target
coordinates not defined.")
        self._offsets = self.compute_offsets(coordinates,self.
target_coordinates)
    @property
    def offsets(self):
        str_out = f"""{self._offsets[0].value:.1f}'' N, {self.
_offsets[1].value:.1f}'' E"""
        return str_out
```

A few things happened here. First, we add a new method, `compute_offsets`, which does *not* take the `self` argument, and has a `@staticmethod` decorator to tell Python the following method will be a static method. It takes two coordinates, computes the offsets, and returns them as a list with the values in arcseconds. We also modified our `add_offset_star()` method to simply call the static method internally when setting the offsets after a coordinate has been added. So now, we have two ways to use the same method. We can set offset stars as before:

```
m81 = Target('M81')
m81.add_offset_star(SkyCoord(ra=148.8282194, dec=69.0529514,unit='deg'))
print(m81.offsets)
```

```
77.2'' N, 44.5'' E
```

Or, now, if we wished, we could also compute some offsets on the fly:

```
c1 = SkyCoord(ra=148.8282194, dec=69.0529514,unit='deg')
c2 = SkyCoord(ra=148.832, dec=69.052613,unit='deg')
print(Target.compute_offsets(c1,c2))
```

```
[<Angle 4.86577537 arcsec>, <Angle -1.21809008 arcsec>]
```

In this particular example, using our static method is only slightly more convenient than calling the spherical offset method itself (the only extra step is the conversion to arcseconds). But hopefully it is clear that if you have a static method which carries out several more operations than this, the pattern becomes increasingly useful.

10.7 Abstract Base Classes

Let's assume for a moment that we have built up our `Target` class into a true powerhouse of a code—maybe it would be useful for other people as well. We could put the code up on Github for people to download and use. But we'll quickly run into a problem: everyone around the world is using different telescopes and instruments. We may have defined some useful subclasses for ourselves, but other users might have different needs.

Sometimes, code we distribute is meant to be used, but not modified (i.e., we don't want users actually adding code to the "under the hood" codebase). Here, we might actually *want* users to be able to define their own special classes. To this end, *Abstract Base Classes* come to the rescue.

These classes (also called abc's) take the idea of a class as a scaffold even further. We can use them to create classes where we explicitly expect the class to be subclassed, and can specify methods that are required for use.

In our above IrisTarget example, we began adding methods directly related to one instrument. One could imagine that instead of subclassing a Target, we had a whole new class, called BaseSpectrograph. This class's purpose would be to define any aspects and properties of the spectrograph and its setup. We could then have SpectroscopyTarget objects actually read in BaseSpectrograph objects wholesale, and use the attributes and methods within to carry out any necessary computations.

In this scenario, we will set up BaseSpectrograph as an abstract base class, creating a scaffold on which *users* define their *own* spectrographs.

```python
from abc import ABC,abstractmethod

class BaseSpectrograph(ABC):
    def __init__(self,name):
        self.name = name

    @abstractmethod
    def get_detector_ndims(self):
        pass
    @abstractmethod
    def get_dichroic_list(self):
        pass
```

Our class definition for BaseSpectrograph looks similar to the classes we have created thus far, but because it inherits from the ABC superclass,[8] it has a few special properties. One of those is that when we use the @abstractmethod decorator, we are telling Python that whenever someone tries to create a Spectrograph object (subclassing this BaseSpectrograph), they *must define* methods with these names. For example:

---

[8] Technically, the ABCMeta metaclass, but the distinction is not critical here.

```
class LRIS(BaseSpectrograph):
    def __init__(self,name):
        super().__init__(name)

spectrograph = LRIS('lris')
```

```
TypeError  Traceback (most recent call last)
Cell In[40], line 5
      2     def __init__(self,name):
      3         super().__init__(name)
----> 5 spectrograph = LRIS('lris')

TypeError: Can't instantiate abstract class LRIS with abstract methods
  get_detector_ndims, get_dichroic_list
```

When I try to make a class using `BaseSpectrograph` as its superclass and define no methods, my attempt to instantiate an object of that class fails, and the warning indicates that the abstract methods `get_detector_dims()` and `get_dichroic_list()` aren't available. We can fix this by adding methods with those names to our subclass:

```
class LRIS(BaseSpectrograph):
    def __init__(self,name):
        super().__init__(name)
    def get_detector_ndims(self):
        return (2,4000,4000)
    def get_dichroic_list(self):
        return ['D460','D500','D560','D680']

spectrograph = LRIS('lris')
```

By adding these methods to our class, Python can create it successfully.

This level of abstraction becomes very useful when we go to add our `Spectrograph`-type objects into, e.g., the `Target` class. By using this formalism, we *guarantee* that every object entered will have a certain set of methods we can call. So we simultaneously increase the flexibility of our codebase (by letting users write their own spectrograph classes) while also maintaining regularity, so that we can treat all spectrograph-objects in the same manner within our codebase, ensuring the necessary methods and attributes will exist when we need them for a computation.

10.8 Summary

There is much, much more we could explore when it comes to classes — as the fundamental building blocks of Python, there is obviously a huge degree of functionality we have not covered. For example, you could define operation methods, e.g., __add__() or __lt__(), to control what happens when users apply operations (+,-,/,*, <,>) between instances of your class.[9] You can play games with inheritance, or create complex composite classes in which attributes are, in fact, other classes. You can enable "method chaining" (e.g., `object.method1().method2().method3()`) by having methods return `self`. We discussed setters and getters; one can use `@attribute.setter` decorators to directly control how attributes are set (similar to our discussion of the `@property` decorator for attribute retrieval). The list goes on, but for the purposes of an introductory text, we will conclude here.

When learning Python for the first time, nearly all learners struggle with classes. They are the nuts and bolts of the language, at some level, and can often seem overly complex and full of boilerplate. The gradual transition from fear of classes to their daily use tends to track with your progress with the language overall, and the rate at which your code is changing from simple line-by-line scripts to sets of functions to something that can reasonably be called "software." Do not be discouraged if for some time you struggle to see where classes would be most useful in your code, or struggle with implementing them. My suggestions for a steady practice with classes are to

1. Remember that at their simplest, classes define a container for some attributes (much like a dictionary) and some custom functions (methods),
2. Remember to add an __init__(self,…) method to the class, so that instances of it can be instantiated, and
3. Remember to start each (normal) method with self as the first argument, and refer to attributes and methods within the class code with `self.XXX`.

Ultimately, classes help us define an API (application program interface) that makes code cleaner and easier to use (by ourselves and others) by moving many of the complex calculations under a layer of abstraction. It is increasingly true that nearly every time you install a Python package, it will have been created using classes, and you will be using those objects and their methods when interacting with the package. So understanding how classes work is a useful exercise when reading or trying to fix others' code, even if you do not use them extensively yourself.

---

[9] This is known as *operator overloading*.

Chapter 11

Data Science with Astronomical Catalogs

11.1 Introduction

Catalogs are an invaluable resource for carrying out science on many objects at once. Catalogs are most often released for large surveys that were always designed to produce data for widespread use (e.g., SDSS, 3D-HST). As opposed to the more "raw" data formats discussed thus far in this text (e.g., images, spectra), catalogs tend to be tabular, containing one or more tables in which each row corresponds to an object (star, galaxy, etc.) while each column is a measured or known property (e.g., RA, DEC, magnitude, redshift, flux in some band).

Working with catalogs requires, in some ways, a very different skillset from working directly with raw astronomical data. In particular, the focus shifts to tasks like filtering data on values in different columns, joining tables on shared (unique) keywords, cleaning bad values, and visualizing trends across large samples (e.g., via density estimation). The skills garnered while working with catalogs are some of the most directly transferable outside of astronomy—tabular data science (using tools like pandas and SQL, which often then leads to the training of machine learning models on said data) is a mainstay of industry.

As an example: let's say you've found a research mentor—graduate student, postdoc,[1] or faculty—and they have agreed to help you get started with research. Generally, this will involve two things:—they will provide a (possibly overwhelming) list of papers and review articles for you to read, and they will provide some data related to the project.

Then they might say something like "As you skim through these papers to get a feel for what you will be doing, try playing around with these data a bit and see if you can make a plot of X vs Y." Or, they might say "Go online and download the

---

[1] A postdoc (postdoctoral researcher) is someone who has earned their PhD and is holding a research-focused position that is usually seen as a stepping stone toward becoming a faculty memember.

NASA Sloan Atlas and see if you can make a BPT diagram for all the galaxies with non-zero line flux."[2]

11.2 Filetypes and Reading in Data

This is where file I/O comes in. Even if you have taken a general introductory Python course online or even in the computer science department of a university, those courses will likely never have had cause to teach you how to read in a FITS file, a specialized filetype which can only be interacted with via special tools. Even if your mentor provides something as simple as a text file with some columns of numbers written out (known as ASCII), it is likely that a function you may have learned such as numpy.loadtxt() or pandas.read_csv() won't work on that data right "out of the box" because astronomical data in ASCII files often
- has missing entries
- have a mix of delimeters
- are not even formatted into clean columns.

If that sounds frustrating, it is—and it's why more constrained data formats (like FITS, asdf, or hdf5) are now preferred. Generally, if you can successfully *write* one of these files, adhering to their internal rules, you can *read* them back in simply and without hassle.

In the following examples, I am going to demonstrate the *general* method for reading several common data formats in astronomy.

11.2.1 ASCII (Text Files)

The simplest (in some sense) format for data storage is the humble text file. The primary advantage of an ASCII file as a data storage format is portability; no special software is needed to read these files (in fact, any computer can do so via one of several programs, including the shell). On many computers, you can preview the contents of a text file without even opening it.

Text files are also very flexible insofar is they impose no rules on how you format the information placed inside. As mentioned above, however, this can lead to headaches later, because decisions that make a text file read-friendly to the eye might make it impossible to parse (easily) via computer.

ASCII files are so basic that even Python itself has an I/O operation that can read them in. Here is an example of that (*Note, this is almost never the recommended method*):

```
with open('../Imaging/relano2016_m33_apertures.txt','r') as f:
    data = f.read()
```

Above, I use a Python-structure called a *context manager* (the "with open" formatting) simply because it allows a file to be opened, extracted, and then

---

[2] Hopefully, that dataset is easy to find online, and definitions for BPT and line fluxes are plentiful in the papers they provided.

automatically closed again when the requisite data has been loaded. The verbose equivalent would be:

```
f = open('../Imaging/relano2016_m33_apertures.txt','r')
data = f.read()
f.close()
```

Let's look at what our **data** variable looks like:

```
data
```

```
'ID, RA,              DEC,             ap_radius, type\n1\t,01:32:32.
   330,+30:35:03.
97,48.5,Mixed\n2\t,01:32:34.685,+30:30:27.45,40.7,Mixed\n3\t,01:32:34.
   687,+30:27
:29.01,44.8,Mixed\n4\t,01:32:37.566,+30:40:08.76,53.
   2,ClearShell\n5\t,01:32:44.8
23,+30:34:58.75,29.5,Filled\n6\t,01:32:44.903,+30:25:10.88,60.
   0,ClearShell\n7\t,
01:32:45.500,+30:38:55.18,47.6,Mixed\n8\t,01:32:46.135,+30:20:25.64,30.
   3,ClearSh
...'
```

```
type(data)
```

```
str
```

As we can see, it's read the file as a single, long string. That's not very helpful. Additionally, the string contains many so-called formatting tags (e.g., \t, \n), which indicate things like tabs and new lines, but from a data-parsing perspective, would have to be removed from this string before we could extract the actual values.

While we could use those formatting tags to construct a set of code that parses this string into actual data we care about, it is instead better to use one of several libraries that have functions directly related to this purpose.

The Numpy library has two functions useful for this: np.loadtxt() and np.genfromtxt(). If you have a very simple text file with columns of numbers and no mix of data types, loadtxt() is probably fine. But in many cases, genfromtxt() is more flexible. Let's try it out:

```
import numpy as np
data = np.genfromtxt('../Imaging/relano2016_m33_apertures.txt')
```

```
-----------------------------------------------------------------------
ValueError                                Traceback (most recent call
   last)
Cell In[5], line 2
      1 import numpy as np
----> 2 data = np.genfromtxt('../Imaging/relano2016_m33_apertures.txt')
```

```
File ~/miniconda3/envs/theia/lib/python3.10/site-packages/numpy/lib/
  ↪npyio.py:2291, in genfromtxt(fname, dtype, comments, delimiter,␣
  ↪skip_header, skip_footer, converters, missing_values, filling_values,␣
  ↪usecols, names, excludelist, deletechars, replace_space, autostrip,␣
  ↪case_sensitive, defaultfmt, unpack, usemask, loose, invalid_raise,␣
  ↪max_rows, encoding, ndmin, like)
   2289 # Raise an exception ?
   2290 if invalid_raise:
-> 2291     raise ValueError(errmsg)
   2292 # Issue a warning ?
   2293 else:
   2294     warnings.warn(errmsg, ConversionWarning, stacklevel=2)


ValueError: Some errors were detected !
    Line #2 (got 2 columns instead of 5)
    Line #3 (got 2 columns instead of 5)
    Line #4 (got 2 columns instead of 5)
    Line #5 (got 2 columns instead of 5)
    Line #6 (got 2 columns instead of 5)
    Line #7 (got 2 columns instead of 5)
    Line #8 (got 2 columns instead of 5)
    Line #9 (got 2 columns instead of 5)
    Line #10 (got 2 columns instead of 5)
    ...
```

This has not worked. This is because genfromtxt() doesn't know the format of the file and doesn't have a strong engine for making guesses. We have to instead provide several extra arguments to the function to indicate the formatting of our data:

```
data = np.genfromtxt('../Imaging/relano2016_m33_apertures.
  ↪txt',skip_header=1,delimiter=',')
```

If you look directly at the text file we're working with, you'll notice that the first row contains names for each column—we want to skip this row when reading in the data because it will (usually) be a different data type (str) than the data in the column below it. Additionally, we had to specify that the delimeter between entries in a row is a comma. But now we should have a reasonable array of data:

```
data[:5]
```

```
array([[ 1. ,  nan,  nan, 48.5,  nan],
       [ 2. ,  nan,  nan, 40.7,  nan],
       [ 3. ,  nan,  nan, 44.8,  nan],
       [ 4. ,  nan,  nan, 53.2,  nan],
       [ 5. ,  nan,  nan, 29.5,  nan]])
```

Unfortunately, most of the data has been read in as a nan — "Not a Number". The reason for this is that we have a mix of data types in our file—the RA, DEC, and "type" columns are all technically strings, and only the ap_radius column is a decimal valued number, which is what Numpy expects by default.

As it turns out, we can also specify the data types for each individual column upon loading the file:

```
data = np.genfromtxt('../Imaging/relano2016_m33_apertures.txt',
                     skip_header=1,
                     delimiter=',',
                     dtype=(int,'<U25','<U25',float,'<U25'))
data[:5]
```

```
array([(1, '01:32:32.330', '+30:35:03.97', 48.5, 'Mixed'),
       (2, '01:32:34.685', '+30:30:27.45', 40.7, 'Mixed'),
       (3, '01:32:34.687', '+30:27:29.01', 44.8, 'Mixed'),
       (4, '01:32:37.566', '+30:40:08.76', 53.2, 'ClearShell'),
       (5, '01:32:44.823', '+30:34:58.75', 29.5, 'Filled')],
      dtype=[('f0', '<i8'), ('f1', '<U25'), ('f2', '<U25'), ('f3',⌐
  ⌐'<f8'),
('f4', '<U25')])
```

Hooray! We finally got the data into Python (in at least what appears to be a sensible way). Above, I specified that columns 1 and 4 were int and float, while the "<U25" option specifies "A string less than 25 characters".

Are we finished? Unfortunately, no. The Numpy library is not well-designed to handle this type of heterogeneous data, and is currently storing it as a structured array, for which we cannot easily access columns or perform operations (like choosing just rows with radii greater than some value).

We could loop through this array and ultimately place each column into a different list, then start indexing. But it turns out there was (all along) a best tool for the job: pandas.

Pandas is a data-science oriented library with tools well-suited to the (often) heterogeneous datasets seen in industry. It also has some logic built in to try to figure out the structure of a dataset.

```
import pandas as pd
data = pd.read_csv('../Imaging/relano2016_m33_apertures.txt')
data[:5]
```

	ID	RA	DEC	ap_radius	type
0	1	01:32:32.330	+30:35:03.97	48.5	Mixed
1	2	01:32:34.685	+30:30:27.45	40.7	Mixed
2	3	01:32:34.687	+30:27:29.01	44.8	Mixed
3	4	01:32:37.566	+30:40:08.76	53.2	ClearShell
4	5	01:32:44.823	+30:34:58.75	29.5	Filled

As we can see, with almost no effort at all, **pandas** read our data into a table, deftly figuring out the column headings and parsing the heterogeneous data types. We can now get the radii out easily, or indeed, find all rows with a radius bigger than some value. There is only one thing that pandas did *incorrectly* which we do have to fix: our delimeter in our file is a comma, but there are many spaces in the header row designed, it seems, to make it easier to see which column heading goes with each column. Pandas didn't know to strip those spaces, so if we actually print the column names, we'll see the following:

```
data.columns
```

```
Index(['ID', ' RA', '          DEC', '          ap_radius', '␣
 →type'],
dtype='object')
```

We need to remove these spaces or we won't be able to easily index out the columns.

```
data = pd.read_csv('../Imaging/relano2016_m33_apertures.
 →txt',skipinitialspace=True)
data.columns
```

```
Index(['ID', 'RA', 'DEC', 'ap_radius', 'type'], dtype='object')
```

The **pandas.read_csv()** function has *many* optional arguments that lets us set up for different data. Here, I've used the "skipinitialspace" argument to tell it to strip out those leading spaces. Now, with this (still relatively easy) line of code, we can perform our data analysis:

```
data.loc[data.ap_radius>55]
```

	ID	RA	DEC	ap_radius	type
5	6	01:32:44.903	+30:25:10.88	60.0	ClearShell
27	28	01:33:15.673	+30:56:40.94	60.0	Mixed
28	29	01:33:15.870	+30:53:24.88	67.5	Mixed
32	33	01:33:23.026	+30:50:23.31	60.0	Mixed
39	40	01:33:28.248	+30:52:49.29	60.0	Shell
47	48	01:33:35.110	+31:00:54.44	60.0	ClearShell
55	56	01:33:44.499	+31:02:04.51	60.0	Shell
82	83	01:34:10.505	+30:21:52.11	59.2	ClearShell
86	87	01:34:13.301	+31:09:14.58	60.0	ClearShell

97	98	01:34:33.060	+30:47:01.71	74.2	Mixed
98	99	01:34:33.726	+31:00:31.15	61.1	ClearShell
102	103	01:34:38.019	+30:57:28.13	60.0	ClearShell
109	110	01:34:49.060	+31:07:47.26	68.5	ClearShell

Above, for example, I've *filtered* the DataFrame for only rows where the aperture radius is greater than 55 arcseconds.

When it comes to ASCII text files, this is only the beginning. These read-text functions have numerous extra optional parameters because many text files are in fact much more complicated and headache-inducing than this one. What I want you to takeaway is:

- Sometimes, one tool/reader will be better suited to your data than another and can help reduce the time it takes to get data imported,
- Indeed, sometimes using one tool to *load* the data, then converting it to another data container of choice (say, from `pandas` to `Numpy`) can be a good trick,
- Sometimes, but not always, simply modifying the underlying text file in a notepad-like program first can make it easier to read (for example, adding a comma delimiter).

11.2.2 Reading Tabular Data with Astropy

Beyond text files, the next most common data format you will find tabular data stored in is the FITS file. We've worked extensively with FITS already in the context of images. The format can also be used to store tables, with the primary difference being that FITS tables are by convention stored in the 1st (or greater) extension of the file.

Tables are thus stored after the image, or if there is no image, the 0th extension is empty.[3] Tabular data, if parsed directly using the format we've used for images, will be a special FITS record array. Let's as an example load the NASA Sloan Atlas, a catalog of hundreds of thousands of galaxies in the SDSS.

```
with fits.open('../../BookDatasets/catalogs/nsa_v0_1_2.fits') as hdu:
    table = hdu[1].data
```

Unfortunately, there's no easy way to display this table in a view-friendly format. Here's a truncated peek at *just the first row*:

```
table[0]

('J094651.40-010228.5', '09h/m00/J094651.40-010228.5', 146.
 71420878660933,
-1.0412815695754145, 0, 72212, 21157, -1, -1, -1, 15.178774, 0.
 021222278,
'sdss', 0.07, 756, 1, 206, '301', 136.29353, 1095.152, 0.020597626, 0.
 020687785,
0.00044536332, 0, array([ 29.552263,    53.198177,   175.86322 ,   819.
 325    ,
```

---

[3] This may not alway be the case.

```
1793.9138 ,
        2480.876    , 3251.03    ], dtype=float32), array([3.0069932e-01,
1.9650683e-01, 1.4631173e-02, 4.3475279e-03,
        9.1012771e-04, 4.7585298e-04, 1.2297294e-04], dtype=float32), 1,␣
    ␣array([
31.202734,   49.386097,  199.0066  ,  824.0366  , 1712.252   ,
        2462.11   , 3454.6162 ], dtype=float32), array([-15.1673565,␣
    ␣-15.816648
, -17.190884 , -18.824028 , -19.66704  ,
        -20.00378   , -20.296247 ], dtype=float32), array([ 222.77438,  ␣
    ␣471.76144,
383.8668 , 2475.7463 , 2484.602  ,

        2484.4727 , 1102.5615 ], dtype=float32), array([0.4536473 , 0.
    ␣44762787,
 0.2820931 , 0.20756142, 0.15054086,
        0.11415057, 0.08093417], dtype=float32), array([-0.00607499,  0.
    ␣00494833,
 0.08098889,  0.0434305 ,  0.03558178,
        0.02031052,  0.01923862], dtype=float32), array([1.7146798e-04,
1.0891152e-08, 1.3611655e-05, 2.1125154e-04,...
        ]...)
```

A lot of this issue here is that this table has many columns and rows. A better way to load a table that has been stored in a `fits` file is via the `astropy` table object:

```
from astropy.table import Table
table = Table.read('../../BookDatasets/catalogs/nsa_v0_1_2.fits')
```

```
table[:2]
```

```
<Table length=2>
      IAUNAME                SUBDIR              ...  PLUG_RA    PLUG_DEC
      bytes19                bytes27             ...  float64    float64
------------------- --------------------------- ... ---------␣
    ␣-----------
J094651.40-010228.5 09h/m00/J094651.40-010228.5 ... 146.71421  -1.
    ␣0413043
J094631.60-005917.7 09h/m00/J094631.60-005917.7 ... 146.63167 -0.
    ␣98827781
```

As we can see, the `Table` read method has parsed that record array and created something similar to the `pandas` display we saw above; column headers with names, and a cleaner interface for viewing the data. We can now filter, if needed. Let's find all sources within a narrow redshift (z) range.

```
filtered = table[(table['Z']>0.02)&(table['Z']<0.021)]
len(filtered)
```

```
2070
```

You might be wondering when to use `pandas` vs `astropy.table`s to work with your data—in truth, you can convert between them fairly easily, so whichever you are more comfortable with will ultimately be fine.

One caveat to this is that `pandas` cannot handle "cells" with multi-dimensional values, e.g., a row-column position can't have a value of a whole array. Some astronomical tables, *because* `astropy.tables` can handle this, will store data that way. Often, however, we only need a few columns, so whichever method we use to read the data in, we can extract those columns, create our own table, and move on. An additional scenario in which the `astropy.Table` shines over `pandas` is in reading in the "machine-readable" tables hosted by some astronomical journals to accompany publications.

If we examine the head of one of these files, we can see an immediate problem:

```
!head -60 ../../BookDatasets/catalogs/Carlsten.txt
```

```
Title: The Exploration of Local Volume Satellites (ELVES) Survey: A
  Nearly
       Volume-Limited Sample of Nearby Dwarf Satellite Systems
Authors: Carlsten S.G., Greene J.E., Beaton R.L., Danieli S., Greco J.P.
Table: Confirmed dwarf satellite distance results
================================================================================
Byte-by-byte Description of file: apjac6fd7t6_mrt.txt
--------------------------------------------------------------------------------
   Bytes Format Units   Label      Explanations
--------------------------------------------------------------------------------
   1- 14 A14     ---     Name       Dwarf galaxy name
  16- 22 F7.3    deg     RAdeg      Right Ascension, decimal degrees
  (J2000)
  24- 30 F7.3    deg     DEdeg      Declination, decimal degrees (J2000)
  32- 38 A7      ---     Host       Host galaxy name
  40- 44 F5.2    Mpc     Dist-host  Host Distance, Mpc
  46- 50 F5.2    Mpc     Dist       ?="" SBF Distance
  52- 56 F5.2    Mpc     lb2Dist    ?="" Lower 2{sigma} bound on Dist
  58- 62 F5.2    Mpc     lb1Dist    ?="" Lower 1{sigma} bound on Dist
  64- 68 F5.2    Mpc     ub1Dist    ?="" Upper 1{sigma} bound on Dist
  70- 74 F5.2    Mpc     ub2Dist    ?="" Upper 2{sigma} bound on Dist
  76- 80 F5.2    ---     SNR        ?="" SBF Signal-to-noise ratio
  82- 89 A8      ---     Source     SBF Source
  91- 95 A5      ---     Confirm    Confirmation criteria exception (1)
  97-100 I4      km/s    vrec       ?="" Recessional velocity
 102-105 A4      ---     r_vrec     Source for vrec (2)
 107-111 F5.2    Mpc     Dtrgb      ?="" Distance, tip of the red giant
  branch
 113-116 A4      ---     r_Dtrgb    Source for Dtrgb (2)
--------------------------------------------------------------------------------
Note (1): Galaxies marked "True" by this flag are exceptions to the usual
    confirmation criteria, see text for details.
Note (2): Sources --
    a = Karachentsev et al. (2013) [2013AJ...145..101K];
    b = Sand et al. (2014) [2014ApJ...793L...7S];
    c = Toloba et al. (2016) [2016ApJ...816L...5T];
    d = Muller et al. (2019a) [2019A&A...629L...2M];
    e = Meyer et al. (2004) [2004MNRAS.350.1195M];
    f = Irwin et al. (2009) [2009ApJ...692.1447I];
    g = Karachentsev et al. (2022) [2022arXiv220301700K];
    h = Cohen et al. (2018) [2018ApJ...868...96C];
    i = Sabbi et al. (2018) [2018ApJS..235...23S];
    j = Tully et al. (2016) [2016AJ...152...50T];
```

```
k = Kim et al. (2020) [2020ApJ...905..104K];
l = Haynes et al. (2018) [2018ApJ...861...49H];
m = Karunakaran et al. (2020b) [2020arXiv200514202K];
n = Rines et al. (2003) [2003AJ...126.2152R];
o = Smercina et al. (2018) [2018ApJ...863..152S];
p = Tikhonov et al. (2015) [2015AstL...41..239T];
q = Danieli et al. (2017) [2017ApJ...837..136D];
r = Bennet et al. (2019) [2019arXiv190603230B];
S = Simbad.

--------------------------------------------------------------------------------
NGC247          11.783 -20.757 NGC253    3.56
159 S      3.72 a
dw0047m2623     11.894 -26.390 NGC253    3.56
3.90 b
dw0049m2100     12.454 -21.017 NGC253    3.56
295 S      3.44 a
dw0050m2444     12.575 -24.737 NGC253    3.56
3.12 c
dw0055m2309     13.754 -23.169 NGC253    3.56
250 a
dw0132p1422     23.249  14.374 NGC628    9.77
669 S
dw0133p1543     23.484  15.731 NGC628    9.77  9.18  7.28  8.24 10.16 11.
 22  8.29
GEMINI   False
dw0134p1544     23.554  15.746 NGC628    9.77 10.50  8.76  9.64 11.42 12.
 36  9.63
GEMINI   False
dw0134p1438     23.674  14.644 NGC628    9.77
731 S
dw0136p1628     24.084  16.470 NGC628    9.77  9.45  7.88  8.71 10.17 10.
 87 22.76
GEMINI   False
```

The whole top portion of this file contains un-commented lines that will seriously confuse Numpy or Pandas. Moreoever, the column names are not, as pandas expects, in a single row above all the data—they are in a column of the header part of the file. We could manually remove these header lines from the file, but we would have to also copy that row of labels and transpose it into a horizontal row before we could even attempt to read this file.

Instead of wrangling this data file significantly before passing it to pandas.read_csv(), we can instead use the fact that Astropy has designed its Table.read() to explicitly handle these types of files. The view of the table output from this simple read operation is in Figure 11.1.

```
from astropy.table import Table
table = Table.read('../../BookDatasets/catalogs/Carlsten.txt',
format="ascii.cds")
```

If we wish to use pandas for our analysis or data storage, we can still easily convert this table to pandas via

```
table.to_pandas()
```

| Name | RAdeg | DEdeg | Host | Dist-host | Dist | lb2Dist | lb1Dist | ub1Dist | ub2Dist | SNR | Source | Confirm | vrec | r_vrec | Dtrgb | r_Dtrgb |
| | deg | deg | | Mpc | Mpc | Mpc | Mpc | Mpc | Mpc | | | | km/s | | Mpc | |
str14	float64	float64	str7	float64	float64	float64	float64	float64	float64	float64	str8	str5	int64	str1	float64	str1
NGC247	11.783	-20.757	NGC253	3.56	--	--	--	--	--	--	--	--	159	s	3.72	a
dw0047m2623	11.894	-26.39	NGC253	3.56	--	--	--	--	--	--	--	--	--	--	3.9	b
dw0049m2100	12.454	-21.017	NGC253	3.56	--	--	--	--	--	--	--	--	295	s	3.44	a
dw0050m2444	12.575	-24.737	NGC253	3.56	--	--	--	--	--	--	--	--	--	--	3.12	c
dw0055m2309	13.754	-23.169	NGC253	3.56	--	--	--	--	--	--	--	--	250	a	--	--
dw0132p1422	23.249	14.374	NGC628	9.77	--	--	--	--	--	--	--	--	669	s	--	--
dw0133p1543	23.484	15.731	NGC628	9.77	9.18	7.28	8.24	10.16	11.22	8.29	GEMINI	False	--	--	--	--
dw0134p1544	23.554	15.746	NGC628	9.77	10.5	8.76	9.64	11.42	12.36	9.63	GEMINI	False	--	--	--	--
dw0134p1438	23.674	14.844	NGC628	9.77	--	--	--	--	--	--	--	--	731	s	--	--
...	...	...	...	...	...	...	...	...	...	...	...	...	...	...	...	...
dw1403p5356	210.937	53.944	NGC5457	6.5	6.02	4.61	5.32	6.79	7.64	7.43	CFHT	False	--	--	6.37	q
NGC5474	211.256	53.662	NGC5457	6.5	--	--	--	--	--	--	--	--	294	s	6.98	a
NGC5477	211.386	54.461	NGC5457	6.5	--	--	--	--	--	--	--	--	317	s	6.76	a
dw1406p5344	211.708	53.741	NGC5457	6.5	6.17	5.0	5.56	6.78	7.42	11.04	CFHT	False	--	--	6.83	r
dw1408p5419	212.156	54.325	NGC5457	6.5	7.49	6.06	6.74	8.37	9.37	7.51	CFHT	False	--	--	6.87	q
dw1905m6316	286.483	-63.272	NGC6744	8.95	7.61	5.57	6.55	8.77	10.26	5.57	DECAM	False	--	--	--	--
dw1906m6357	286.742	-63.964	NGC6744	8.95	7.31	5.55	6.44	8.21	9.17	9.68	DECAM	False	--	--	--	--
dw1907m6342	286.844	-63.706	NGC6744	8.95	7.99	6.26	7.12	8.94	10.01	11.51	DECAM	False	--	--	--	--
dw1908m6343	287.183	-63.73	NGC6744	8.95	5.33	4.35	4.8	5.96	6.73	43.93	DECAM	True	--	--	--	--
dw1911m6413	287.844	-64.223	NGC6744	8.95	--	--	--	--	--	--	--	--	779	j	--	--

Table length=251

Figure 11.1. View returned by `astropy.Table.read()` when loading a machine-readable ASCII datafile associated with a scientific publication.

Once again, we've seen that *which* library you use to read in a data file can dramatically affect the amount of work needed to get it imported.

11.2.3 ASDF (Advanced Science Data Format)

One new and up-and-coming data storage option is the ASDF format. It is convenient because it stores data in a tree-like structure that, from the outside, looks exactly like a Python dictionary. This means that when we want to take some data we've been working with and save it to a file, instead of trying to format everything into columns and writing an ASCII file, we can simply store things by column name in a dictionary and then drop it straight into the file:

```python
import asdf
from asdf import AsdfFile
import numpy as np

tree = {
    'a': np.arange(0, 10),
    'b': np.arange(10, 20)
}

target = AsdfFile(tree)
target.write_to('target.asdf')
```

As we can see, any arrays we are working with we can name, drop into the tree dictionary, and ultimately write to the file. We can also add metadata—either directly into the tree, or into a metadata dictionary which then gets added to the tree. Opening an ASDF file is also simple:

```python
with asdf.open("example.asdf") as af:
    tree = af.tree
```

This `tree` attribute contains the dictionary we saved above, so everything pops right back out.

There's a lot more to `ASDF` in terms of motivation and capability, but for our purposes at present, the important part is knowing how to pack and save, then open and unpack, data with this format.

11.2.4 HDF5 (Hierarchical Data Format 5)

The `HDF5` format is another new, semi-popular file standard. It structures its interior like a full file directory system, with "folders" and "files". To deal with `HDF5` files in Python, we need to use the `h5py` package, which handles the I/O operation for us.

```
import h5py

f = h5py.File('example.hdf5', 'r')
data_keys = list(f.keys())
```

From here, things will change depending on the file—HDF5 stores things in "datasets," whose names will appear in that list of keys we just accessed. Assuming our example file key list had "dataset1" as one of the keys, we would then have

```
dataset1 = f['dataset1']
```

This dataset would be an object similar to, but not exactly, a `Numpy` array. This object can be indexed and sliced like a `Numpy` array. For many frameworks, however, it is beneficial to extract and convert these datasets into `Numpy` before continuing further.[4]

11.3 Working with Tabular Data in Pandas

In this section, we will be using a set of catalogs released by the 3D-HST collaboration to explore the usage of the `pandas` library beyond the reading in of data. Extremely popular in the data science industry (indeed, moreso than in astronomy), `pandas` was designed to enable work on very large tables, which may have hundreds of columns and thousands of rows. It has functionality, unlike `Numpy`, to index by *either* row name *or* column name, and has built-in methods to group by certain parameters or join and split tables in a SQL-style manner. As we saw above, it also has a robust `read_csv()` function which often handles the importing of tabular data more elegantly than other solutions. Often, when loading an astrophysical catalog, storing it in a `pandas.DataFrame` will make it easier to work with than storing it in a `Numpy` array or even record array.

---

[4] More on datasets can be found here: https://docs.h5py.org/en/stable/high/dataset.html#dataset.

Whether you use `pandas` or `Astropy` tables is primarily a matter of personal preference. I will focus on `pandas` in this text, primarily because a familiarity with the library is also highly marketable outside of astronomy. But as we've already seen, there are times when temporarily using one or the other becomes much more efficient.

The data we will be working with is the 3D-HST survey (Brammer et al. 2012, van Dokkum et al. 2013; Skelton et al. 2014; Momcheva et al. 2016). The survey, carried out with the Hubble Space Telescope, focused on several extragalactic fields which have been extensively observed. Using 248 orbits of HST, the survey added to a decade of imaging in those fields by taking *grism spectroscopy*—a special form of slitless spectroscopy that disperses everything in the field of view (making for a challenging data reduction). This new dataset allowed for accurate redshifts (distances) to be measured to many of these galaxies for the first time, via the measuring of spectral features (as we did in Exercise 7.1).

We will be using the catalogs for one of the several fields observed, which is split into multiple files.

- One file contains the redshifts (where available) and star formation rates derived from the rest-frame UV+IR emission
- One file contains the output of FAST (Kriek et al. 2009, 2018), which measures properties about the galaxies (e.g., stellar mass, star formation rate, and metallicity) based on the photometry and redshift,
- One file contains the *rest frame* colors—that is, magnitude differences between bands, de-redshifted.
- One file contains the coordinates and all the photometry and errors for the many bands the galaxies have been observed in.

Each of these files has anywhere from 11 to 155 columns. We won't need every column from every file, but we *will* need *some* columns from each file. You can imagine that working with four Numpy arrays, each with shapes like (155,3000), independently, attempting to extract the same source from each or produce plots of one property versus another, might be somewhat inefficient. This is where `pandas` will be most useful.

To begin, we need to read in the files. Two of our files are ASCII text files which we can read in with `pd.read_csv()`, two are FITS files for which I'll use the `astropy.Table.read()` method, which can then convert to `pandas`.

```
from astropy.table import Table
import pandas as pd

sfr_file = '../../BookDatasets/catalogs/cosmos_3dhst.v4.1.5.zbest.sfr'
fast_file = '../../BookDatasets/catalogs/cosmos_3dhst.v4.1.5.zbest.fout'
```

```
rf_colors_file = '../../BookDatasets/catalogs/cosmos_3dhst.v4.1.master.
  ↪RF.FITS'
photometry_file = '../../BookDatasets/catalogs/cosmos_3dhst.v4.1.cat.
  ↪FITS'

sfr_df = pd.read_csv(sfr_file,delim_whitespace=True)
fast_df = pd.read_csv(fast_file,delim_whitespace=True)
rf_colors_df = Table.read(rf_colors_file)
phot_df = Table.read(photometry_file).to_pandas()
```

We can examine the head (first several lines) of one of our tables to see what we are working with:

```
fast_df.head()
```

```
   id    z  ltau  metal  lage   Av  lmass    lsfr    lssfr  la2t    chi2
0   1  0.89   7.0   0.02   9.7  2.1  10.46  -99.00  -99.00   2.7  56.700
1   2  0.89   7.0   0.02   9.7  2.0  10.41  -99.00  -99.00   2.7  43.400
2   3  0.81   8.4   0.02   9.2  0.3  10.45   -0.45  -10.90   0.8   1.090
3   4  3.10   8.8   0.02   9.3  0.2  10.16    0.23   -9.93   0.5   0.948
4   5  0.51   7.0   0.02   9.9  0.6   8.55  -99.00  -99.00   2.9   7.010
```

Note that the pandas DataFrame has an extra "column" out to the left of the named columns, which here is similar (off by one) to the id column. This is known as the Index, and it is what allows us to index by row name as well as column name. The index need not be integers, but rather can be arbitrary (unordered) values or even strings, so long as each row has a unique key. You could imagine the index here being "galaxy alpha", "galaxy beta", "galaxy gamma", etc.

11.3.1 Indexing Columns

Indexing columns in pandas works in one of two ways: dictionary style, in which we pass a string column name to postpended square brackets, and attribute style, in which we access the column via dot notation of the column name. The former always works, the latter only works if the column name does not have a space in it. As a note, we can also pass a *list* of column names to our indexing, e.g.,

```
(fast_df[['id','metal','lmass']])
```

```
     id  metal  lmass
0     1   0.02  10.46
1     2   0.02  10.41
2     3   0.02  10.45
3     4   0.02  10.16
4     5   0.02   8.55
...   ...   ...    ...
```

```
33874   33875   0.02    9.86
33875   33876   0.02    9.13
33876   33877   0.02    6.28
33877   33878   0.02   12.97
33878   33879   0.02   11.17

[33879 rows x 3 columns]
```

11.3.2 Indexing Rows with .loc

Being able to set keys for each row via the Index, or indeed, using the default index of counting numbers, we can slice a dataframe by row *or* column. To retrieve a row, we use the df.loc[index] format. When we do so, we actually get a fairly useful structure:

```
fast_df.loc[0]
```

```
id         1.00
z          0.89
ltau       7.00
metal      0.02
lage       9.70
Av         2.10
lmass     10.46
lsfr     -99.00
lssfr    -99.00
la2t       2.70
chi2      56.70
Name: 0, dtype: float64
```

Here, we can see that for just this object (id==1), we can read off the interesting parameters, such as metallicity (metal) or log star formation rate (lsfr). We can retrive those values in one of three ways. We can index this object now by name:

```
fast_df.loc[0]['metal']
```

```
0.02
```

Or, if the name does not contain any spaces, we can even directly dot-access the property

```
fast_df.loc[0].metal
```

```
0.02
```

This is convenient especially when dealing with a lot of queries, dot-accessing the name has fewer characters than square brackets and string quotes. Finally, we could have retrieved this value immediately in our call to .loc, by adding in the column(s) to retrieve after the row, as follows:

```
print(fast_df.loc[0,'metal'])
print(fast_df.loc[0,['metal','lsfr']])
```

```
0.02
metal     0.02
lsfr     -99.00
Name: 0, dtype: float64
```

When we ask for a single "cell" (row column pair), we get the value out directly. When we pass a list of column names to that second argument of the indexing, we get out a mini-version of what we got above — which could then be indexed in the same ways we just showed.

11.3.3 Filtering Dataframes

When we want to filter a data frame on some condition, we will use the .loc operation and add our conditions into the brackets much like we would when making a mask on a list or array. For example, let's find all galaxies with log star formation rate greater than 2 (i.e., SFR >100 $M_\odot$ yr^{-1}).

```
df_high_sf = sfr_df.loc[sfr_df.sfr>100]
df_high_sf
```

	id	z_best	z_type	sfr	flag
0	1	3.067	3	214.34	0
1	2	3.227	3	191.43	0
3	4	3.088	3	306.65	0
73	74	2.090	3	147.31	0
81	82	4.840	3	482.09	0
...	...	...	...	...	...
33807	33808	2.904	3	141.18	0
33824	33825	4.739	3	276.00	1
33861	33862	5.034	3	163.43	1
33869	33870	5.961	3	460.38	1
33872	33873	5.598	3	1206.70	0

```
[1285 rows x 5 columns]
```

As we can see, 1285 galaxies in the overall sample have measured star formation rates of more than 100 solar masses per year![5]

However, some of these galaxies might not be worth including: in the flag column, flags of 1 and 2 indicate a low signal-to-noise measurement and a bad redshift, respectively. So let's index for just those galaxies with flag 0.

---

[5] That is a brisk rate of star formation. For reference, the Milky Way forms roughly 1 solar mass per year, and the nearest starbust, M82, forms around 10 solar masses per year.

```
df_high_sf = sfr_df.loc[(sfr_df.sfr>100)&(sfr_df.flag==0)]
print(df_high_sf.head())
print(f'{len(df_high_sf)} galaxies meet criteria.')
```

```
     id  z_best  z_type      sfr  flag
0     1   3.067       3   214.34     0
1     2   3.227       3   191.43     0
3     4   3.088       3   306.65     0
73   74   2.090       3   147.31     0
81   82   4.840       3   482.09     0
918 galaxies meet criteria.
```

As we can see, we are down from 1285 to 918 galaxies after removing those with flags. To accomplish this double-condition, we add parentheses to each, and add the "&" symbol between them. As you might expect, we can combine arbitrary and/or conditionals here. We can also subselect only certain columns during this .loc call, just as we did above. In this example, I will use the sfr from our sfr_df, but masses from the fast_df. These are the same length, so this masking should still work. Let's grab all galaxies with high star formation rates OR those with very high mass (keeping our flag condition intact).

```
df_m_sf = fast_df.loc[((sfr_df.sfr>100)|(fast_df.lmass>12))
                       &(sfr_df.flag==0),
                       ['id','lmass','metal','lage']]
```

```
print(df_m_sf)
```

```
          id  lmass  metal  lage
0          1  10.46   0.02   9.7
1          2  10.41   0.02   9.7
3          4  10.16   0.02   9.3
73        74  10.74   0.02   9.4
81        82  10.37   0.02   9.0
...      ...    ...    ...   ...
33649  33650   9.05   0.02   7.6
33670  33671   9.42   0.02   8.2
33717  33718   9.83   0.02   8.3
33807  33808  10.70   0.02   9.3
33872  33873  11.41   0.02   8.7
```

```
[921 rows x 4 columns]
```

Here, we use the "|" (or) operator to indicate that *either* galaxies with high log mass, or those with high SFR can be accepted. Those two possibilities are in their own parenthesis, and are evaluated first. We also still check that flag is 0, and this is joined to the output of the other filtering. And finally, we elect to extract only two columns: id and lmass. This example demonstrates the care that needs to be taken when combining conditionals; if we had input "a or b and c", that means something different than "(a or b) and c".

Sometimes, particularly when using some other tool, we want to strip away all the pandas wrapper and retrieve a straightforward Numpy array. To do so, we can take any pandas Series (every column of a data frame is a series) and access the .values attribute to output a Numpy array.

```
df_m_sf['lmass'].values[:10]
```

```
array([10.46, 10.41, 10.16, 10.74, 10.37, 10.89,  7.96,  8.32, 10.  ,
        8.15])
```

As we can see, the .values attribute gives us a standard array, which from then on we can treat with all the Numpy formalism we have learned.

11.3.4 Merging Dataframes

Let's now join together our four DataFrames, selecting only the columns we'll need for the rest of this chapter. I'll demonstrate the usage for two of the DataFrames, and in Exercise 11.2, you will combine the rest in.

In order to combine multiple tables together, we need a criteria that can be used to establish the merge. The terms we'll use here (inner, outer, left, and right join) come from a SQL/database background, where multiple tables with common keys are used to store data.

We have a relatively easy task here: every table has a column called id, which refer to the same object in all the tables. Thus, we can have pandas merge our frames rather seamlessly.

```
sfr_df = sfr_df[['id','z_best','z_type','sfr','flag']]
fast_use = fast_df[['id','metal','lage','lmass']]
merged_df = pd.merge(sfr_df,fast_use,how='outer',on='id')
print(merged_df.head())
```

	id	z_best	z_type	sfr	flag	metal	lage	lmass
0	1	3.0670	3	214.3400	0	0.02	9.7	10.46
1	2	3.2270	3	191.4300	0	0.02	9.7	10.41
2	3	0.8131	3	2.4763	0	0.02	9.2	10.45
3	4	3.0880	3	306.6500	0	0.02	9.3	10.16
4	5	0.5092	3	-99.0000	2	0.02	9.9	8.55

Now we have a merged frame with the two redshift columns and the four selected stellar population columns (along with id and a flag that we'll get to in a moment). Here, the z_best refers to the adopted redshift, and the z_type column is whether the redshift is from full spectroscopy (best, 1), grism spectroscopy (less good, 2), or just photometry (worst, 3).

The first two arguments to pd.merge() are the frames to use, with the first being deemed "left" and the second being deemed "right". The how argument specifies how to join the tables, in cases where the left hand table has more or less rows than the right hand table.

- Using 'left' means all rows from the left table are kept; if the right table has fewer rows, the final row will have the left hand values, and then NaN for the columns from the right hand table, and extra rows in the right hand table *not* in the left table are dropped,

- Using 'right' is the same, but for the right hand table,
- Using "inner" means only rows present in *both* tables are kept (so there should be no NaN sections), and
- Using 'outer' means all rows in both tables are kept, so there may be swaths of NaN in both the columns from the left table and the right table.

In our current example (I used outer), it did not matter which method we used, because all of these tables were the same length and had the full id column. And finally, the on argument tells the function which column of each frame is the "key" which should refer to the same object in each table.

Let's examine these different merge types in a bit more detail to get a feel for them.

To begin, let's return to our frame from the SFR catalog that had only very star-forming objects. Let's similarly make a cut on the redshift catalog for only sources with spectroscopic redshift (i.e., those with the z_type==1).

```
z_spec = sfr_df.loc[sfr_df.z_type==1]
print(f'{len(z_spec)}/{len(sfr_df)} galaxies have spectroscopic
    redshifts.')
```

```
308/33879 galaxies have spectroscopic redshifts.
```

Of the roughly 34,000 galaxies total, 308 of them have quoted spectroscopic redshifts.[6]

We now have two subsamples of the total datasets, each with different selections of galaxies. If we now try our merging again, the mode we choose will matter significantly.

```
merged_df = pd.merge(z_spec,df_m_sf,how='outer',on='id')
print(merged_df)
```

	id	z_best	z_type	sfr	flag	lmass	metal	lage
0	457	0.5336	1.0	2.0949	1.0	NaN	NaN	NaN
1	937	0.8783	1.0	20.8620	0.0	NaN	NaN	NaN
2	1324	0.6666	1.0	3.0822	0.0	NaN	NaN	NaN
3	1424	0.6810	1.0	5.2284	0.0	NaN	NaN	NaN
4	1592	0.5325	1.0	-99.0000	2.0	NaN	NaN	NaN
...	...	...	...	...	...	...	...	...
1223	33650	NaN	NaN	NaN	NaN	9.05	0.02	7.6
1224	33671	NaN	NaN	NaN	NaN	9.42	0.02	8.2
1225	33718	NaN	NaN	NaN	NaN	9.83	0.02	8.3
1226	33808	NaN	NaN	NaN	NaN	10.70	0.02	9.3
1227	33873	NaN	NaN	NaN	NaN	11.41	0.02	8.7

```
[1228 rows x 8 columns]
```

---

[6] Extragalactic galaxy spectroscopy at intermediate to high redshift is a challenging and time consuming (and therefore expensive) proposition, often requiring many hours of observing time per galaxy. It is not surprising that only a few hundred galaxies in a sample like this would have full spectroscopic redshifts.

As expected, the outer join keeps all rows from both tables—its length (1228) is thus the length of each table summed, minus any overlapping rows. We can see in the example here that the first few items were in our spectroscopic sample, but did not make it into our highly-star-forming sample (so those columns are NaN). Similarly, we have some rows at the end which are in our highly star-forming sample but were not in our redshift-selected sample. In both cases, non-present entries are filled with NaN. Let's do the opposite, and find only the galaxies which *are* in both samples.

Exercise 11.1: Joint Sample.
Exercise: which join will give us just the galaxies in both samples? Try it on the frames and see which, if any, galaxies are in both.

In Exercise 11.1, you should find that a single galaxy, ID 27 916, is present in both of our samples—i.e., it has high SFR or mass, and it has a spectroscopic redshift.

With our current datasets, it makes more sense to join the full tables together with the columns we need, and perform filtering like this after the fact. But when working with catalogs from *different* surveys, with wildly different selections, you may need to think carefully about how to merge them, and on what column. In Exercise 11.2, you'll merge the other two tables, following the syntax I've used above.

Exercise 11.2: Merging Tables.
Join the rest of the tables (all rows) to our merged frame.
- From the photometry frame, keep only the x, y, ra, and dec columns.
- From the restframe colors frame, keep only filters U,B,V,J,H,K, which have column names 1153, 1154, 1155, 1161, 1162, and 1163, respectively. As a bonus, look up how to rename the columns of a dataframe, and update the non-descriptive restframe columns to the filternames mentioned above.

It does not matter what join you select here, since all frames will have all rows.
Your final dataframe should have the full 33,879 rows, and 18 columns.

Exercise 11.3: Initial Cleaning.
Now that we have our full frame, it's time to cull out some bad values. Using the filtering mechanism we covered above, index only the rows for which
- the lmass column is not -1.0,
- the lage column is not -1.0,
- the metal column is not -1.0,
- the flag column is not -99.0 or 2,
- none of the filters (UBVJHK) are -99.0.

You can add all these conditions to a single `.loc[]` call, using the syntax we learned above.

Finally, there are some NaN values scattered around the file. Drop any row which has a NaN for one of the columns. You can use the `df.dropna()` method to do this. You should have 18,170 galaxies left after this cleaning. I suggest calling this frame `clean_df`, as it is the variable name I will be using in the example.

As a note: if you can't find a README or other file for your catalog that lays out what the flag values are for a dataset you are using, it is often useful to create histograms of the values, or of the log values, of that column. Generally, the values will have some distribution, and flag values are chosen to be far away from that distribution (and/or a sign the values can't be, e.g., negative). Seeing a distribution and a narrow spike separated might mean the narrow spike is a flag value.

11.3.5 Saving Dataframes

In the parlance of data science, we have now successfully "wrangled" our dataset—that is, read it into Python in a usable fashion and cleaned it up to only contain the values we wish to use for science. Of course, this wrangling was on the easy end of the spectrum; these files contained flags to indicate where we should filter data out. The more typical case involves the ingesting of much larger datasets (millions of rows, hundreds of columns, possibly from many sources with different file formats and structures), with no meaningful *a priori* knowledge of how to isolate the elements of the dataset worth using. An old adage goes, "A data scientist's job is 95% data wrangling and cleaning, and 5% running the actual machine learning pipeline."

Nevertheless, it took the reading in of several files, and several steps of cleaning to get to our final, cleaned `DataFrame`. It might be worthwile, then, to save this specific frame to disk, so that we can load it back into Python in the future without having to repeat the steps we've done so far.

There are a few ways to write to disk. The easiest is the `df.to_csv()` method, which will simply write out a `csv` (comma separated) file with all our rows and columns in such a way that it can be easily read back in.

```
clean_df.to_csv('3DHST_Cleaned.csv',sep='\t',index=False)
```

After providing the desired filename, I've utilized two optional arguments, `sep` (delimiter), for which I indicate to use tabs instead of commas between values, and `index`, which I set to false so that the first column of the file is not the dataframe index (which is not meaningful in this context).

This dataset is relatively small by data science standards. At only 2.3 MB, the csv file created is lightweight and easily portable from system to system. We could also use the FITS library, by first converting our frame to a record array and feeding it into the FITS file creation process:

```
from astropy.io import fits

rec_array = clean_df.to_records()
header = fits.Header()
header['DESC'] = 'cleaned 3dhst frame'
header['AUTH'] = 'Imad Pasha'
header['DATE'] = '01/01/2024'
primary = fits.PrimaryHDU(header=header)
tab = fits.TableHDU(rec_array)
hdulist = fits.HDUList([primary,tab])
hdulist.writeto('3DHST_Cleaned.fits')
```

Now that we have our cleaned DataFrame, we can start exploring the data within. How you approach this stage depends on the data and your goals: in one case, you might not even know what relationships exist between columns of your files, in another, you might have taken the data that went into this catalog specifically to measure a specific relation.

11.4 Research Example: Analysis with 3DHST

11.4.1 Star-forming Sequence

Let's use our frame to make a few key scientific figures based on this data. We'll start with the Star-forming Sequence (SFS), a well known relation in which the star formation rate of a galaxy correlates strongly with its stellar mass. The figure (as for many in astronomy) is typically plotted in log units. Thanks to our cleaning, we simply need to call the lmass and sfr columns of our frame to create this plot (Figure 11.2)

```
import pandas as pd
clean_df = pd.read_csv('3DHST_Cleaned.csv',delim_whitespace=True)
```

```
import matplotlib.pyplot as plt
import numpy as np

sf_ms_df = clean_df.loc[(np.log10(clean_df.sfr)>-5)
                        &(clean_df.lmass>1.0)]
fig, ax = plt.subplots(figsize=(6,6),constrained_layout=True)

ax.plot(sf_ms_df.lmass,np.log10(sf_ms_df.sfr),',',)
ax.set_xlabel(r'log Stellar Mass [$M_{\odot}$]',fontsize=15)
ax.set_ylabel(r'log SFR [$M_{\odot}$ $yr^{-1}$]',fontsize=15)
ax.tick_params(direction='in',length=5,top=True,right=True)
```

You may have noticed a new Matplotlib argument used here: rasterized=True. By default, every point from our dataset would be plotted as its own vector—so if we saved our figure as a PDF, the filesize might be large. By rasterizing the points, they become "pixel" representations, which is much lighter on memory (but which won't scale to arbitrary zoom like a vector). As an experiment, try out saving this

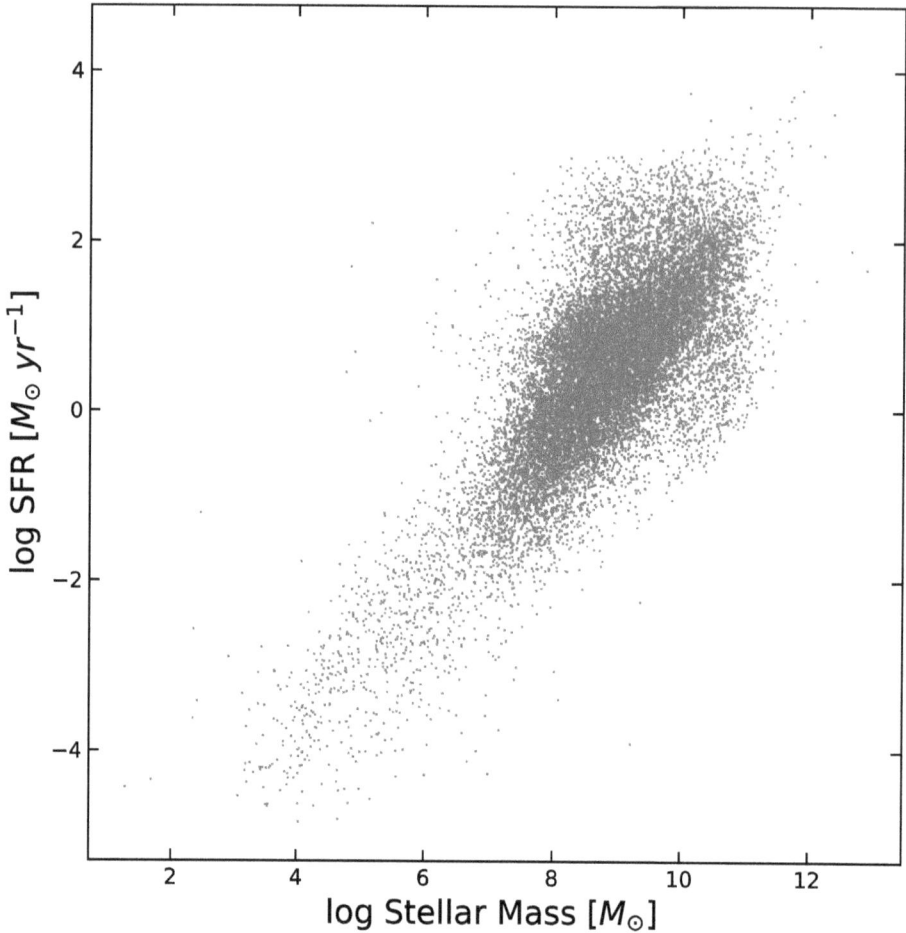

Figure 11.2. The star-forming sequence: log stellar mass versus log star formation rate. The majority of the points lie above log mass of 8 and log SFR of −2, but it is hard to make out the exact density structure of this data because many points overlap each other.

figure both ways! You should find that the PDF from the non-rasterized plot is roughly a factor of 10 larger in filesize (though in this instance, both are so small that using the non-rasterized version is fine.)

A point-cloud plot is not the only way we can visualize these data. Let's zoom in on the region of higher density, between log mass 7 and 11.5. To better visualize the distribution of data, we can use a 2D histogram to bin the points, giving a measure of the density at each location in parameter space.

Exercise 11.4:
Use pandas .loc indexing to create a new frame, df_zoom, containing only the rows with log masses in the range 7 to 11.5. Also cut on SFR: take those with log SFR greater than −2.5.

Assuming you have made your df_zoom frame, the following code will produce a 2D histogram of the data, binning into larger pixels the count of galaxies lying in that region (Figure 11.3).

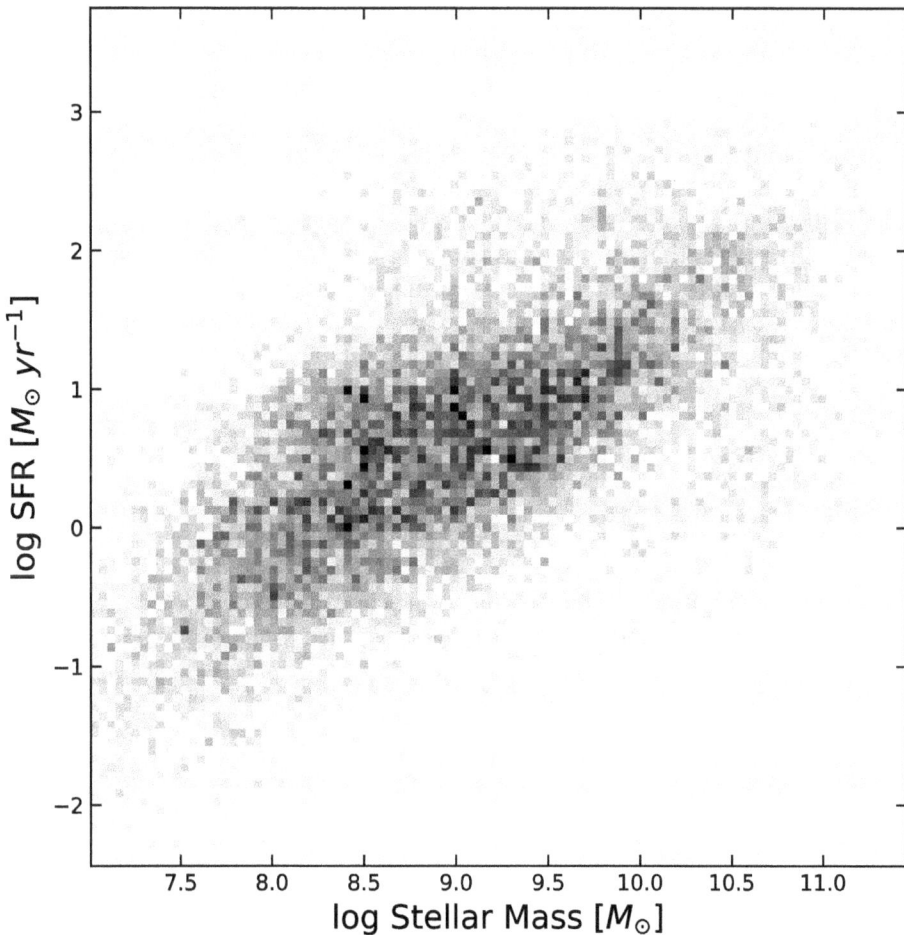

Figure 11.3. Two-dimensional histogram of the star-forming sequence, using a dataframe filtered on a mass and SFR range. The histogram bins nearby points and is colored by the relative density from bin to bin, so the darkest pixels here contain the most galaxies.

```
fig, ax = plt.subplots(figsize=(6,6),
                       constrained_layout=True)

ax.hist2d(df_zoom.lmass,
          np.log10(df_zoom.sfr),
          bins=100,
          cmap='magma_r')

ax.set_xlabel(r'log Stellar Mass [$M_{\odot}$]',fontsize=15)
ax.set_ylabel(r'log SFR [$M_{\odot}$ $yr^{-1}$]',fontsize=15)
ax.tick_params(direction='in',length=5,top=True,right=True)
```

In Figure 11.3, we can really clearly see the relation and the locus where most of these galaxies fall. How would we parameterize the ridge of high-density that characterizes the shape of this distribution of galaxies? By eye, you can see where this line should go. But think for a moment: how would you calculate or fit it?

One possibility is to create a histogram that, rather than displaying the *counts* in each bin, computes some other statistic. For example, we could compute the median SFR in bins of stellar mass. The `scipy.stats` module has a function, `binned_statistic`, that allows us to compute a histogram using a list of possible statistical measures (mean, median, etc), or indeed supply our own function handle to compute a custom statistic. Here, we can just use the median (it will be less sensitive to outliers).

```
from scipy.stats import binned_statistic

bin_means,bin_edges,bin_number = binned_statistic(x=df_zoom.lmass.
    values,values=np.log10(df_zoom.sfr),bins=15,statistic='median')
```

Annoyingly, this function returns the values (in this case, means) for each bin along with the bin *edges* and bin number. We need the centers of the bins, so we will have to compute those ourselves:

```
bin_width = (bin_edges[1] - bin_edges[0])
bin_centers = bin_edges[1:] - bin_width/2
```

Now we can plot our binned values over our 2D histogram, and see if the two agree (Figure 11.4)

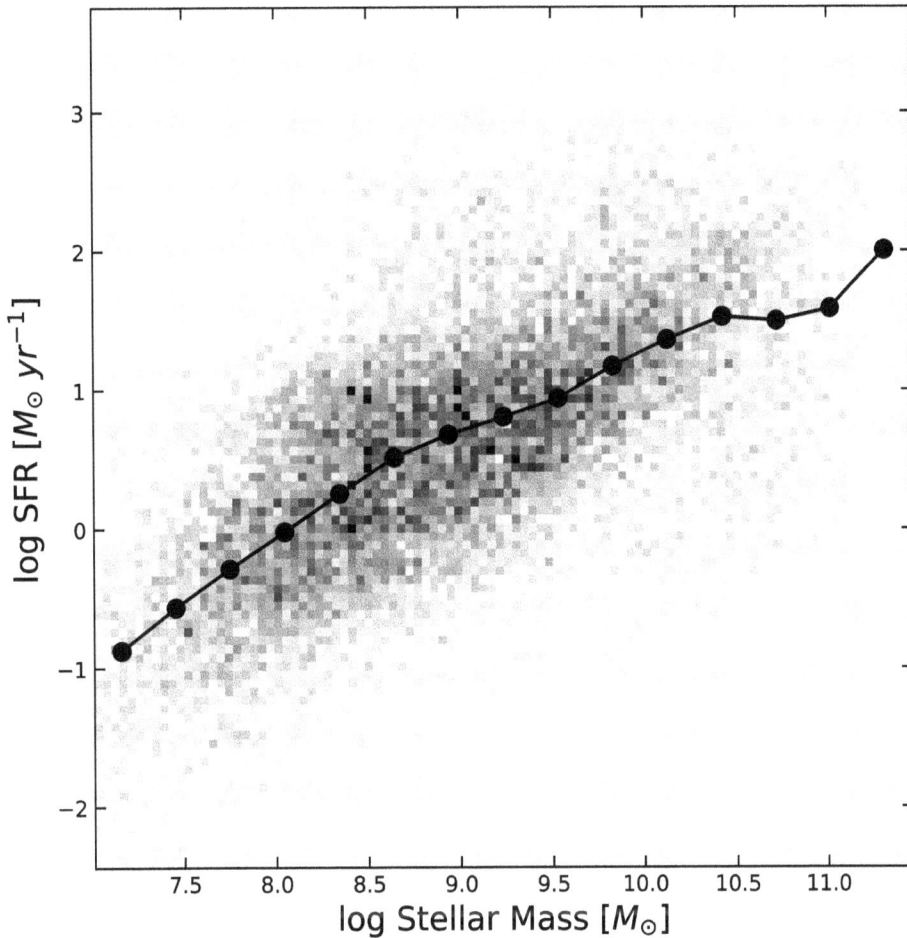

Figure 11.4. Two-dimensional histogram of the SFS with the median SFR in 15 mass bins computed with `binned_statistic()` overplotted.

```python
fig, ax = plt.subplots(figsize=(6,6),
                       constrained_layout=True)

ax.hist2d(df_zoom.lmass,
          np.log10(df_zoom.sfr),
          bins=100,
          cmap='magma_r')
ax.plot(bin_centers,bin_means,'k')
ax.plot(bin_centers,bin_means,'o',ms=8,color='k')
ax.set_xlabel(r'log Stellar Mass [$M_{\odot}$]',fontsize=15)
ax.set_ylabel(r'log SFR [$M_{\odot}$ $yr^{-1}$]',fontsize=15)
ax.tick_params(direction='in',length=5,top=True,right=True)
```

As we can see, this measure appears to describe the relation well.

Another way we could compute binned values is directly within our pandas DataFrame.

The pd.cut() function allows us to assign rows of our frame into categories or bins based on the value of a column.

Let's try it out, splitting our dataset into bins of stellar mass. For convenience, we'll also create a new version of the frame in which the SFR value has also been logged, and the mass bins are added as an explicit column of the new frame:

```
smass_bins=pd.cut(df_zoom.lmass,bins=12,retbins=True)


logdata={'log_smass':df_zoom.lmass,'log_sfr':np.log10(df_zoom.
  ⌐sfr),'smass_bin':smass_bins[0]}
log_sfr_df=pd.DataFrame(data=logdata)
log_sfr_df
```

```
       log_smass   log_sfr         smass_bin
0          10.46  2.331103  (10.348, 10.718]
1          10.41  2.282010  (10.348, 10.718]
2          10.45  0.393803  (10.348, 10.718]
3          10.16  2.486643   (9.977, 10.348]
4           8.81  0.760762    (8.493, 8.864]
...          ...       ...               ...
18162       8.38  2.213332    (8.123, 8.493]
18166       9.33 -0.092992    (9.235, 9.606]
18167      11.41  3.081599   (11.089, 11.46]
18168       9.86  0.401194    (9.606, 9.977]
18169       9.13  0.555578    (8.864, 9.235]

[17094 rows x 3 columns]
```

We can see that this new column, smass_bin, shows the particular bin that each row has been assigned to (the log_smass value for each row should be inside the range shown in the bin column). There are only 12 unique values of the bin (since we chose to cut with 12 bins).

Why was assigning one of twelve bins to each galaxy in a new column helpful? We can now compute statistical measures of all the galaxies in each bin thanks to a powerful DataFrame method known as groupby(). Group-by, as it sounds, allows us to group and fold a dataframe by some quantity. Ideally, it is a quantity that repeats throughout a dataset, on which you want to collapse. In our example, we can group by the value of the smass_bin to end up with twelve abstract objects, which have methods attached to compute statistics. This makes a bit more sense seen in action:

```
mean_grouped_df=log_sfr_df.groupby(['smass_bin']).mean()
med_grouped_df=log_sfr_df.groupby(['smass_bin']).median()
std_grouped_df=log_sfr_df.groupby(['smass_bin']).std()

mean_grouped_df
```

```
                 log_smass    log_sfr
smass_bin
(7.006, 7.381]    7.221843  -0.772704
(7.381, 7.752]    7.593970  -0.394160
(7.752, 8.123]    7.950556  -0.075036
(8.123, 8.493]    8.315394   0.275864


(8.493, 8.864]     8.677456   0.558410
(8.864, 9.235]     9.050799   0.770245
(9.235, 9.606]     9.414558   0.907579
(9.606, 9.977]     9.781272   1.189815
(9.977, 10.348]   10.146080   1.341965
(10.348, 10.718]  10.517585   1.396597
(10.718, 11.089]  10.873117   1.464846
(11.089, 11.46]   11.230541   1.735623
```

Notice how our dataframe has "pivoted" such that the unique values of that bin column are now our index, while the remaining columns no longer correspond to individual galaxies, but rather to the mean value of all rows within each bin (for that column).

Put another way, this overall set of **pandas** operations allowed us to compute the mean mass and SFR in 12 bins of mass. We also computed the median, and standard deviation for those bins—and other methods could be run as well. Let's assume the standard deviation in each bin is the uncertainty on the median SFR in each bin, and plot an error-bar plot over our histogram (Figure 11.5)

```
fig, ax = plt.subplots(figsize=(6,6),
                       constrained_layout=True)

ax.hist2d(df_zoom.lmass,
          np.log10(df_zoom.sfr),
          bins=100,
          cmap='magma_r')
```

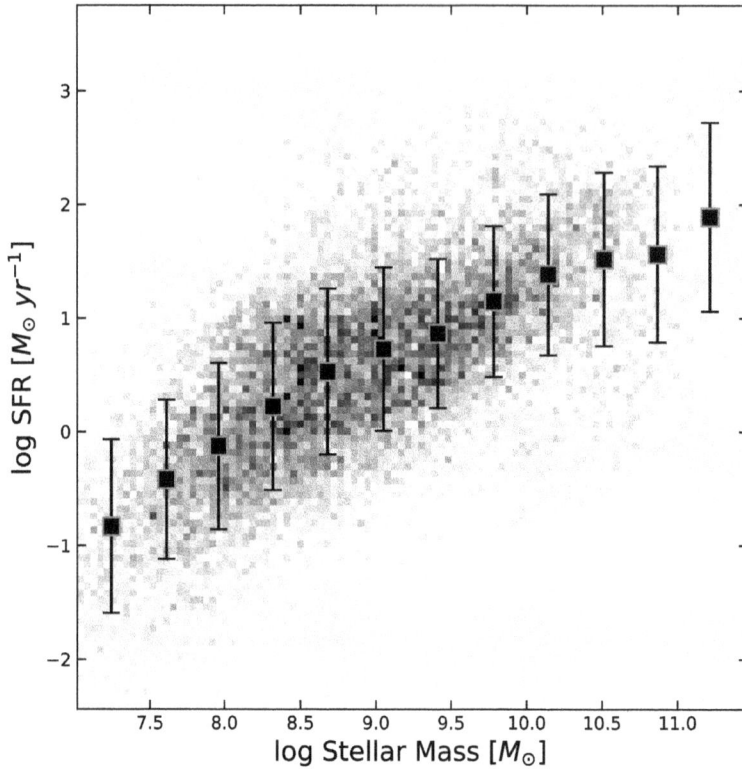

Figure 11.5. SFS with the median SFR in 12 bins overlaid, this time calculated via **pandas** operations. The error bars are computed by taking the standard deviation of values in the bin, and seem to more or less encompass the spread in the relation.

```
ax.errorbar(med_grouped_df['log_smass'],
            med_grouped_df['log_sfr'],
            yerr=std_grouped_df['log_sfr'],
            c='k',
            fmt='s',
            capsize=5,
            ecolor='k',
            markersize=10,
            label='binned median SFR')
ax.set_xlabel(r'log Stellar Mass [$M_{\odot}$]',
            fontsize=15)
ax.set_ylabel(r'log SFR [$M_{\odot}$ $yr^{-1}$]',
            fontsize=15)
ax.tick_params(direction='in',
            length=5,
            top=True,
            right=True)
```

Exercise 11.5: Redshift dependence.

Add a column to your frame called `ssfr`, computed as log(SFR)—log(Mass) (so technically, log sSFR). This quantity is known as the specific star formation rate, and by normalizing the star formation rate by stellar mass, we divide out the star forming sequence relation between mass and SFR. With this quantity, you can ask the question: does the specific star formation rate of galaxies in this sample vary with redshift?

Using the `pandas` cut technique we learned above, divide the 3DHST sample into 6 bins of redshift z (using `z_best`), adding the bins as a column to your dataframe. Then group-by the redshift bins to determine the median specific star formation rate in each bin, and plot them versus your redshift bin centers. Is there a relation? Note that you will need to compute the bin centers from the bin edges.

Did you find that the specific star formation rate of galaxies increases the further back in time (higher redshift)? While it is true that earlier in the universe's history (especially around "cosmic noon", from redshifts 2–4) the average specific star formation rates was much higher than today, it doesn't increase monotonically back to the beginning of the universe. As data scientists looking at this result, we should question if we have considered all factors. For example: if you plot a histogram of all the galaxy redshifts in the sample, you will find that there are very few in the high redshift bins compared to cosmic noon—it is thus likely that the uncertainty in those bins is large. On top of this, a survey of this nature will pick up the light from only the brightest galaxies in the early universe, which are also the most star-forming. This selection effect must be taken into account when attempting to interpret these kinds of plots.

11.4.2 UVJ Diagram

Another useful diagram one can create from these data (in particular, the rest-frame colors) is the UVJ diagram. Here, we plot the $U - V$ colors on the y-axis, and the $V - J$ colors on the x-axis. This parametrized space is useful because star-forming and quiescent (or quenched) galaxies separate themselves into clusters, and confounding effects (such as the fact that dust can block light from star formation and make a galaxy appear redder and more quiescent) tend to push galaxies along axes that don't lead to overlaps.

We can extract our U, V and J fluxes from our full `DataFrame` and convert them to magnitudes following instructions in the 3DHST readme, which indicates to compute

$$U - V = -2.5 \log\left(\frac{F_U}{F_V}\right)$$

and so on.

```
import matplotlib.pyplot as plt
import numpy as np

#sf_ms_df = clean_df.loc[(clean_df.lsfr>-5)&(clean_df.lmass>1.0)]
u_v = -2.5*np.log10(clean_df.U.values / clean_df.V.values)
v_j = -2.5*np.log10(clean_df.V.values/clean_df.J.values)
fig, ax = plt.subplots(figsize=(5,5),constrained_layout=True)
```

```
ax.plot(v_j,u_v,'.',alpha=0.1,rasterized=True)
ax.plot([0,0.8],[1.3,1.3],lw=4,color='k')
ax.plot([0.8,1.6],[1.3,2.05],lw=4,color='k')
ax.plot([1.6,1.6],[2.05,2.5],lw=4,color='k')
ax.set_xlim(0,2.5)
ax.set_ylim(0,2.5)
ax.set_xlabel(r'$V-J$',fontsize=14)
ax.set_ylabel(r'$U-V$',fontsize=14)
```

In Figure 11.6, I have added the UVJ "bounding box" (by eye), which separates a population of quiescent galaxies (within the box) from star-forming galaxies (rest of

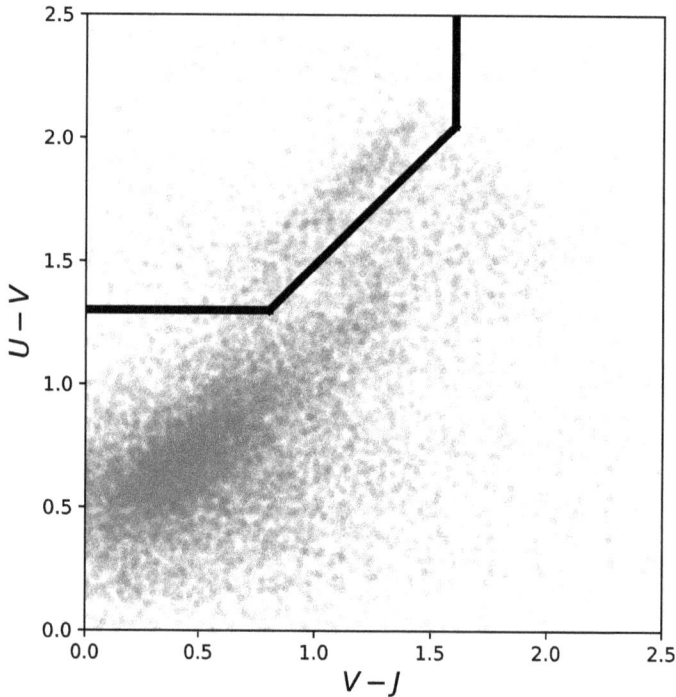

Figure 11.6. Diagnostic UVJ diagram, which uses restframe colors to separate galaxies into quiescent (found in the upper left box) or star-forming.

the plot). Dust tends to move galaxies up and to the right at a 45 degree angle, helping to keep from confusing dust obscured star formers with quiescent galaxies.[7]

Exercise 11.6:
Construct a conditional to feed to `.loc` to return only the quiescent galaxies based on their position in UVJ space.

Let's see if the measured star formation rates correctly fall in the right parts of this diagram. Using the `ax.scatter()` command, we can color these points by their measured sSFR (Figure 11.7).

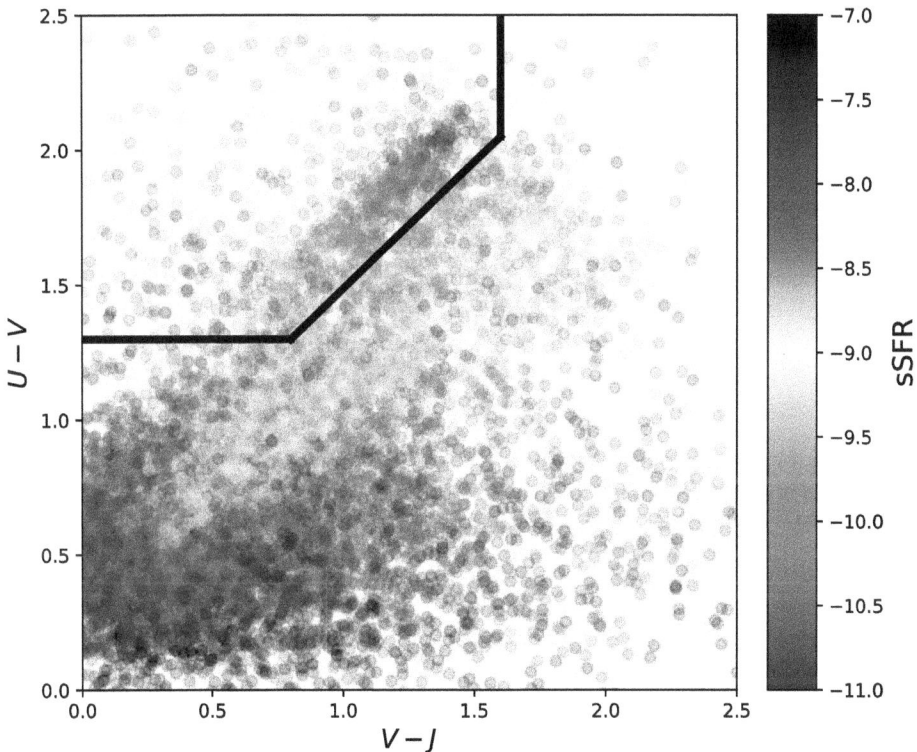

Figure 11.7. UVJ diagram updated such that points are colored by their log sSFR. As the SFR is computed using different data than the UVJ colors, it is a secondary affirmation that the plot generally works; red and dead galaxies live in the bounded box, while blue, star-forming galaxies live outside it.

---

[7] The diagram isn't perfect, of course. But it has been used in hundreds of papers, and has been updated over the years for different redshifts, and more recently, other "versions" using other filters are being explored.

```
from mpl_toolkits.axes_grid1.axes_divider import make_axes_locatable
fig, ax = plt.subplots(figsize=(7,6.5))
mapp = ax.scatter(v_j,u_v,c=clean_df.ssfr,marker='o',alpha=0.
   ↪2,rasterized=True,cmap='seismic_r',vmin=-11,vmax=-7)
ax_divider = make_axes_locatable(ax)
cax = ax_divider.append_axes('right', size='7%', pad='2%')
cbar =plt.colorbar(mapp,cax=cax,label='sSFR')
cbar.set_label('sSFR',size=15)
cbar.solids.set(alpha=1)
ax.plot([0,0.8],[1.3,1.3],lw=4,color='k')
ax.plot([0.8,1.6],[1.3,2.05],lw=4,color='k')
ax.plot([1.6,1.6],[2.05,2.5],lw=4,color='k')
ax.set_xlim(0,2.5)
ax.set_ylim(0,2.5)
ax.set_xlabel(r'$V-J$',fontsize=14)
ax.set_ylabel(r'$U-V$',fontsize=14)
```

As one would hope, if we color by the specific star formation rate, quenched galaxies (observers often use $sSFR < -11$ as a threshold), which are redder in this plot, live in the part of UVJ space in he bounding box, while blue galaxies live outside. Based on this coloring, my bounding box might need to expand slightly to catch more of the quiescent population! You might notice a few Matplotlib tricks played in the above code, in order to modify the color bar's opacity (since the points are low in opacity, the color bar, by default, is the same `alpha` value) and text fontsize.

Exercise 11.7: Size-mass relation.
Does a galaxy's size correlate with its mass? For this exercise, you will need to return to the `photometry` catalog and extract the `flux_radius` column to add to your frame. This quantity is the effective radius (the radius which contains half the total light from the galaxy). We use this as a measure of galaxy size. Again using the six redshift bins (but ignoring the highest-redshift few, which do not have many galaxies), plot the galaxy sizes versus their log stellar mass. Are the two correlated? Does the relation shift with redshift? Replot, but split or color your sample by whether galaxies are quiescent or star forming (use either your UVJ or the sSFR value to make this split). You can check your work by reading Miller et al. (2019).

In this chapter, we discussed the data formats commonly used for tabular, catalog-style data in astronomy. We covered several libraries useful for the reading in and parsing of such data, as well as the Pandas package as (in addition to file reading) a tool for the analysis, filtering, and management of large amounts of data within Python. Using Pandas, we explored a real tabular catalog of galaxy data, and saw how to clean, modify, filter, and combine multiple tables into one useful catalog for doing science. Working with "big data" involves scaling up these kinds of skills, and is a valuable tool in a modern astronomer's arsenal.

11.5 Summary

In this chapter, we have discussed how tabular data can be ingested and analyzed in Python. Working with large catalogs is a common aspect of the data scientist's day-to-day operations, requiring us to load, clean, combine, and filter massive datasets for the key elements we need for a calculation. Whole textbooks exist on how to get fancier with Pandas (or other tabular libraries) when it comes to these sorts of techniques, but hopefully this introduction has illustrated the core requirements of working with catalog data.

References

Brammer, G. B., van Dokkum, P. G., Franx, M., et al. 2012, ApJS, 200, 13

Kriek, M., van Dokkum, P. G., Labbé, I., et al. 2009, ApJ, 700, 221

Kriek, M., van Dokkum, P. G., Labbé, I., et al. 2018, FAST: Fitting and Assessment of Synthetic Templates, Astrophysics Source Code Library, record ascl:1803.008

Miller, T. B., van Dokkum, P., Mowla, L., & van der Wel, A. 2019, ApL, 872, L14

Momcheva, I. G., Brammer, G. B., van Dokkum, P. G., et al. 2016, ApJS, 225, 27

Skelton, R. E., Whitaker, K. E., Momcheva, I. G., et al. 2014, ApJS, 214, 24

van Dokkum, P., Brammer, G., Momcheva, I., Skelton, R. E., & Whitaker, K. E. 2013, arXiv e-prints, arXiv:1305.2140

Astronomical Python
An introduction to modern scientific programming
Imad Pasha

Chapter 12

Vectorization and Runtime Improvements

12.1 Introduction

In this chapter, we'll discuss how to turbo charge our code to get it running quickly. Whenever we code solutions to problems we are facing, there are (inevitably) multiple ways to accomplish the same task. Some solutions are more or less efficient in terms of number of lines of code. Some are more or less efficient in how long they take us, the programmer, to write and develop. And some are more or less efficient at runtime—does your code do the job in several seconds, several minutes, several hours, or several days?

Across all these different axes, we have to make choices about where to spend our time. If we need to run some code *once*, and we know it will take roughly six hours to finish, and it would take us two days of effort to craft a more elegant, runtime-friendly solution… perhaps it is better to just run the slow (but working!) method. On the other hand, if we need to run that same code a hundred times, two days spent optimizing it up front would be well spent.[1] Generally, when you are first getting started with astronomical research, a relatively inefficient solution to a problem is fine, as you are often working with small amounts of data. A thousand-times difference in runtime might be the difference between several milliseconds and several minutes—pair this with the fact that optimizing often requires learning some new coding techniques which might take considerable time and practice and we arrive at the conclusion that it would be a waste of time.

That is why this chapter comes late in this text. Nevertheless, there comes a point in every astrophysics researcher's career in which some inefficient code will be dragging down your day to day progress. That point is usually fairly early. At that point, it's time to sit down and learn to write vectorized, optimized code.

---

[1] Of course, the "trap" is we can't *know* how long optimizing and refactoring our code to run faster will take.

12.2 Identifying Bottlenecks

Compared to "human time," computer code in general executes very quickly. Thus far in this text, most blocks of code (or even full programs) we have written can likely execute in a second or less on a modern laptop. For these cases, spending time optimizing performance is not generally a necessity. The primary situation in which speed becomes a paramount concern is when we have to complete *many* calculations *in a row*.[2] In these instances, even single-computation runtimes of a fraction of a second quickly add up to minutes, hours, or days when we have many, many calculations to carry out.

If the above process sounds familiar, it should: what I've described more or less encompasses the idea of a Python for-loop or while-loop.

You may be wondering, why are Python loops so slow? Other (generally compiled) languages can run loops considerably faster than Python. Indeed, when we use something like Numpy, *somewhere* on our computer, a loop is still being carried out—it is just being carried out in a compiled language, e.g., C. There are several reasons why these loops are faster. Of particular importance is Python's dynamic typing; unlike other languages where every variable is assigned a type at allocation, and an array-like object will have a specific type associated with it, Python allows types to be defined dynamically. This is convenient for us as programmers, but it means that when Python runs a loop, it has to carry out a multitude of checks on the type of the current loop elements, even when those elements are the same types each time.

When we first learned about Numpy, I mentioned that arrays, unlike lists, can only support one type at a time. Now we learn why; when Numpy goes to carry out a mathematical operation, because the array knows all of its elements are of that one type, the operation can be efficiently downcompiled to the C code underlying Numpy and be run as a loop there. There is of course overhead associated with passing the arrays back and forth between Python and C, but almost always, that overhead is significantly smaller than the time running that same operation as a loop in pure-Python would take. Not having to query each loop object to retrieve its type is one area the loop speeds up, but not the only one; compilers over the years have employed many tricks which work in that framework and can speed up loops.

All this is to say that the first and easiest way to identify bottlenecks in your code is to find every loop. Some of those loops may only iterate over a few elements for a very simple task (e.g., a loop to read in five files). These are likely not worth fixing (and might not be possible to improve). But it is likely that any other loops in your code are your greatest source of slowdowns. Finding refactorings[3] that remove these loops is the first priority.

---

[2] In programming parlance, commands carried out one after the other on one processor are said to be run *in serial*.

[3] *Refactoring* is the process of rewriting code such that it accomplishes the same task and behaves the same externally, but which via the rewriting has been improved in simplicity, efficiency, or other benchmark metrics.

That said, there are several ways we can quantitatively assess the speed of our code, ranging from simple timing checks to entire programs. For most day-to-day research, the following type of check may be sufficient.

Let's say we have a research script that is executing too slowly. A few print statements or logging statements scattered through the code may give us a qualitative idea of which stage of the script is taking the longest. We can then pull the offending section into a few cells of a notebook, and use the magic command %%timeit to run a speed test.

```
%%timeit
import numpy as np
for i in np.arange(1000):
    arr = np.random.random(size=(1000,1000))
```

7.32 s ± 46.7 ms per loop (mean ± std. dev. of 7 runs, 1 loop each)

The advantage of using the timeit command is that it will actually run the cell several times, to average over the fact that a computer processor, as it handles different tasks, may perform the calculation with slightly different efficiencies each time you run it. If we wanted an even simpler, one-execution estimate of the time, we could use the time module:

```
import time
import numpy as np
time_start = time.time()
for i in np.arange(1000):
    arr = np.random.random(size=(1000,1000))
time_total = time.time() - time_start
print(time_total)
```

7.439732074737549

We can see that this time is similar to the average time returned by the multi-run test. In this example, all we are attempting is to create 1000 instances of a 1000×1000 array of random numbers (though we don't do anything with them here). Random number generation is not particularly efficient in Python, scaling poorly with the number of elements needing to be created, so this loop overall is taking roughly 7.3 seconds to complete. In computer terms, that is quite slow. Depending on our circumstances, 7.3 seconds might be okay for a one-time calculation. However, now imagine this loop being a part of a larger program which needs to be run on hundreds, or thousands of objects. Factors of 10 hit quickly, and our code would soon become unusable.

Unfortunately for our small example here, there is not (to my knowledge) an easy refactoring of this loop in pure-Python or Numpy terms which would provide dramatic improvements; the offending code is a single line of one Numpy function, and the runtime will scale with however many times that one slow line needs to be repeated. As it turns out, random number generation of this kind can be computed with much higher efficiency for large arrays using a Graphical Processing Unit (GPU), as opposed to the standard CPU that we have been using throughout this text. New Python packages, e.g., jax (which we will discuss more in a moment) allow for several different operations to be carried out on GPUs, which in this case can lead to large speed improvements.

Yet another way to decrease the runtime of this code is to *parallelize* it. Notice here that each of these calls to create a 1000 × 1000 random number array are independent of one another—put another way, subsequent steps of the loop do not rely on previous steps. This means that if we have multiple CPU processors at our disposal, we could consider splitting the computations of this loop across those CPUs. There are special modules, e.g., `multiprocessing`, which facilitate this task. If we had 10 CPU cores, for example, we could have each of those cores run a loop of only length 100 (rather than 1000), and collect all the responses.[4]

Thankfully, most slow loops in our Python code do *not* require moving to a GPU or multiprocessing to solve, and instead can be approached via some clever applications of Numpy and its ability to broadcast arrays.

12.3 Fast Array Operations with Numpy

We can take advantage of Numpy's speed by carrying out calculations in a *vectorized* manner. That's just a fancy way of saying the following: "*When there exists a Numpy function or operation that acts on arrays, use it rather than any form of loop*". The simplest example of this can be summarized as follows:

Let's say we have `array_a` and `array_b`, and we want to generate `array_c` as the multiplication of each element of `array_a`, pairwise, with the value in `array_b` at the same index. We could write a `for-loop` over the indices, or over the `zip()` of the two arrays, and fill a new array with the result of each multiplication, like this:

```
array_c = []
for i in range(len(array_a):
    array_c.append(array_a[i]*array_b[i])
```

In the above, we iterate over a set of indices which can be used to pull each value from the two arrays. We could get them directly with `zip`:

```
array_c = []
for a,b in zip(array_a,array_b):
    array_c.append(a*b)
```

As you probably already know from various examples in this text, there is a much better way to compute the multiplication of two arrays: simply use the muliplication operator on them.

```
array_c = array_a * array_b
```

---

[4] There is of course overhead here as well, and Python is generally not a language for which multiprocessing helps dramatically, for subtle reasons which may be changing in upcoming years. You can read more about that in a recent proposed change to the language: https://peps.python.org/pep-0703/.

In this case, knowing the "vectorized solution" simply meant knowing the behavior resultant from multiplying two arrays together, and is usually fairly easy to spot by eye.

That's not always the case. Let's take an example from a real research application. In astronomy, we often spend time comparing a *model* to *data*, trying to determine which model parameters produces something which best matches our data (we'll discuss this further in Chapter 13).

Let's say we have one real spectrum of a star—in our example, these measurements are the *data*. We also have 5,000 *models*— essentially, 5,000 spectra derived synthetically using some code, which have different input parameters. Our task is to determine which of these 5,000 models is the "best-fit" to our data.[5] An easy metric to assess the goodness of fit between each model and our data is the χ^2. This is defined as

$$\chi^2 = \sum_{i}^{N} \frac{(d_i - m_i)^2}{\sigma_i^2} \tag{12.1}$$

(with some constants we won't worry about for now). You can read this as: for each corresponding data point and model point, take their difference squared and normalize by the uncertainty on the data point squared. When we do this for a spectrum of several thousand points, we'll get several thousand of these values, which we sum over to get some global estimate of how different that particular model is from the data. If we don't have any uncertainties, we could leave off the denominator (which I will do for the remainder of this example). The actual *value* of this χ^2 doesn't matter, but what does matter is that if one model is closer to the data, its χ^2 will be smaller in magnitude than a worse model. Thus, we wish to find the model that has the minimum value of this metric.

How would you approach this task? A natural thought is to break the problem into units. We know (now) how to compute the χ^2 between *one* model and the data. So what if we did the following:

```
chi2_array = []
for model in model_spectra:
    model_chi2 = 0 #
    for i in range(len(model)):
        chi2_point = (data[i] - model[i])**2
        model_chi2 += chi2_point
    chi2_array.append(model_chi2)
best_model_ind = np.argmin(chi2_array)
best_model = model_spectra[best_model_ind]
```

---

[5] In this example, we assume the model spectra are defined on a grid of wavelengths identical to our real spectrum.

This isn't the simplest code, so I'll also write it out in pseudo-code:

```
initialize empty container to hold all model chi2s
for model in model_spectra:
    initialize that model's chi2 as 0
    for each data point in the model and data:
        compute chi2 of that data point and model point
        add to the running sum for this model
    append the total chi2 for this model to the container
find the loc of model with min chi2
index that model
```

If the solution you were imagining looked something like the above, you're not alone! You could consider this a rather "faithful" adaptation of the mathematical expression above. Unfortunately, this solution uses *two* loops—one over all our 5000 models, and one over each value in those models (which for our spectrum is a few thousand). This is going to be *extremely slow*.

Stop for a moment now and think about how we can eliminate one (or more) of those loops, using the vectorization method.

Hopefully, the first improvement is immediately apparent. Just like with `array_a` and `array_b` above, our operation of subtracting one spectrum, pairwise, with another can be accomplished with one `Numpy` operation. This resolves our inner loop:

```
chi2_array = []
for model in model_spectra:
    model_chi2 = np.sum((data-model)**2)
    chi2_array.append(model_chi2)
best_model_ind = np.argmin(chi2_array)
best_model = model_spectra[best_model_ind]
```

Notice that the `np.sum()` function is taking in a single array (the residuals between the model and data, squared) and summing it across that long axis.

This code will run roughly a thousand times faster than the first solution. Can we eliminate the second loop too?

It turns out we can.

Using the same logic, we want to consider if we can make this inner comparison (subtracting one spectrum from another) happen for all our models at once:

```
chi2 = np.sum((data - model_spectra)**2,axis=1)
best_model_ind = np.argmin(chi2_array)
best_model = model_spectra[best_model_ind]
```

Okay, so what's going on here? First, our `data` spectrum is a 1D array, so its shape is (2000,1) while `model_spectra` is an array of all our models, which has a shape of (2000,5000). How can we simply subtract one from the other? It turns out, `Numpy` is appropriately *broadcasting* these shapes, extending that length-1 axis of the first array to match the 5000. So the output of a subtraction here is an array, (2000,5000) in shape, which is the data subtracting each model. I can square every value in this array easily, then sum over the columns to get a vertical array with one χ^2 for each model-data comparison. Then finding the minimum happens the same way.

This solution will run *much, much faster* than our original solution.

Let's now examine an example where we need to be even more clever with Numpy to avoid looping. I'll load up some data here (available in the textbook data sets) and then explain what we are looking at:

```python
import pandas as pd

draco = pd.read_csv('draco.dat',
            delim_whitespace=True,
            names=['ra','dec','sdss',
                    'g','u-g','g-r',
                    'r-i','i-z','E'])
iso = pd.read_csv('iso2.txt',delim_whitespace=True)
```

```python
import matplotlib.pyplot as plt

x = iso.g_f0-iso.r_f0 - 0.05
y = iso.g_f0+18

fig, ax = plt.subplots(figsize=(8,8))
ax.plot(draco['g-r'],draco['g'],'.',alpha=0.2)
ax.set_ylim(ax.get_ylim()[::-1])
ax.set_xlabel(r'SDSS $g-r$',fontsize=20)
ax.set_ylabel(r'SDSS $g$',fontsize=20)
ax.plot(x[:-750],y[:-750],'k.')
```

What we are looking in Figure 12.1 is a color–magnitude diagram (CMD). A CMD compares the difference in magnitude (flux ratio) between two filters (in this case, SDSS g and r) to the magnitude of one of those filters (in this case SDSS g). This turns out to be a useful exercise because stars of a given age, drawn from an initial mass function, tend to lie on "tracks." Using physical models, we can predict what these tracks look like for a population of stars of some age and metallicity (among other parameters). These tracks are known as *isochrones*. Isochrones can be used to find the age of astronomical stellar populations (if the population is thought to have all formed at roughly the same time, e.g., a globular cluster), and can be used to find the distance to astronomical objects, based on how much one needs to shift an intrinsic isochrone to match a set of data.

In this example, we will be using photometry of stars in the Draco Dwarf galaxy (Rave et al. 2003). Draco is an old (~10 Gyr) dwarf, and while it has a multiple-age stellar population, *most* of the stars are in a very old component. Using the stated age and metallicity quoted in Aparicio et al. (2001); I computed an isochrone using the Padova library of tracks. The exact details are not important, but after correcting

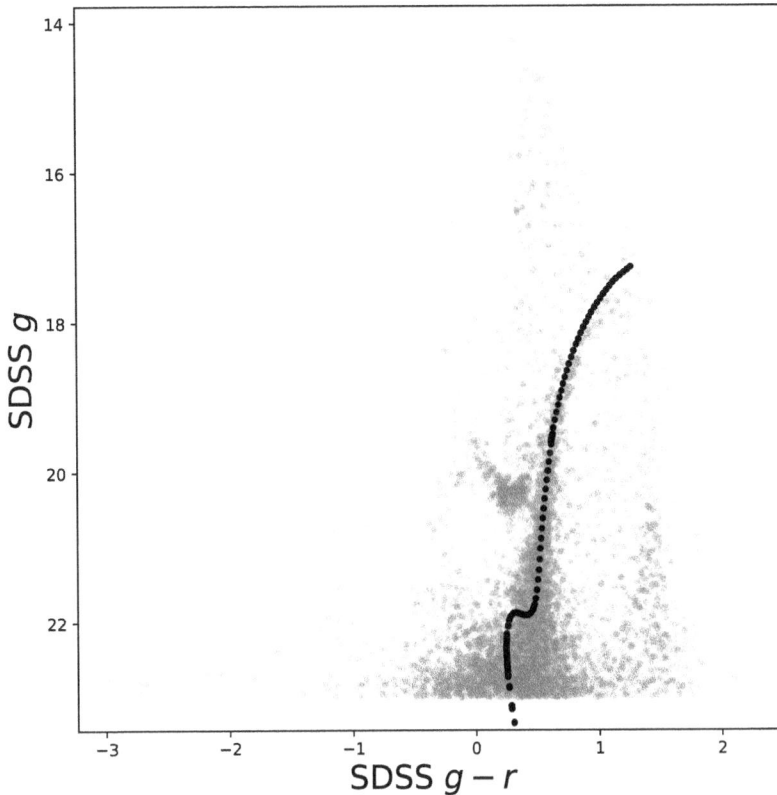

Figure 12.1. Color-magnitude diagram of the Draco Dwarf galaxy, using SDSS photometry (Rave et al. 2003). Overlaid is a single isochrone track from the Padova isochrone library.

for the distance effect, we can see that the computed isochrone (black dots) is a reasonable fit to the data (blue dots), showing the red giant branch rising above the main sequence (Figure 12.1).

A common astronomical task at this stage is to determine membership probability. The stars in this plot were all stars which spatially (on the sky) appeared to be within the bounds of the galaxy. But without distance measurements to each star, it is challenging to know if they are in the foreground, or actually part of Draco. If we assume that the stars actually associated with Draco lie close to this theoretical isochrone, we can assign some type of probability, giving stars far away from the isochrone low probability of being in Draco, and stars directly on the isochrone a high probability. Note that most stars in the clump to the left of the main near-vertical track (known as the red giant branch) are likely variable stars in Draco, such as RR Lyrae. This particular exercise, using distance from the isochrone to assign membership, would not be ideal for characterizing this clump.

I selected this problem to use because we can now lay out a simple programming task: for each star in the Draco sample, we need to determine both which point on the isochrone track it is closest to, and determine its distance to that point. You can attempt this problem in Exercise 12.1.

Exercise 12.1: Distances from Isochrone.

Before we learn the Numpy solution, carry out this task yourself, using loops if necessary. As a reminder, your goal is:

1. For each data point in the Draco sample (i.e., the stars spatially coincident with Draco on sky), determine *which data point* on the isochrone track is the closest to it (using Euclidean distance).

2. For each star in the Draco sample, keep track of that minimum distance, such that you can assess using a single value how distant from the isochrone a star is.

3. Plot your results using the `ax.scatter()` function in Matplotlib, feeding your distance array in as the `c` parameter, which sets how the points are colored. You should find that there is a gradient of increasing distance as you move away from the isochrone normal to its orientation at any given position.

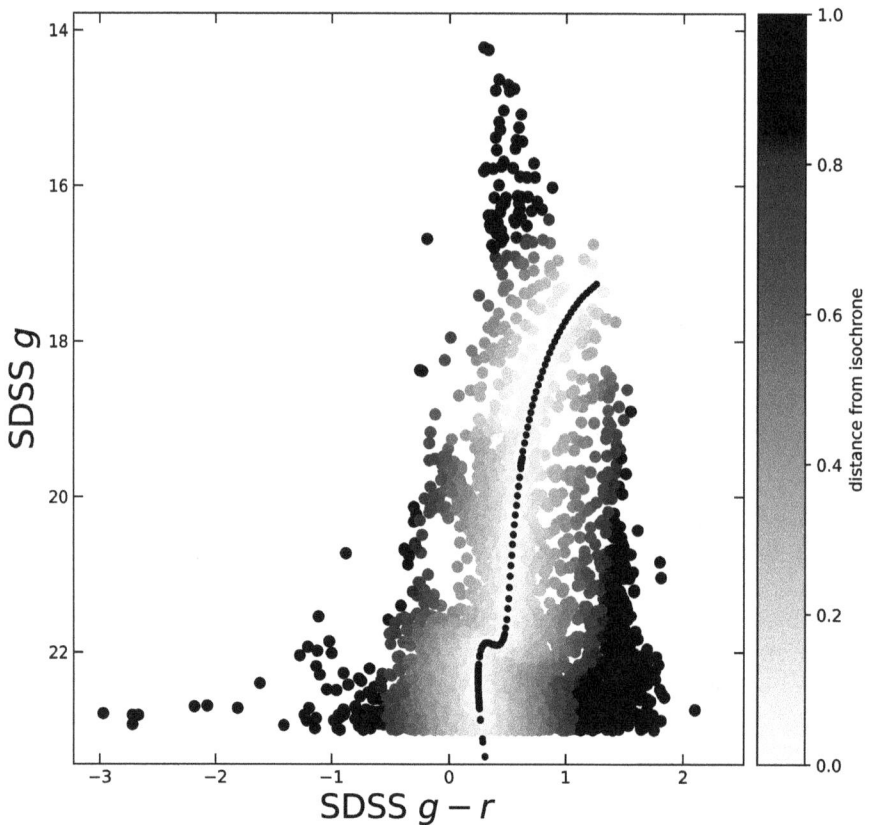

Figure 12.2. Color-magnitude diagram of stars in the Draco Dwarf galaxy, now colored by their computed euclidean distance from the isochrone track itself.

Hint: You can peek at Figure 12.2 to see what this should look like, or glance at the pseudo-code below for a starting scaffold.

You may have found that in your solution to Exercise 12.1, you may have used one, or even two loops. A possible solution might have pseudo-code that looks like the following:

```
final_distances = []
for star_i in draco_sample:
    iso_distances = []
    for iso_j in isochrone:
        compute distance between star_i and iso_j
        append to iso_distances
    take minimum of iso_distances
    append to final_distances
```

You may have also tried to replicate the algorithm we discussed above for the model spectra example, in which you subtract the isochrone track's positions from the full array of data positions... but if you did, you likely ran into errors about shapes—unlike our spectral library example above, the isochrone track does not have the same number of points as the data set, and both are two dimensional. Thankfully, this is not an insurmountable obstacle. Instead, it is an opportunity to talk about array broadcasting in the context of mismatched array sizes.

Let's examine what shapes we are working with in this example. I am going to composite together the positions (x,y) for both data sets:

```
D = np.column_stack([draco['g-r'].values,draco['g'].values])
M = np.column_stack([x,y])

print(D.shape)
print(M.shape)
```

```
(5766, 2)
(980, 2)
```

The length-2 axis in both cases represents the spatial dimensions we are working with (each data point has an x and y value). The differing lengths are the number of points in the Draco data set and the model isochrone track, respectively.

We'll begin with the solution (a two-liner which could be a one-liner) and then subsequently unpack what is going on.

```
dist2 = np.sum(np.abs(D[:,np.newaxis,:]-M[np.newaxis,:,:])**2, axis=-1)
min_dist = np.min(dist2**0.5,axis=1)
```

What happened here? Let's first examine the shapes of our two output arrays:

```
print(dist2.shape)
print(min_dist.shape)
```

```
(5766, 980)
 (5766,)
```

As we might expect, the dist2 array (the array containing all the squared distances) has a shape 5766 × 980. This matrix represents the pairwise distances between each point in D and each point in M. So each row of this matrix is one of our Draco stars, and each column is its distance to a different point on the isochrone track. We only care about each star's distance to the *closest* of the points on the isochrone to it; i.e., for each row, we only need the minimum value of all the columns. So the second line of code should be more straightforward; we take the minimum of the distance array (square rooted to get from squared distance to distance) along axis 1 (the columns).

What about the first line? How did dist2 get constructed? We can separate this composite operation into parts. The inner most computation is the key one:

```
pairwise = D[:,np.newaxis,:] - M[np.newaxis,:,:]
pairwise.shape
```

```
(5766, 980, 2)
```

This bit is where the magic happens. We know that we need to compute all the $x_i - x_j$ terms and all the $y_i - y_j$ terms for each point (which will then get summed together as part of the distance calculation). Both D and M have 2 axes, but we will need a third axis to hold both an x difference and y difference for each position on a grid of (draco point, model point) combinations. The np.newaxis call tells Numpy to increase the dimension of an existing array by one, along the axis specified by where the np.newaxis is placed in the array slice.

D is shaped (5766,2). We are fine with leaving the first axis alone, but we want to grow our array from 5766 × 2 to 5766 × 980 × 2, making the room needed to accommodate the isochrone array. We can leave our length 2 axis at the end. Hence, we'll use our commas to assert an additional axis, in the middle, to grow our array into via np.newaxis.

M, meanwhile, is shaped (980,2). Because our subtraction operator is going to need both these arrays to have compatible shapes, we don't actually want our first axis to be 980 anymore—we want the first axis to grow to match the 5766. If we then keep the existing two axes as is, as the second and third now rather than the first and second, we will end up with an array compatible with D.

This is because by default, the length assigned to the new axes created by the slicing will be 1:

```
print(D[:,np.newaxis,:].shape)
print(M[np.newaxis,:,:].shape)
```

```
(5766, 1, 2)
(1, 980, 2)
```

But when we go to subtract them, Numpy can *extend* each of those length 1 empty axes to match the corresponding length in the other array. This extension is what Numpy was doing in our more simple spectral library fit above—one array was (2000,1) (the real fluxes), and the other was (2000,5000) (the model fluxes for 5000 models). Because of that size 1 axis, Numpy extended it to match the shape of the 5000, allowing for that subtraction.

Thus, subtracting these two quantities leads to our $(5766,980,2)$ array. Put another way, we have 2 (5766,980) arrays, one with all the pairwise x-differences and one with all the pairwise y-differences. The rest of the operations (the absolute value and the sum) are simply responsible for combining our x-differences and y-differences appropriately; notice the sum is over axis -1, as expected since our length 2, final axis is the one handling the x versus y data. (which get summed in the distance formula).

Let's examine what our distance metric gave us, by coloring our Draco data points by their distance value determined using this method. The code below will produce the plot shown in Figure 12.2.

```
from mpl_toolkits.axes_grid1.axes_divider import make_axes_locatable
fig, ax = plt.subplots(figsize=(8,8))
im = ax.scatter(draco['g-r'],
                draco['g'],
                c=min_dist,
                alpha=1,
                vmin=0,
                vmax=1,
                cmap='magma_r')
ax.set_ylim(ax.get_ylim()[::-1])
ax.plot(x[:-750],y[:-750],'k.')
ax.set_xlabel(r'SDSS $g-r$',fontsize=20)
ax.set_ylabel(r'SDSS $g$',fontsize=20)
ax_divider = make_axes_locatable(ax)
cax = ax_divider.append_axes('right', size='7%', pad='2%')
plt.colorbar(im,cax=cax,label='distance from isochrone')
ax.tick_params(top=True,right=True,direction='in',length=6)
```

As we can see, our calculation has successfully completed the task we set out to do. The distance (as indicated by color) for each Draco sample star increases roughly perpendicular to the isochrone track, as we would expect when the distance was properly computed to the closest point on said track.

Exercise 12.2: Membership Probability. Using the following formula (Equation (12.2) from Collins et al. (2020)), compute the membership probability for each star in the sample. Then, use `np.where()` to extract only stars with $>60\%$ probability of being members, and recreate the above figure using just those points, colored by their probability.

$$p(m) = e^{-(d_{\min})^2/2(0.1)^2} \tag{12.2}$$

Hopefully, these two examples have illustrated one key takeaway: if you have loops in your code, but in some way shape or form, all you are doing is computations with multiple arrays, then there is likely a one- or several-line solution invoking just Numpy operations and broadcasts which can, if you have many computations to make, speed up your code (and make it shorter, as well).

12.4 Jax

The Jax library (James et al. 2018) is rapidly growing in popularity as a tool for enabling just-in-time (JIT) compilation of functions, tracking of derivatives (needed for some gradient-based algorithms; see Chapter 13), and seamless execution on GPUs in addition to CPUs. As mentioned at the top of this chapter, GPUs are highly efficient at certain mathematical operations, because unlike a CPU (~1 process), a single GPU device has hundreds of parallel threads available. It has taken some time for computational tools to successfully wrangle data such that passing it back and forth from CPU to GPU is relatively easy: Jax is one such tool.

In the introduction to this chapter, I asserted that the scenarios in which bottlenecks are the worst are ones in which we need to repeat some operation *many* times. In your own standard research code, it is rare to need to repeat an operation more than a few hundred times (e.g., smoothing a few hundred images, or fitting 100 isochrone tracks to several CMDs). There are two tasks, however, which *demand* not hundreds, but tens of thousands or even millions of repetitions.

The first is Markov Chain Monte Carlo (MCMC) algorithms. We'll learn much more extensively about these in Chapter 13, but the takeaway here is that where above, we compare 5000 model spectra to our data, in an MCMC chain, we might produce hundreds of thousands of models and compare them to our data. You can see then, that the model creation time—the function that spits out a new model for a different set of parameters—matters greatly. If it takes 1 second per model, a hundred thousand model creations will take around 27 hours to run. If we can pare down the individual model creation time to just 0.05 seconds, that's less than two hours of runtime (ignoring other overheads).

The second task that requires many repetitions is machine learning. When we train a machine learning algorithm on some data, it needs to see the data (and sometimes augmented versions of it) millions of times before it can converge to a predictive model.

As a result, the community (both within and beyond astronomy) is rapidly adopting tools like Jax, which help facilitate both of these tasks. It was not at all unheard of within the decade between 2010 and 2020 to be running MCMCs that *did* take 27 hours to finish. Now, new packages are even *combining* machine learning with MCMC to attempt inference problems at lightning speed.

To see a real example of this type of exercise in action, you can check out the pysersic (Pasha & Miller 2023) or AutoPhot (Stone et al. 2023) codes. Both are concerned with fitting the structural parameters of galaxies based on images—namely, the size and flux of these systems. The pysersic code uses an MCMC to compute Bayesian posteriors on the fit parameters (which define the Sersic profile often fit to galaxies), but does so many orders of magnitudes faster than traditional MCMCs by compiling functions, carrying out computations (like convolutions) on a GPU, and leveraging more efficient MCMC algorithms that require derivative information (which Jax tracks).

All this is to say, you likely will not need to invoke Jax as you begin writing your first research code. But being aware and abreast of its development and usage is a useful exercise for when you eventually find an opportunity within your research code to leverage it, and you may quickly find yourself *using* code that leverages it, so understanding some of the basic principles is always wise.

12.5 Summary

In this chapter, we examined how to approach the task of improving the runtime of our code. While not always a requirement, some astronomical tasks benefit greatly from optimizations of this kind. The simplest (and usually best) form of optimization one can perform in Python is simply to ensure vectorized, Numpy approaches are used in place of any loops over data. A code with no loops will likely run relatively quickly, and even oddly-shaped problems can generally be accommodated into Numpy's broadcasting framework.

References

Aparicio, A., Carrera, R., & Martínez-Delgado, D. 2001, ApJ, 122, 2524

Collins, M. L. M., Tollerud, E. J., Rich, R. M., et al. 2020, MNRAS, 491, 3496

James, B., Roy, F., Peter, H., et al. 2018, JAX: composable transformations of Python+NumPy programs http://github.com/google/jax

Pasha, I., & Miller, T. B. 2023, JOSS, 8, 5703

Rave, H. A., Zhao, C., Newberg, H. J., et al. 2003, ApJS, 145, 245

Stone, C. J., Courteau, S., Cuillandre, J.-C., et al. 2023, MNRAS, 525, 6377

Astronomical Python
An introduction to modern scientific programming
Imad Pasha

Chapter 13

Astronomical Inference

13.1 Introduction

Thus far, we have built up our familiarity with the astrophysics-oriented Python ecosystem of packages and tools primarily via examples of *measurement*. Tasks like measuring the flux in an aperture placed on an astronomical image (Chapter 8) do not involve the explicit creation of a model. As it turns out, however, very few things can be learned about the universe purely by making a measurement—what we can *see* (photons, gravitational waves, neutrinos) are only a noisy, biased, sometimes non-linear *tracers* of the processes we actually care about (such as the mass of a galaxy, supermassive black hole, or progenitor of a supernova). The process of turning our measurements of observables into knowledge about intrinsic properties almost always involves the creation of a model which we *think*[1] encapsulates the necessary physics, observational biases, etc. to represent the data. We then compare many realizations of that model (varying its parameters) to our data set to determine which set of parameters "best fits." I quote "best fits" because establishing the best fitting model is non-trivial—we will discuss this more throughout the chapter.

This process of model-data comparison is known as *inference*, and colloquially is often referred to as model fitting. Examples of inference in astronomy range from "fitting a straight line to some points" to "attempts at simultaneously constraining all of cosmology and galaxy evolution at once." The former has two parameters (m and b), while the latter may have hundreds of free parameters. Commensurately, the frameworks for carrying out inference vary in scale and complexity by several orders of magnitude.

This is not a textbook on inference, and several other textbooks cover the in-depth process of fitting complex and highly-dimensional problems. However, one cannot progress far into astronomy on the basis of direct measurement alone. While taking the sum of some pixels doesn't require the definition of an explicit model, even the

---

[1] and hope...

process of determining a redshift (velocity) from a spectrum will necessitate the adoption of a model to fit, e.g., the position of an emission or absorption line. During that process, assumptions are made (for example, the functional form chosen to approximate the shape of the emission or absorption line).[2] Additionally, the inference task is responsible not just for picking an answer out, but in estimating the uncertainty on that answer. Uncertainties in the results of an inference depend not only on the uncertainties in the data (which may, in turn, also have to be estimated via some model) but also on the degree to which the model chosen for fitting is actually an accurate representation of the underlying data.[3]

That sounds like a lot! And it is. But we will focus in this chapter on exploring a series of relatively straightforward inference problems, in order to practice using the set of packages and tools now ubiquitous (or growing in popularity) in the field of astronomy.

13.2 Fitting a Line to Data

Let's begin by looking at fitting a line to data. This is not as trivial as it sounds, as we will see, but can (crudely) be done very simply.

For this example, we will be examining the so-called $M - \sigma$ relation. This empirical relation has emerged from dynamical measurements of the supermassive black hole masses in a set of nearby galaxies. It turns out that there is a strong correlation between the mass of a galaxy's supermassive black hole and the measured velocity dispersion (non circular motions) measured for those galaxies.

I have provided a file, m-sigma.txt, which denotes a set of galaxies with these measurements taken from Ferrarese & Merritt (2000), if you would like to follow along.

```
import numpy as np
import pandas as pd

df = pd.read_csv('m-sigma.txt',delim_whitespace=True)
df.head()
```

	Galaxy	M_bh	M_bh_err_up	M_bh_err_down	sigma	sigma_err
0	MW	0.0295	0.0035	0.0035	100	20
1	I1459	4.6000	2.8000	2.8000	312	41
2	N221	0.0390	0.0090	0.0090	76	10
3	N3115	9.2000	3.0000	3.0000	278	36
4	N3379	1.3500	0.7300	0.7300	201	26

---

[2] It is critical to understand the ways our model assumptions affect the answers we get out, particularly when characterizing uncertainty.

[3] It is perhaps worth noting that even the summing of pixels in an image is not model-free, the model just happens to be more implicit. For example, the counts in a pixel are a function of the detector properties, which are not exactly known—we assume (or indeed, attempt to measure) these properties, and implicitly adopt those choices when making measurements on the data.

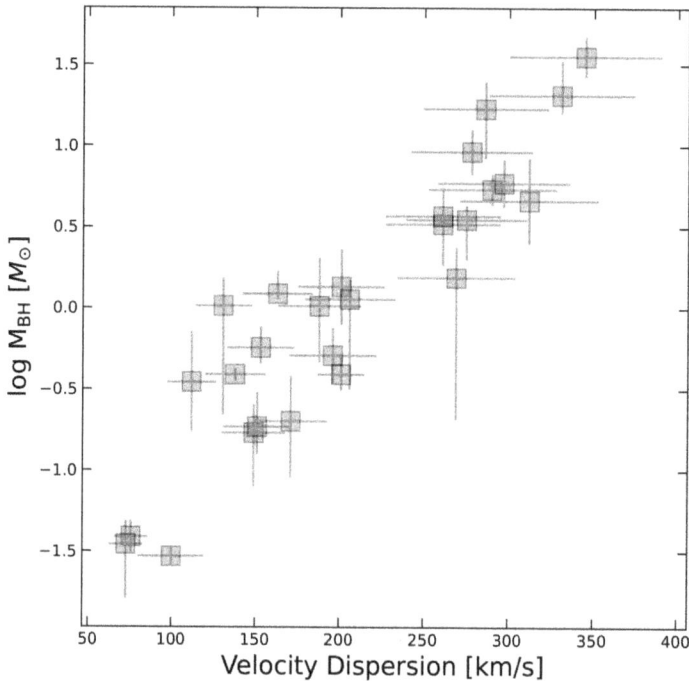

Figure 13.1. M-σ relation as published in Ferrarese & Merritt (2000).

Our first step when looking at any data should always be to plot it, to see what we are working with (Figure 13.1).

```
import matplotlib.pyplot as plt

log_BH = np.log10(df.M_bh)
BH_err = 0.434*np.array([df.M_bh_err_up,df.M_bh_err_down]) / np.
    array(df.M_bh)[np.newaxis,:]


fig, ax = plt.subplots(figsize=(7,7))
ax.errorbar(df.sigma,log_BH,yerr=BH_err,xerr=df.
    sigma_err,fmt='None',color='gray')
ax.plot(df.sigma,log_BH,'s',color='C0',ms=12,alpha=0.5,mec='k')
ax.set_xlabel('Velocity Dispersion [km/s]',fontsize=16)
ax.tick_params(which='both',top=True,right=True,direction='in',length=6)
ax.set_ylabel(r'log M$_{\rm BH}$ [$M_{\odot}$]',fontsize=16);
```

As we can see, there is a positive, reasonably tightly correlated relationship between velocity dispersion and (log) black hole mass.

To quantify this relationship (or, say, to create a predictive model for black hole mass based on a velocity dispersion), we need to *fit* the data.

As mentioned in the introduction, this necessitates the adoption of a model. In this case, we are going to assume a linear model—one could chose many other models, based on many other motivations.

With a linear model, we look to determine (α, β) such that

$$\log M_{\rm BH}[M_\odot] = \alpha V_\sigma + \beta$$

I will use V_σ to denote velocity dispersion in order to avoid confusion with σ, the uncertainties on the data. Perhaps the most basic method of determining these parameters would be via the Linear Least Squares (LLS) approach. This is a mathematical formalism which asserts that the best fit model is the one which minimizes the squared residuals between the model and the data—that is,

$$\chi^2 = \sum \frac{(d_i - m_i)^2}{\sigma_i^2},$$

where d_i are the data points and m_i are the model predictions for that data point given some (α, β). The weighting by the data uncertainty, σ_i, in the denominator, is optional (it will not be in a moment, when we discuss bayesian sampling).

LLS is not appropriate for all situations—it assumes, for example, that the data are drawn from a population that follows the best-fit relation exactly and are perturbed only by noise that is Gaussian in nature, which is not often the case in astronomy. Another problem you may have already noticed is that if we want weight by the uncertainty on each data point, LLS alone does not have the ability to fit data with uncertainties in two dimensions. Furthermore, because the fit is attempting to minimize the square of the residuals, outliers (data points far from the true relation) are heavily weighted in the fit, which we often want to avoid.

Let's (at least initially) ignore the obvious problems with LLS for this data set and fit a line anyway. For this, I am going to use the simple np.polyfit function:

```
fit = np.polyfit(df.sigma,log_BH,deg=1,w=1/np.max(BH_err,axis=0))
xx = np.linspace(50,400,100)
yy = np.polyval(fit,xx)
```

```
fig, ax = plt.subplots(figsize=(7,7))
ax.errorbar(df.sigma,log_BH,yerr=BH_err,xerr=df.
    sigma_err,fmt='None',color='gray')
ax.plot(df.sigma,log_BH,'s',color='C0',ms=12,alpha=0.5,mec='k')
ax.plot(xx,yy,'k',lw=3)
ax.set_xlabel('Velocity Dispersion [km/s]',fontsize=16)
ax.tick_params(which='both',top=True,right=True,direction='in',length=6)
ax.set_ylabel(r'log M$_{\rm BH}$ [$M_{\odot}$]',fontsize=16);
```

The resulting fit is shown in Figure 13.2. Let's look at what actually happened here. The np.polyfit used linear least squares to determine the best (α, β) (these were stored in the variable fit). Notice that the weight parameter, w, cannot handle

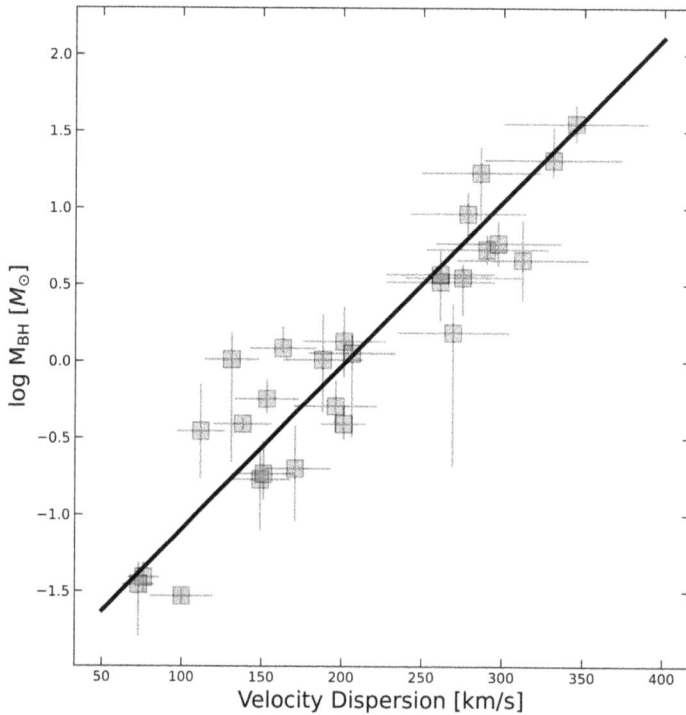

Figure 13.2. Linear Least Squares fit to the relation, in which only the uncertainties on log mass were treated in the fit.

uncertainties that are different in the positive and negative direction, so I had to make a choice about how to insert them (I chose to keep the larger error of the two, which is not a particularly well-motivated choice).

We can look at our fit:

```
fit
```

array([0.01067229, -2.16101042])

We can see that the measured intercept is −2.16, and the slope is 0.01.

Are we done? We know that various questionable assumptions and choices went into our fit. I mentioned above that errors were only included in one direction. We can illustrate the effect of this to ourselves by re-fitting for $V_\sigma(M_{BH})$, i.e., the inverse fit (Figure 13.3).

```
fit2 = np.polyfit(log_BH,df.sigma,deg=1,w=1/df.sigma_err)
in_masses = np.linspace(-1.5,2,100)
out_velocities = np.polyval(fit2,in_masses)
```

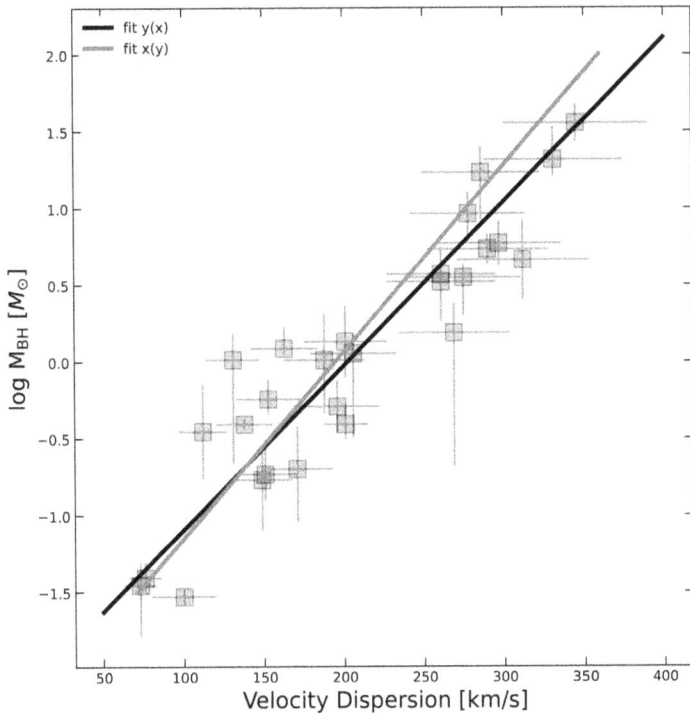

Figure 13.3. Same as Figure 13.2, but with an added trendline determined by fitting velocity dispersion as a function of mass (weighted by the velocity, rather than mass, uncertainties).

```
fig, ax = plt.subplots(figsize=(7,7))
ax.errorbar(df.sigma,log_BH,yerr=BH_err,xerr=df.
   sigma_err,fmt='None',color='gray')
ax.plot(df.sigma,log_BH,'s',color='C0',ms=12,alpha=0.5,mec='k')
ax.plot(xx,yy,'k',lw=3,label='fit y(x)')
ax.plot(out_velocities,in_masses,'r',lw=3,label='fit x(y)')
ax.set_xlabel('Velocity Dispersion [km/s]',fontsize=16)
ax.tick_params(which='both',top=True,right=True,direction='in',length=6)
ax.legend()
ax.set_ylabel(r'log M$_{\rm BH}$ [$M_{\odot}$]',fontsize=16);
```

As we can see, fitting the dispersion as a function of black hole mass, weighting by the uncertainties in velocity dispersion (and ignoring uncertainties in $M_{\rm BH}$) produced a different fit!

There are many ways to deal with this problem, including higher-complexity fits that include uncertainties in both x and y, though these are beyond the scope of this text (see, e.g., Hogg et al. (2010)). One different, but conceptually simpler way we can estimate our uncertainty in (α, β) is via *bootstrap resampling* or *perturbative re-fitting*.

With bootstrap resampling, we re-fit the data many times after creating subsamples of randomly drawn data sets (with replacement) from our actual data.

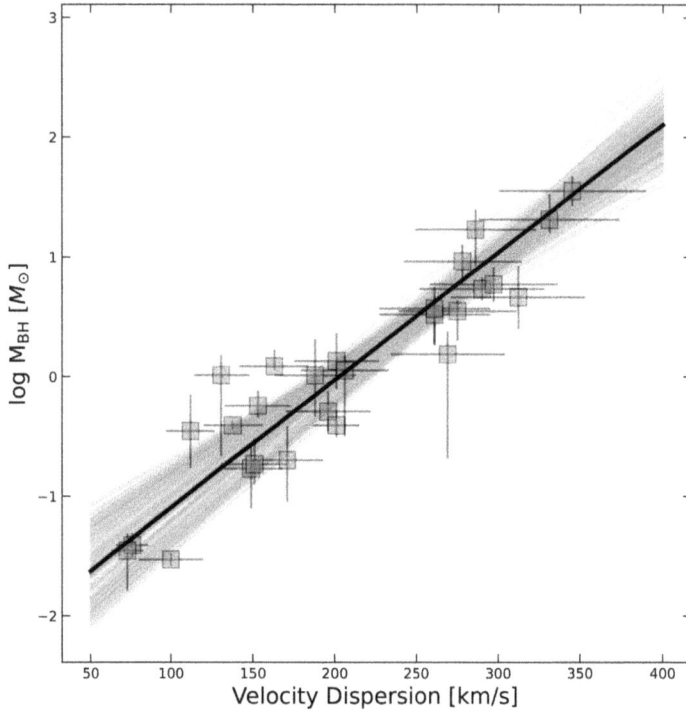

Figure 13.4. Fits to the relation carried out via bootstrap resampling, in which the data set is resampled with replacement, leading to scenarios where sometimes a datum is not included, and sometimes it is included more than once in the same fit.

This is not (strictly) an inference technique, but it is very useful, so let's try it out here. For now, we will go back to our initial (implicit) assumption that uncertainty in the x direction is negligible (Figure 13.4).

```python
def bootstrap_resample(x:np.array,y:np.array,yerr:np.array,n:int=500):
    data_ind = np.arange(len(x))
    fits = []
    for i in range(n):
        ind = np.random.choice(data_ind,size=len(x),replace=True)
        resampled_x = x[ind]
        resampled_y = y[ind]
        resampled_yerr = yerr[ind]
        fit = np.polyfit(resampled_x,resampled_y,
                        deg=1,
                        w=1/resampled_yerr)
        fits.append(fit)
    fits = np.array(fits)

    return fits
```

```
fits = bootstrap_resample(np.array(df.sigma),
        log_BH,
        np.max(BH_err,axis=0))
```

```
fig, ax = plt.subplots(figsize=(7,7))
for i in fits:
    ax.plot(xx,np.polyval(i,xx),color='gray',alpha=0.1)
ax.errorbar(df.sigma,
            log_BH,
            yerr=BH_err,
            xerr=df.sigma_err,
            fmt='None',
            color='k',
            alpha=0.5)
ax.plot(df.sigma,log_BH,'s',color='C0',ms=12,alpha=0.5,mec='k')
ax.plot(xx,yy,'k',lw=3,label='fit y(x)')
ax.set_xlabel('Velocity Dispersion [km/s]',fontsize=16)
ax.tick_params(which='both',top=True,right=True,direction='in',length=6)
ax.set_ylabel(r'log M$_{\rm BH}$ [$M_{\odot}$]',fontsize=16);
```

Meanwhile, we can also perform perturbative re-fitting. In this technique, we don't remove or double weight individual points in our sample, but rather perturb the data points repeatedly by an amount commensurate with each point's uncertainty in *both* the x and y direction. We will *still* make the assumption that our uncertainties are Gaussian, but we will assume that each Gaussian is centered at the center of the uncertainty region for each measured point.

13.3 χ^2 Fitting

When our models grow more complicated, we have to take things into our own hands. If we have some set of data, and a model with some parameters, and we wish to find the set of parameters that best fits the data, we can (brute-force) perform a fit. If the model is one-dimensional and can be encapsulated in a Python function, we can use the `scipy.optimize.curve_fit()` routine.

If it is not (e.g., we need to call an external program to generate our model), then we need to create a loop in which we

1. Select a set of model parameters (α_i, β_j, γ_k,...),
2. Construct a model using those parameters,
3. Compare that model to the data and compute a quantitative measure of how well the two agree,
4. Repeat, for every combination of every model parameter, and
5. Find the set of model parameters in that multidimensional grid which had the closest agreement with the data, and take that to be our best fit.

A standard metric used to carry out step 3 above is the χ^2 (chi-squared) goodness of fit metric. It should look familiar:

$$\chi^2_{m,d} = \sum \frac{(m_i - d_i)^2}{\sigma_i^2}$$

Here, our measure is the squared residuals between each model point and corresponding data point squared (normalized by each data point's uncertainty), all summed up. As a note, this differs slightly from something called χ^2_ν, which is the chi-squared per degree of freedom. This is also called the reduced chi-squared, and is helpful as the unreduced chi-squared will have a different value depending on how many data points are being fit (due to the sum). To compute the reduced version, we will simply divide by $(N - 1)$, where N is the number of points being fit. In this reduced sense, a value of ~1 is considered a good fit.

As a note, you can see that when using χ^2 as our metric, outliers—single points d_i that are very far from our model whose uncertainties are not also large—will be *very heavily weighted*, contributing to high values of χ^2. This is an unavoidable feature of our choice of χ^2; there exist other metrics which do not so heavily weight outliers. In general, we usually attempt to prune outliers before fitting, or create a generative model that assigns a probability of being an outlier on each point (see, e.g., Hogg et al. (2010)).

13.4 Bayesian Inference

Understanding and characterizing uncertainties is one of the major motivators for the use of *Bayesian Inference*. This technique involves, generally, the sampling of a distribution of model parameter values and assessing their likelihoods with respect to the data. In short, instead of simply optimizing to the best-fit (something which in Bayesian parlance would be called the *maximum a posteriori*, or MAP value), we actually estimate the fit quality (encapsulated in the likelihood) in a region *around* the MAP.

In concert with a technique known as *forward modeling*, in which we create "intrinsic" models and then pass them through the transformations needed to make them "look like" the data (e.g., adding noise, etc.), Bayesian inference can produce best-fit results which contain a physically-motivated, realistic estimate of the uncertainties.

Bayesian inference is a topic which can fill multiple textbooks. Here, we will engage in what is more akin to a crash course, learning about the basics of sampling (and why that gets us parameter estimates). I highly recommenddiving deeper into this topic, as it underpins much of modern astronomy.[4]

I want to emphasize that while this topic is extensive, and can be intimidating, the basic execution of performing Bayesian inference on data using a simple, physically motivated model is *not that challenging*, in part in thanks to community-developed tools. By the end of this section, you *will be carrying out Bayesian inference*. From there, it is a matter of adding complexity to your models, fitting more complex data, and learning more advanced and efficient samplers to make both of those tasks tenable.

---

[4] Additionally, I will use the phrasing and terminology currently used in the field, but I'll emphasize these terms are often qualitative and not formally correct in some sense.

13.4.1 Bayes Theorem

Every chapter on Bayesian statistics starts with Bayes theorem, which states that

$$P(A|B) = \frac{P(B|A)P(A)}{P(B)}.$$

This is read in English as "the probability of A given that B is true is equal to the probability of B given A is true, times the probability of A, divided by the probability of B."

In the context of model fitting, we can write it this way:

$$P(m|d) = \frac{P(d|m)P(m)}{P(d)}$$

where d is the *data* and m is our *model*. So, in English, the thing we care about—the probability of our model given some data—is equal to the probability of the data given some model times the probability of the model, all divided by the probability of the data. More specifically, we have one framework of a model, and we want to know the probability of certain *parameters* of that model given the data we have collected. What does that mean?

Each of these terms has a name in the Bayesian formalism.

1. The lefthand side, $P(m|d)$, is called the *posterior* or *posterior distribution*, and it represents the probability distribution function that describes the space of model parameters and how likely those models are, given the data we have.
2. The term $P(d|m)$, the probability of the data given the model, is called the *Likelihood*, and describes our objective function that determines whether our data is a good fit to a particular model. For example, a common likelihood has a Gaussian form, which (we'll see below) translates into the same χ^2 squared-residuals based metric from above. It is *our job* to write down the likelihood function when carrying out Bayesian fits.
3. The term $P(m)$ is the *prior*, or the intrinsic probability distribution of the model parameters. It is also our job to set priors—these encode our prior knowledge of the parameters (e.g., a parameter can't be less than 0, or is Gaussian distributed around some value). If we know nothing about the value of some parameter, we can set an *uninformative* prior, which is simply a flat equal-probability distribution over some large range.
4. Last but not least, we have $P(d)$, the probability distribution of the *data*, which is known as the *evidence*. We will ignore the evidence in this text; it is complex to compute and many sampling algorithms do not even make use of it (nested sampling is one exception). Why can it usually be ignored? We would need the evidence if we wished to produce a likelihood that is correct in an absolute sense. But a standard Bayesian framework needs only the *relative* likelihoods, i.e., how much more likely one set of model parameters is than another. We'll see more on this below.

Our goal is to find the most likely set of model parameters given our data. Because the probability distribution of the priors does not change, this is accomplished by maximizing the likelihood, which will be different for every set of model parameters we try.

13.4.2 Why Does This Help with Uncertainties?

Thus far, we have introduced a mathematical way of describing a distribution. When we introduced the topic, I stated that this technique would allow us to get a handle on our parameter uncertainties, we are interested in as scientists. Why is that the case?

Above, when we discussed basic χ^2 fitting, we were inherently limited by our resolution in the parameters, i.e., by the grid. Our grid was uniformly spaced (or at least somehow parametrized) meaning that we do not capture any regions in which the gradient in likelihood for a parameter is steeper than our grid size. In higher dimensions, the number of grid points needed to accurately sample the posterior grows exponentially, and doing so well becomes impossible. This is a reflection of the fact that χ^2 is fundamentally a *parameter search* tool, not a sampling tool.

In a Bayesian framework, we dynamically sample the posterior (more specifically, we sample the likelihood times the prior) in a way that much better traces the regions of high probability. By sampling these regions well, we can compute integral statistics such as quantiles (e.g., 16-84 percentile). These integrals are what sampling is actually estimating, and in the limit of enough independent samples, these integrals are correct, leading to accurate estimates of parameter uncertainties. We also gain the benefits of *marginalization* — that is, the ability to know the posterior distributions for parameters of interest, having accounted for the distributions of our nuisance parameters.

So what is sampling, how does sampling solve integrals, and which integrals are we interested in when using sampling for inference? These are the questions that underpin the use of Markov Chain Monte Carlo (MCMC) methods in astronomy. A canonincal recent overview of this topic is presented in Hogg & Foreman-Mackey (2018).

13.4.3 Estimating Integrals

Let's take the following integral.

$$I = \int_a^b f(x)dx$$

If I asked you to integrate $f(x)$, what would you do? Well, it depends on what $f(x)$ is, right? If

$$f(x) = x^2,$$

you would simply tell me that

$$I = \int_a^b f(x)dx = \int_a^b x^2 dx = \frac{b^3}{3} - \frac{a^3}{3}$$

Now let's imagine that $f(x)$ is *ugly*—so ugly that we have no idea how to integrate it. What then?

Well, usually the answer would be either "Wolfram Alpha" for an attempt at symbolic integration, or, more generally, "Numerical Integration". Numerical integration says, "I can estimate the area under this complex curve by chunking it into finite rectangles/trapazoids/etc. and then calculate a sum". You've probably heard of (or used some of these methods): Midpoint rule, Trapezoidal rule, Simpsons rule, Gaussian quadrature... (many of these are implemented in the `scipy.integrate` module).

When you're dealing with a (relatively) well behaved function in one dimension, those methods are often the way to go (and the first thing we jump to in our code). But what happens if our problem is not one dimensional? What if, for example, f is a function of three spatial quantities and three additional parameters,

$$f(\theta) = f(x, y, z, a, b, c)$$

We now have θ as a vector of six parameters, meaning our integral looks more like

$$I = \int \int \int \int \int \int f(x, y, z, a, b, c)dx\,dy\,dz\,da\,db\,dc$$

We can now ask ourselves,

Can our above numerical integration schemes handle this?

Each scheme above has an associated error, which comes from how the scheme is integrated. From Calculus, you probably remember that the trapezoid rule usually produces smaller errors than the midpoint rule, as it better approximates the curve being traced. We can actually write down how the error of each of these scales. I'll use the Trapezoid rule here.

$$\epsilon \propto \frac{1}{N^{2/d}}$$

where N is the number of sample points (i.e., how fine our grid where we evaluate our trapezoid) and d is the number of dimensions being integrated over. This is a big problem. The error in our numerical solution to the integral scales to a power of the dimensions being integrated over, which requires us to have intractably large values of N to get accurate results. This is often referred to as "the curse of dimensionality".

So how do we get around this?

What if instead of trying to "grid up" this multidimensional space and evaluate our function at each location, I simply "threw a dart" at a random location and evaluated it there? It turns out, you can show that the error in such a sampling method has an error of

$$\epsilon \propto \frac{1}{N^{1/2}}$$

Amazingly, this does not have any dependence on dimensionality! So doing a little math with the trapezoid-rule error above, we can see that for problems with dimensionality greater than $\sim d = 4$ (for this rule, and closer to $d = 6 - 8$ for, e.g., Simpson's rule), the error properties of an MCMC algorithm win out, and make the integration tractable.

13.4.4 Sampling

Let's back up for a moment to the 1D case of $f(x)$ to aid in our visualization. If I draw some arbitrary function $f(x)$ across my plot, I can evaluate the integral (area) by any of the tools above. I could also sample, which in the absolute first order case means choosing random values from $U(a, b)$ (uniformly drawn) values over the bounds of the integrand (i.e., in the 1D case here, values of x between a and b), and then evaluate $f(x)$ at those values. This is, quite literally, throwing darts to pick values (and where the method gets the Monte Carlo part of it's name). Imagine I have my function $f(x)$ that looks like Figure 13.5.

My sampling, as I described it above, corresponds to something like the points in Figure 13.6, where the four points x_i are presumed to have been drawn from some random uniform distribution. (so more likely, they will not be in ascending order of x).

To estimate the area under the curve, I create a rectangle for each sample $f(x_i)$ with a width of $(b - a)$ and a height of $f(x_i)$. For example, $f(x_1)$ in (Figure 13.6) would be estimated by the rectangle in Figure 13.7, while $f(x_3)$ would be estimated by the rectangle in Figure 13.8. We can see that sometimes, we will overestimate the integral, and sometimes under estimate it. However, I will claim here that the *expectation value* (i.e., the *average*) of all of these rectangles represents an accurate estimate of the integral of the function $f(x)$. In short, I'm asserting for the moment that the expectation value by the normal integral, i.e.,

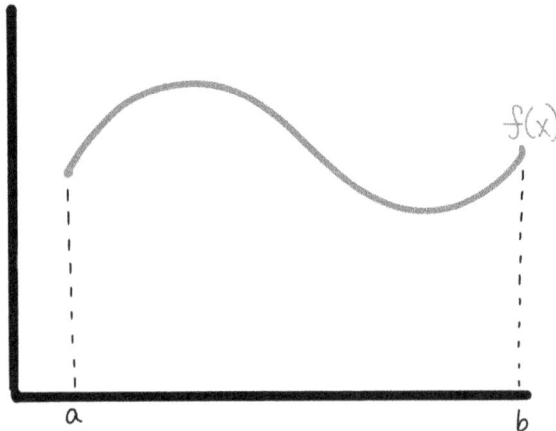

Figure 13.5. A function $f(x)$.

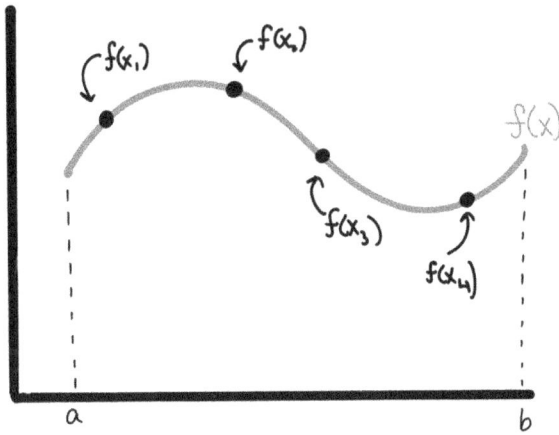

Figure 13.6. A function $f(x)$, sampled at four points.

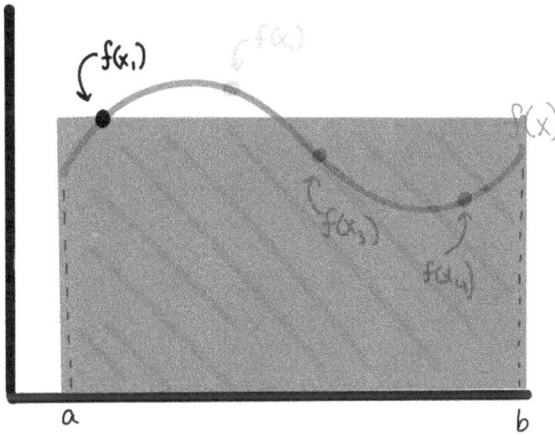

Figure 13.7. Area assuming a rectangle with height $f(x_1)$.

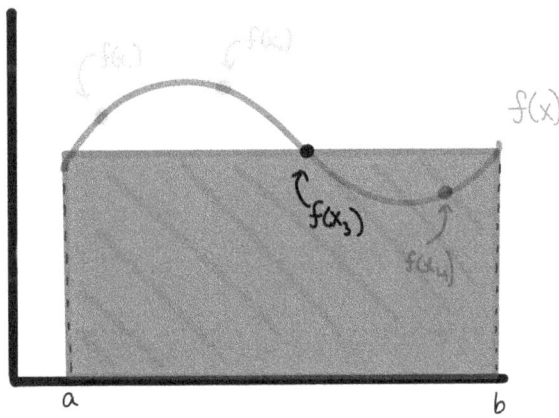

Figure 13.8. Area assuming a rectangle with height $f(x_3)$.

$$\int f(x)p(x)dx.$$

where $p(x)$ is the probability density, is going to be approximated by

$$E_{p(\theta)}[f(\theta)] = \int f(\theta)p(\theta)d\theta \approx \frac{1}{K}\sum_{k=1}^{K}\frac{f(\theta_k)}{p(\theta_k)}$$

Let's explain why. In the case of a uniform distribution, we know that our $p(\theta_k)$ is given by, simply

$$p(\theta_k) = \frac{1}{b-a}$$

That is, a uniform over some bounds is normalized at the level $1/(b-a)$ such that the area of the distribution is properly normalized to 1.

Recall I computed my rectangle areas as the width $(b-a)$ times the height of the function at different sample locations. I'm thus approximating my integral as

$$I \approx \frac{1}{K}\sum_{k=1}^{K}f(x_k)(b-a)$$

Notice though, that $(b-a)$ is just $1/p(x)$ as we've defined it.

Now that we have seen (at least, in crash-course style!) how we can estimate an integral by sampling, we can turn our attention back to the integral we want to estimate — the posterior distribution of a set of parameters which define our model (which we are comparing to data). Our integrand will now look like that of the likelihood times the prior (and for many fitting scenarios, we will ignore the normalizing term of the evidence).

The suite of techniques we will use to sample that posterior is known as Monte Carlo techniques — so before we dive in, let's use them to estimate some more simple integrals.

13.4.5 Simple Monte Carlo

In the simplest case (Monte Carlo) we simply draw random (uniform) values of θ and compute the expectation value using the sum. We then use that expectation value, and the bounds of our integral, to solve for the area.

For example, let's take

$$f(x) = x^2$$

and I want to integrate from 1 to 2,

$$I = \int_{1}^{2} x^2 dx$$

Obviously we know the answer to this is $8/3 - 1/3 = 7/3$. Let's solve it using Monte Carlo:

```
import numpy as np
from scipy import stats
import matplotlib.pyplot as plt

unif = stats.uniform(1,1)  #this creates a uniform over the range [1,2]

def f(x):
    return x**2

sample_sum = 0
N = 1000
for i in range(N):
    s = unif.rvs()
    call = f(s)
    sample_sum += call

sample_sum /= N
print("integral as estimated from 1000 Samples: {}".format(sample_sum))
```

```
integral as estimated from 1000 Samples: 2.3503764887062295
```

We know that the true value is 2.33 repeating; here we can see that with 1000 samples, we estimate the integral to be 2.35. As a note, if you haven't seen it used before, I set up an object called `unif` that is an instance of the `stats.uniform` class, which takes arguments `loc` and `scale` and "sets up" a distribution from `loc` to `loc +scale`. I can then draw random values from that distribution using the `unif.rvs()` method. There are lots of other useful methods too, like `.pdf(x)` to evaluate the PDF, etc. But I could've also simply used `np.random.uniform` to draw values.

We can also try with a much (somewhat absurdly) higher N:

```
N = 100000
for i in range(N):
    s = unif.rvs()
    call = f(s)
    sample_sum += call

sample_sum /= N
print("integral as estimated from 100,000 Samples: {}".
    format(sample_sum))
```

```
integral as estimated from 100,000 Samples: 2.33538901
76720573
```

We can see that in this case we're very close to the true value.

I mentioned above that the error in our estimate of the integral in this Monte Carlo scheme scaled as $\propto N^{-1/2}$. We can write this more formally as

$$\epsilon = kN^{-1/2}$$

where k is a constant that captures the normalization of our scaling relation. Our goal at this point is to bring k down as much as possible, so that our scaling with error has a lower normalization. (In the parlance of the expectation value above, we want to reduce the *variance* in the estimates of I for any given sampling run.)

13.4.6 Importance Sampling

Imagine you have a distribution that looks something like a Gaussian, defined at some range, like below (Figure 13.9):

```python
def f2(x):
    out = 3 * np.exp(-(x-5.)**2/(2*0.5**2))
    return out

xx = np.linspace(0,10,1000)
y = f2(xx)
fig, ax = plt.subplots(figsize=(10,5))
ax.plot(xx,y);
```

I could sample this function using a $\sim U(0, 10)$. But many of my samples would be "wasted" because they would be sampling regions (e.g., between 0 and 3, or 8 and 10) where the value of $f(x)$ is very small, and thus the contribution to the integral is negligible. What if I had a way to throw darts that were more likely to land near 5, where I want to be well-sampled, and not as much near 10?

In order to improve my k value, and assign some *importance* to some values of θ (in this case x) to sample over others, I need a new probability distribution to sample from that isn't just the uniform. Thinking about this for a moment, it would seem

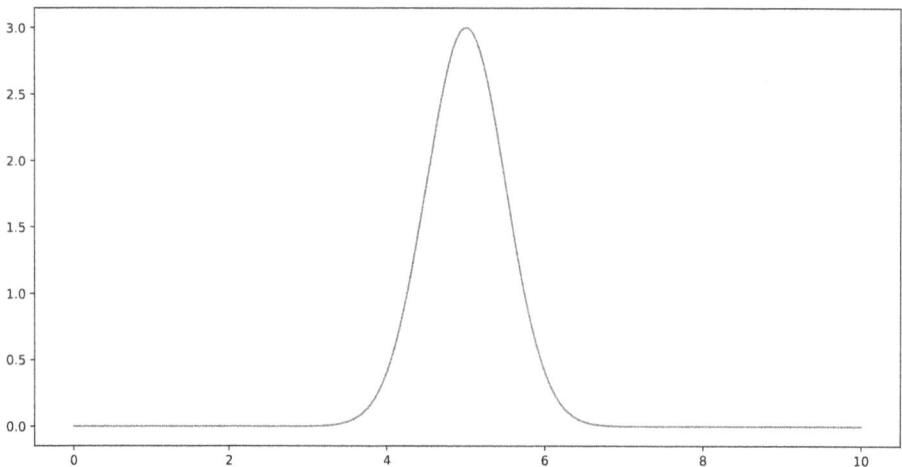

Figure 13.9. Rather than sampling with a uniform PDF over the range from 1 to 10, we can sample from a different distribution—in this case, a Gaussian, in order to sample more often in the region where our actual function $f(x)$ has non-zero weight.

like the obvious choice is in fact, $f(x)$ itself, (or rather, $f(x)$ normalized such that it is a probability density function).

This would naturally capture what I want to do: where $f(x)$ is larger, the PDF will be larger, and the chance of drawing values there will be larger than elsewhere, where $f(x)$ is smaller. In this case, instead of a PDF that is just $1/(b - a)$, we will plug a real PDF into our sampling expression:

$$\int g(\theta)p(\theta)d\theta \approx \frac{1}{K}\sum_{k=1}^{K} \frac{g(\theta_k)}{p(\theta_k)}$$

Let's try setting up a problem using a Gaussian like above, and sample from a PDF that is the Gaussian itself. I'll set up my "arbitrary" function to return something that is Gaussian shaped, but arbitrarily normalized. I then set my "PDF" distribution to be a true, normalized normal distribution at the same (μ, σ) (if we don't know these values, we can approximate them). I repeat the exercise from before, normalizing each evaluation of my function by an evaluation of the proposal PDF at the same value

```
# Note; f(x) is not normalized, it's just something with a gaussian
  →form,
# as I've multiplied by a constant
def f2(x):
    return 3 * np.exp(-(x-5.)**2/(2*0.5**2))

gauss = stats.norm(5,0.5) #this is my new p(theta)
N=100000
area = []
for i in range(N):
    val = gauss.rvs()
    call = f2(val) / gauss.pdf(val)
    area.append(call)

norm_area = np.sum(area) / N

print('Calculated Area: {}'.format(norm_area))
```

Calculated Area: 3.759942411946498

We know analytically that the area we should get is

$$\int_{-\infty}^{\infty} ae^{-(x-b)^2/2c^2}dx = a\sqrt{2\pi c^2}$$

where here, a is 3, b is 5, and c is 0.5. This gives me a computed analytical value of:

```
area_theoretical = np.sqrt(2*np.pi*0.5**2)*3
area_theoretical
```

3.7599424119465006

We can see that once again we've gotten the answer almost exactly right. Note that this didn't only work because both my sampling distribution and PDF were Gaussians with different normalization. Any $f(x)$ that looked roughly like a "bump" could have been estimated this way. I simply chose a Gaussian because we could compare to an analytical solution.

Exercise 13.1: The Value of π.

You might've heard of, or seen an example online, demonstrating that you can estimate the value of π using this method. The idea is to inscribe a circle into a square, and estimate the ratio of their areas by sampling, since these will be

$$R = \frac{\pi a^2}{4a^2} = \frac{\pi}{4}$$

and thus we can get π by calculating $4R$. If we adopt a uniform distribution (good for this problem, we want to sample everywhere in the square with equal probability). For each sample we draw, we just need to know whether that sample fell in the circle inscribed or in the space in between, which we can do by evaluating

$$x^2 + y^2 < a^2$$

for each sample. We then just divide the number of samples that meet this requirement by the total number (as all samples will fall in the square), and multiply by four. Give it a try!

Now rerun your Monte Carlo integrator, but keep track (in arrays) of the N_{circle} (the number in the circle thus far in the loop) and an intermediate calculation of π, and then at the end, make a plot that shows your estimate of pi as a function of `nsteps`. Place an `axhline` at the value of `np.pi`, and from the plot, make a rough estimate of how many steps you would need to hit π (at least to the accuracy that it seems consistent on the plot).

For a bonus problem: try creating an animation similar to the one in the link I showed! Perhaps the easiest way to do this is to use the `matplotlib.widget` library. For this in particular, I recommend the `Slider` module. With it, you can make a plot that allows the user to slide through a number of steps, and see both the square/circle and points, along with the plot we made in the problem above, all updating. (Keep in mind, you'll have to store not only N_{circ} for this to work, but also every x and y value sampled.)

13.4.7 Metropolis MCMC

Now that we understand qualitatively how this process works with some simple 1D integrals, let's go back to thinking about ugly, multidimensional integrals. In the above situation, we were able to set a sampling distribution to be our target distribution because we knew the functional form of $f(x)$ completely. Now, if we knew it was a Gaussian but didn't know (μ, σ) we could have just run an optimizer on $f(x)$ first to find the maximum, and perhaps chosen a reasonably wide spread.

But in more realistic cases, we may know nothing about what $f(x)$ looks like, but can at least *evaluate* $f(\theta)$ for some multidimensional vector θ.

The Metropolis-Hastings algorithm alows you to create a **chain** of evaluations of your function, which don't depend on the initial conditions, but rather only on the evaluation immediately before. This biased "walker" is programmed to move loosely towards areas of higher probability, but occasionally will also move towards lower probability. This walker moves in "steps" that are usually a small sphere around its current location in parameter space. This allows us to very efficiently sample from the high-probability (or in terms of the integral, most important regions) *even if we don't have intimate knowledge of what that region looks like*. Our only requirement at this point is that we can evaluate our function at some position θ, and that our function as usual is positively defined over the bounds. As a note, we are also not sampling in absolute terms; rather, we are at any step comparing two points and calculating the ratio in posterior probability between $f(\theta_1)$ and $f(\theta_2)$.

Exercise 13.2: Write a Sampler.

Write your own sampler which implements the Metropolis algorithm. Here is an outline of how such a sampler is built.

- First, pick an initial value of θ and evaluate it as above. Add this to a stored "chain" of values
- Next, pick a θ' from the *proposal PDF*, a PDF distribution centered on θ (more on this below)
- pick a number r from a $Unif(0, 1)$
- if $f(\theta')/f(\theta) > r$, then move to that position and add it to the chain
- otherwise, the next position in the chain is set to be the current position (and it is added to the chain again).

What do we mean by a proposal PDF? Our walker needs to know how to choose a step to take. The easiest, and most statistically simple, method for doing this is a Gaussian (multivariate if θ is multivariate) with a mean of $\mu = \theta$ and some spread σ that is chosen for each given problem by the amount of parameter space being covered and how sharply $f(\theta)$ varies. The width of σ in each parameter axis is a tunable parameter that can affect how efficient a sampler is.

13.4.8 Emcee: The Workhorse Astronomy Sampler

If you completed Exercise 13.2, you could use your sampler in order to carry out an inference problem. But in all likelihood,[5] it is time to graduate to the emcee package.

There are numerous ways to improve the sampling algorithm presented above to create a more stable, more efficient sampler. Many such improvements have been made in the code emcee (Foreman-Mackey et al. 2013), which uses MCMC to

---

[5] Pun intended.

perform parameter estimation. It is by no means the only sampler out there; in particular, there are samplers which are tuned to handle large numbers of parameters, or to take advantage of differentiable methods in order to carry out really efficient samplings (e.g., Hamiltonian MCMC). Other samplers (e.g., nested samplers) are designed to better catch multimodalities in parameter space (consider: how likely is the MH-MCMC sampler to traverse *down* a peak in likelihood and across a large, low probability trough, to discover a second peak in likelihood in a different region of parameter space?)[6]

However, many introductory problems in astronomy only require a "standard" MCMC, and emcee has become the workhorse sampler for this purpose. It is so popular not only because it is a solid implementation of a sampler, but because it is extremely user friendly. We are thus going to learn about it here, and apply it to some data.

For example: let's begin by returning to our $M - \sigma$ data set — this is a classic example for where emcee can be useful; we would love to have some motivated knowledge of the *distributions* of slopes and intercepts which characterize our data well, in light of the present uncertainties. It is possible, in this framework, to sample in a way which takes into account both the uncertainties in black hole mass and those in velocity dispersion. But for now, let's begin assuming no uncertainty in x, as for typical LLS-style fits.

13.4.9 Priors

Whenever we perform Bayesian inference, we have to write down probability distributions for our parameters (in this case, m and b). We do not need to impose highly restrictive priors on our parameters, if we do not have strong prior knowledge.

For this example, we already have starting estimates on m and b — namely, the results of running he weighted linear least squares. We could (reasonably) decide to choose Gaussian priors centered on these values. However, I'm actually going to select flat priors, allowing uniform probability across the prior range. If we leave this unbounded, it is known as an "improper" prior (as it cannot be normalized), whereas if we chose lower and upper bounds, it is a proper prior. A uniform prior is very easy to implement in the emcee framework, because we simply need to check if a prior is within the bounds. The emcee package asks us to provide a function which, given some vector θ, can return the value of the likelihood times the prior. It is helpful to write our prior and likelihood functions separately, however, and combine them after. Here is an example of a prior function:

---

[6] Traditionally this was approached by running the MCMC several times, each time starting off in a very different region of parameter space.

```
def lnprior(theta):
    m,b = theta
    if ((0<m)&(1>m)&(-3<b)&(-2>b)):
        return 0
    else:
        return -np.inf
```

Here, we return 0 any time a θ vector's parameters are within their prior range, and negative infinity if they are outside. I've chosen arbitrary prior ranges here that encompass where we know the best-fit should lie.

13.4.10 Likelihood

We now need to write down our likelihood function and our wrapper which invokes both.

```
def lnlikelihood(theta,x,y,yerr):
    m,b = theta
    model = m * x + b
    lnlike = (y- model)**2 / yerr**2
    lnlike = - np.sum(lnlike)
    return lnlike

def lnprob(theta, x, y, yerr):
    lp = lnprior(theta)
    if not np.isfinite(lp):
        return -np.inf
    return lp + lnlikelihood(theta, x, y, yerr)
```

We can now evaluate the likelihood for any input θ. We first check if the prior is not negative infinity (if it is, we just pass that through). We then construct a model with this particular set of m, b, compute the likelihood (which here is the same as the χ^2 statistic), and return that value plus the prior (for our flat priors this won't do anything because an accepted prior will add 0). Formally, there are factors missing from our likelihood; they are constant offsets that don't matter for this relative calculation.

Armed with this likelihood function, we are ready to use **emcee**!

To get going, we need to initialize our parameters with some starting values, and decide on the proposal distribution.

```
data = (df.sigma,log_BH,np.max(BH_err,axis=0))
nwalkers = 128
niter = 1000
initial = np.array([0.01,-2.1])
ndim = len(initial)
p0 = [np.array(initial) + 1e-7 * np.random.randn(ndim) for i in np.
    arange(nwalkers)]
```

Above, I've set up a tuple that contains our x, y, and y-uncertainty parameters. I create an initialization, *p0*, by taking Gaussian steps around our initial choice. I've also set some hyper parameters that **emcee** needs: how many steps to take, and how many walkers to use. These can be varied by problem; the documentation for **emcee** and experimentation can help you determine how many walkers and steps are needed to reach a converged, stationary distribution. We'll talk more about convergence below. For now, let's sample!

```
import emcee

sampler = emcee.EnsembleSampler(nwalkers, ndim, lnprob, args=data)

print("Running burn-in...")
p0, _, _ = sampler.run_mcmc(p0, 100)
sampler.reset()

print("Running production...")
pos, prob, state = sampler.run_mcmc(p0, niter)
```

```
Running burn-in...
   Running production...
```

We can extract the flattened chain of θ values from the sampler:

```
flatchain = sampler.flatchain
flatchain.shape
```

```
(128000, 2)
```

This number should make sense; we have two parameters, *m* and *b*, and should have 128*1000 samples. Let's visualize the spread in chain values (Figure 13.10):

```
import corner

fig = corner.corner(sampler,labels=['m','b'],quantiles=[.16,.5,0.
   .84],show_titles=True)
```

This so-called "corner plot" is a useful metric when carrying out Bayesian sampling. It creates 1-D histograms of each parameter (I've elected to show the 16th, 50th, and 84th percentiles as dashed vertical lines), as well as showing the covariance between any pair of parameters. We have only two parameters, so we have 1 correlation plot and two 1-D histograms. This result looks "good", insofar as values of highest posterior probability for both *m* and *b* appear well constrained, within their prior (not bumped up against the edge), and appear well-sampled. We can also see from the covariance 2D plot that the values for *m* and *b* are strongly correlated. This makes sense; if we select an intercept that is smaller, we need a larger slope to fit the data at roughly the same level of fit goodness.

There are several other demonstrative plots we should make to assess the quality of our fit. For example, we can directly visualize the chains, in an attempt to ensure they are "stationary" and "converged."

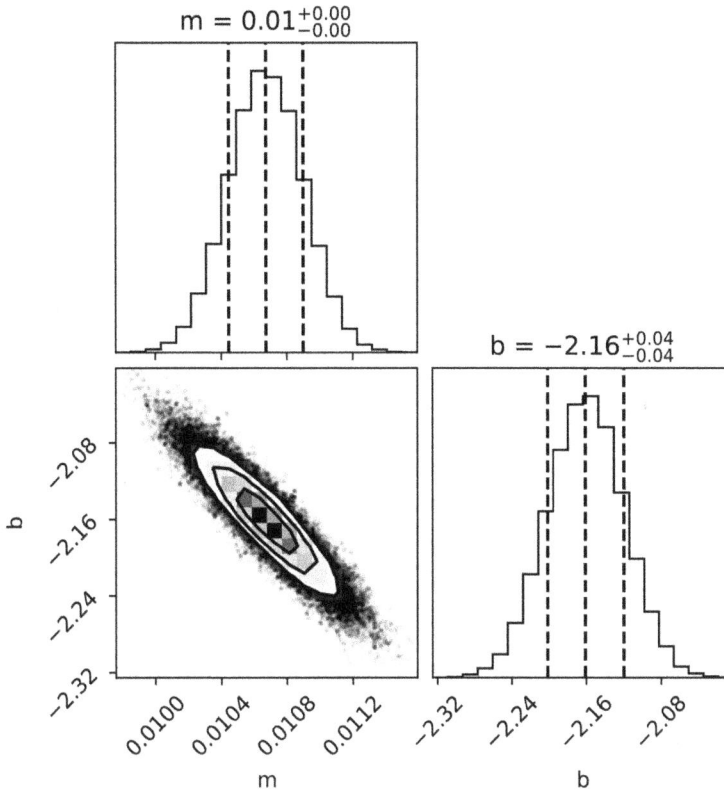

Figure 13.10. A Corner plot showing the marginalized 1D posterior distributions of the slope and intercept as well as the 2D posterior distribution of (m, b), demonstrating the covariance between the two parameters.

```
fig, ax = plt.subplots(2,1,figsize=(20,4))
ax[0].plot(flatchain[:,0::10],lw=0.1)
ax[1].plot(flatchain[:,1::10],lw=0.1)
for i in ax:
    i.set_xlim(0,128000)
ax[0].set_ylabel('m')
ax[1].set_ylabel('b')
ax[1].set_xlabel('Sample Number')
```

In Figure 13.11; I've plotted the chains for m and b, taking every 10th element in the flatchain simply to save on plotting memory. This is an important plot for judging the convergence of our sampling. What does that mean? I mentioned in the introduction to MCMC that it estimates the value of certain integrals in the limit of enough samples. This is where that point is rearing its head. In particular, we need a certain number of *independent* samples in order to say we are reasonably converged (we are never fully converged except in the case of infinite samples).

Recall that each new sample position depends on the position before, so samples of some walker are correlated with each other. But after enough steps, the walker has effectively "forgotten" where it was N steps ago. The number of steps needed before an

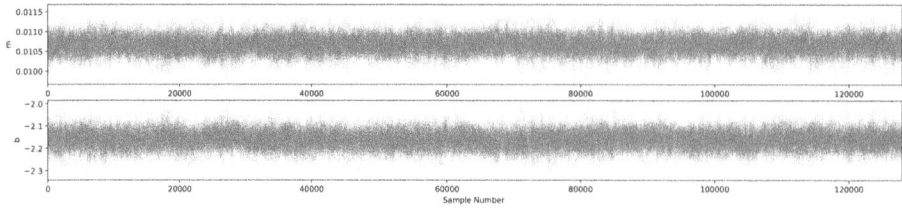

Figure 13.11. Sample chains from the MCMC for the two parameters being fit. This is an example of what we want our chains to look like—stationary (not changing en-masse in value) and not pushing up against one of our prior boundaries.

independent sample is generated in a chain is known as the *autocorrelation time*. This number is often difficult to compute, and will depend on the size of the steps being taken by your sampler. The long and short is there is no analytical, easy way to take a standard MCMC run and definitively state it is converged. But the above plot, which shows us that the chains for *m* and *b* were not changing in width or central location for the length of the chain (i.e., the distribution is stationary), is good heuristic evidence that we have enough samples (and thus enough independent samples) to call our run converged.

Sometimes, when you initialize an MCMC far from the optimal parameter values, the chain distribution takes some time near the beginning of the run to "find" the region of highest probability, where it will then sit and traverse the posterior space many times (ideally). This period when the distribution is not stationary is known colloquially a "burn in," and it is usually recommended to chop these chains off the MCMC before computing summary statistics or parameter uncertainties. As we can see, that didn't happen here (the chain starts off pretty much where it stays the whole time). If it had, we might cut off the first 20,000 samples, for example, and then compute our statistics.

Last but not least, we can of course plot our models directly over the data (Figure 13.12). In the code below, we have chosen a random set of 100 θ values from the flatchain. These should be fairly representative of the distribution of highly probable models, if our sampling has gone well. We plot these directly, and can see visually where most of the models lie as well as their typical spread.

```python
ind = np.random.randint(0,len(flatchain),size=100)
draws = flatchain[ind]
xx = np.linspace(50,400,100)

fig, ax = plt.subplots(figsize=(7,7))
for theta in draws:
    ym = xx* theta[0] + theta[1]
    ax.plot(xx,ym,alpha=0.09,color='k')

ax.errorbar(df.sigma,log_BH,yerr=BH_err,xerr=df.
    sigma_err,fmt='None',color='gray')
ax.plot(df.sigma,log_BH,'s',color='C0',ms=12,alpha=0.5,mec='k')
ax.set_xlabel('Velocity Dispersion [km/s]',fontsize=16)
ax.tick_params(which='both',top=True,right=True,direction='in',length=6)
ax.set_ylabel(r'log M$_{\rm BH}$ [$M_{\odot}$]',fontsize=16);
```

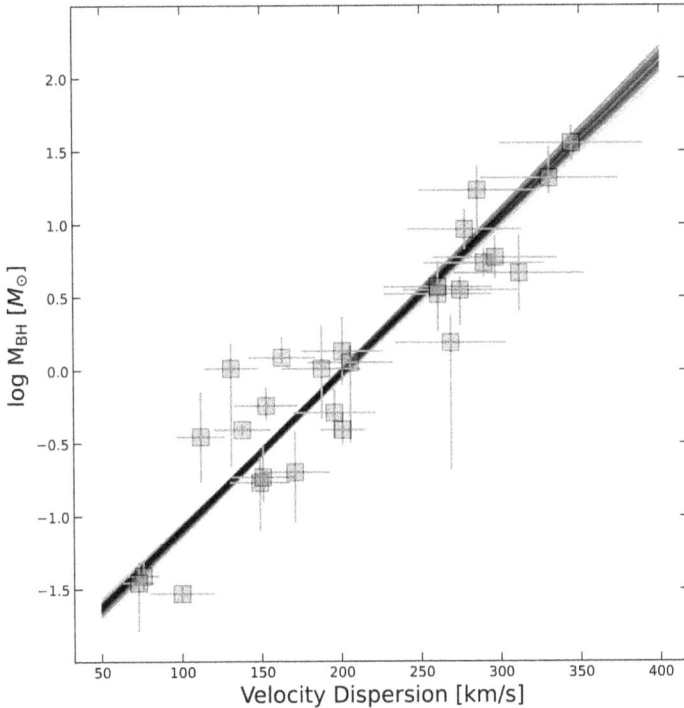

Figure 13.12. Sample models (generated from sampling the posterior chains in slope and intercept of our emcee sampling run), overplotted on the M-σ relation. The stacking of these "draws" naturally shows models of higher and lower likelihood.

In one sense of "being Bayesian," plotting a random subset of posterior draws is the best possible representation of the result of our sampling (e.g., Hogg & Foreman-Mackey 2018). What we should *not* do is ask "What was the model in the chain that had the highest probability? Can we just plot this model?" The answer is no, for two reasons. First, many make the mistake of calling this their MAP (maximum *a posteriori*) value. It is not, because there is no guarantee that the actual MAP (which can be computed directly via optimization) was ever actually sampled in our MCMC. The actual "most likely" model in our MCMC is some random model which will likely be close to the MAP, but otherwise holds no special properties whatsoever. The second reason is that reducing our MCMC to the MAP value, even if we had it, would be sacrificing the *point* of sampling, which is to understand the range of credible models given the data. If all we wanted was the most likely model, we would optimize.

While simply plotting models drawn from the posterior is the best way to represent a sampling run, sometimes this is intractable (or not aesthetically pleasing). An alternative is to characterize the spread in models by, e.g., creating many instantiations of the model from the posterior as we did above (but perhaps more than 100), and then for each position x_i, quantify the spread in model values. This has the upside of allowing us to use a single shaded region to color the credible regions, but has the downside that the upper and lower bounds of that region will not correspond to actual physical models, which can be confusing for those interpreting your plot.

13.4.11 Quoting Parameters and Uncertainties

Last but not least, what shall we write down after this exercise as m and b? Statistics on the chain are the general approach to this, with a common practice being to compute the quantiles containing the 16th, 50th, and 84th percentiles, as follows:

```
m_quant = np.percentile(flatchain[:,0],[16,50,84])
b_quant = np.percentile(flatchain[:,1],[16,50,84])
```

```
m = m_quant[1]
up = m_quant[1] - m_quant[0]
down = m_quant[2]-m_quant[1]
print(f'Best fit m = {m:.4f} (+{up:.6f}) (-{down:.6f})')
```

Best fit m = 0.0107 (+0.000218) (-0.000223)

```
b = b_quant[1]
upb = b_quant[1] - b_quant[0]
downb = b_quant[2]-b_quant[1]
print(f'Best fit b = {b:.4f} (+{upb:.6f}) (-{downb:.6f})')
```

Best fit b = -2.1601 (+0.040704) (-0.039081)

This distribution reflects something akin to 1σ uncertainties. We could repeat the exercise using, e.g., the 5th and 95th percentile as an estimate of the 2σ uncertainties, etc.

13.5 Summary

This completes our introduction to sampling and emcee. Remember that many other sampling tools exist, and they often use different techniques with various improvements and speed and efficiency (e.g., No U-Turn sampling), or have the ability to compute the evidence and assess stopping criteria, while better capturing multi-modal posterior spaces (e.g., nested sampling). There is much more to investigate on the subject, if you are interested!

References

Ferrarese, L., & Merritt, D. 2000, ApJ, 539, L9

Foreman-Mackey, D., Conley, A., Meierjurgen Farr, W., et al. 2013, emcee: The MCMC Hammer, Astrophysics Source Code Library, record ascl:1303.002

Hogg, D. W., Bovy, J., & Lang, D. 2010, Data analysis recipes: Fitting a model to data, arXiv:1008.4686

Hogg, D. W., & Foreman-Mackey, D. 2018, ApJS, 236, 11

Astronomical Python
An introduction to modern scientific programming
Imad Pasha

Chapter 14

Software Development

14.1 Introduction

What is software development, and how does it differ from all of the coding techniques we've discussed in this book so far? When we write code, we are (whether implicitly or explicitly) considering who is going to be the ultimate audience or user interacting with that code. Much of the time, that person is us—and when this is the case, we tend to write code that is not particularly robust. Why? It's not just about laziness; when we are the sole user of a piece of code, and are using it alongside its development, we know the function, and methodology, of each component of that code relatively well. It can thus actually be a waste of time (or at least an inefficient use of time) to write "good" code, when the goal is not the code itself, but the calculation or data that comes out of it.

Of course, that type of code development (a sort of "move fast, break things" approach) breaks down quickly in the real world when either

- Someone else comes along and tries to understand or implement your code,
- Or when *you* come along and try to understand or implement your code six months or six years later,
- Or when the formatting or nature of the data being used in the code changes, but the code is too inflexible to handle those changes (thinking back to "hard-coding").

As soon as we write code with the explicit intent for someone else to use it, we instinctively write better code. We name our variables more descriptively, add little comments throughout to explain what a function does—we know, ultimately, that someone attempting to read or parse our code will benefit from these additions, and make it more likely for them to use the code *correctly*, and thus obtain accurate results within the scope of the code's function.

It does not take long examining this idea to determine that, generally, similar steps would make *all* the code we write better. Any seasoned programmer will attest to the fact that code written as little as a few weeks ago more or less approximates never-before-seen code when you return to it. In that case, every small step we can take to make our code more readable, reliable, and usable (properly) is a step toward helping our future selves.

Software Development is the process of taking this concept to its logical conclusion. Any piece of code—say, a script that reads some data from a file, computes a number, and returns it—constitutes software in an abstract sense. But when a programmer calls something "Software" with a capital "S," they are generally referring to a few benchmarks of code robustness, reliability, and architecture that indicate code is at least able to be easily leveraged by other users than the original writer. These benchmarks include things like

- Code that has its functions and classes well documented in a recognizable style (e.g., Numpy or Google style),
- Code that separates its modular components into separate files and folders in such a way that you can import functions and classes easily within other codes,
- Going one step further, code that conforms to the conditions needed to make it *install-able* (e.g., pip install my_code),
- Code that has unit-tests that check individual functions and classes behave as expected, which is run when changes are made to the code,
- Code that is version-controlled via something like git,
- Going one step further, code that is available on a version-controlled platform such as Github, with ***continuous integration*** (that is, every time changes are made, unit tests are run *automatically*), and, finally,
- code that has a well thought out Application Programming Interface (API), which clearly describes the ways in which a user actually uses a code.

In this chapter, we're going to take a look at some of these benchmarks, and talk about how to implement them in our own code. Not all of those things listed are needed to make something that constitutes software—they simply represent a set of ever-improving steps one can make with their code. Many of these steps are actually easier to implement than they sound, as the open source community has made tools available to automate or partially automate some of these steps as well.

As you approach this chapter, it is important to keep in the back of your mind that not every piece of code you write to carry out your astrophysics research need adhere to these standards—and indeed, much of the code you will write early on wouldn't even benefit that much from them (at the expense of the extra time it would take to implement them). So in this first section, we're going to stop for a moment and ask ourselves: *When should I make my code a Python package?*

14.2 Why (and When) to make a Python Package a Python Package

Throughout this book, we have been making use of Python packages. We have our friendly, ever-present packages like Numpy and Astropy, as well as numerous other smaller packages useful for various parts of the process of data analysis. What makes these packages? In the abstract, they are simply functions someone wrote, which we find useful to use (e.g., `numpy.arange()`). In the technical sense, Numpy is a package because it is install-able. This means we use something like `pip install` or `conda install` to add Numpy to our system, and after that, any code we run or interpreter we open, anywhere on our system, can import Numpy without issue.

This is in contrast to the following scenario: I have two files in a directory, `foo.py` and `bar.py`. The `foo.py` file contains a defined function `spam()`. In this case, I can use code from foo import spam in that `bar.py` code in order to access it. This looks like a package import, and the fundamentals are the same (to Python, *all* .py files are modules), but I couldn't run that same import in any other directory where `foo.py` didn't live—this is the point of "installing" it.

The other things that make Numpy "Software" include qualities on the list above: the fact that Numpy is organized (for example, all the functions for dealing with random numbers are stored in the submodule `np.random`, e.g., `np.random.normal()`, or `np.random.randint()`), the fact that its components are vetted and tested constantly, etc.

So, when should we take our code, and make it behave like Numpy in that way?

To approach this, the first question to ask is *how general is this code?* Sometimes you'll write a function that will only *ever* be used for a highly obscure calculation deep in another code. That function is not very high priority to make a package out of. On the other hand, if you find yourself copying and pasting a function you wrote for one code or project into several others, then it is general enough to be a strong candidate for installation. Doing so would save you the trouble of copying and pasting—in every new code you start to write, you could simply add `import my_useful_function` at the top of your code with your other imports, and be off to the races.

This is good for an additional reason. Let's say somewhere down the line you find a bug in your useful function. If you've been copying and pasting it across different codes, you now need to go through and execute the fix everywhere. On the other hand, if you've installed a package with your function, you need only fix the bug once (in the package code) and reinstall it. You can now simply re-run all the other codes that rely on it and they will be updated accordingly.

A second question we can ask ourselves when mulling over whether to turn code into software is *will other people want to use this code?* As mentioned in the Introduction to this chapter, we tend to write better (or at least more descriptive) code when we expect it to be used by others. There *absolutely* exists good, usable code that is shared (either directly with colleagues, or online via a GitHub type service) which is not in the form of a package. But increasingly, packages are the "unit" by which code is distributed. This is in part because writing code as a package forces a certain amount of organization, and forces the developer to at least consider what the API is going to look like (i.e., when a new user installs your package, what are they going to import and how are they going to use it).

Lastly, a question we can ask is *How rock solid do I want the behavior of this code to be?* We entrust our code to carry out calculations we don't want to do by hand.

At higher levels, we might write code that has several thousand lines and can take a full night's worth of data from a telescope and automatically reduce it. We might run a single command, e.g., `kcwi_reduce *.fits`[1] and then let it carry out the many complex calculations and modifications needed to turn all of the raw frames from a night into wavelength calibrated, science-ready spectra. When we do this, we are *not privy to* the details of the actual lines of code being carried out, with the exception (often) of some summary statistics or plots. We're trusting the code to do things right, and handle aberrant cases appropriately.

For a piece of code we are running and interrogating the output of closely, perhaps a package-style development is not needed. But when we are letting a function (or class, or set of functions and classes) perform many operations on "mission-critical" parts of a process, we want to be *as sure as possible* that those classes and functions are behaving as we expect. To that end, composing them into a package, with strong unit testing, is a valuable enterprise.

14.3 Organizing Packages: Modules and Submodules

Python packages are structured by files and directories. Inside a Python package that is responsible for analyzing and visualizing imaging data from a telescope, we might create directories for, e.g., `io/` (data in-out), `reduction/` (for reducing raw data), `analysis/` (for measuring properties of the reduced data), and `plotting/` (for scripts containing plotting functions tuned to our data). We might also have a Python file with a function called `reduce_data()`, whose job is to call functions from the Python files inside all the other folders described. We might place this file outside those folders. All of this must live somewhere, so let's make that a folder, and give it a name: `myreduce/`. To summarize, our directory structure looks something like this:

```
myreduce
├── reducedata.py
├── io/
│   └── io.py
├── reduction/
│   ├── reduce_darks.py
│   ├── reduce_flats.py
│   └── reduce_science.py
├── analysis/
│   ├── measure_fwhm.py
│   └── measure_flux.py
└── plotting/
    └── plotting.py
```

---

[1] From the KCWI data reduction pipeline: https://github.com/Keck-DataReductionPipelines/KCWI_DRP

Even before we do anything else, the fact that we have *organized* our code this way means that from our `reducedata.py` file (where our `reduce_data()` function lives), we could run the following types of imports:

```
from io.io import some_io_func
from reduction.reduce_science import science_reduce
from plotting.plotting import plot_image
```

Notice that via our dot-notation imports, we specify the top level folder first, then the Python file name (without ".py"), and then the specific function or class *within* that file we want to import—I've made some up above that might reasonably exist inside those files.

By doing so, we are taking advantage of Python's *relative import* syntax. But our code is not yet a *package*. To do that, we need to create special files named __init__.py inside anything that will be a module of our package. In practice, that's all the directories, including `myreduce/`. By default, these files can even be empty. Let's examine our directory structure again, this time assuming we've added __init__.py to the right places:

```
| myreduce
└── __init__.py
    ├── reducedata.py
    ├── io/
    │   ├── __init__.py
    │   └── io.py
    ├── reduction/
    │   ├── __init__.py
    │   ├── reduce_darks.py
    │   ├── reduce_flats.py
    │   └── reduce_science.py
    ├── analysis/
    │   ├── __init__.py
    │   ├── measure_fwhm.py
    │   └── measure_flux.py
    └── plotting/
        ├── __init__.py
        └── plotting.py
```

When our __init__.py files are empty, the imports we can do from `reducedata.py` look identical to those above—we have to specify each module/file down the chain to the function in question. But notice for a moment that some of our modules (directories), such as `io/` and `plotting/`, only contain one Python file. It doesn't make much sense to force the import to be `from plotting.plotting import someplot` since the information in those dot notations is somewhat redundant.

This is where the __init__.py comes in. We can add import statements to __init__.py files, and they get propagated when other imports seek out that module. When we first discussed package imports, there was a method of import that would import *all* functions from a package into the global namespace without being "attached" to a library (e.g., from numpy import *). That method made use of the wildcard, and I warned that it was bad practice to use it, since np.func doesn't add much typing, make it clear which library a function is from, and avoids conflicts (such as when two separate packages have functions with the same name). `

In this case, however, we can take advantage of the wildcard to pull all functions from plotting.py into the namespace of the plotting/ module. To do this, we'll add the line from .plotting import * to the __init__.py sitting inside plotting/. Since there's only one file inside plotting, this is relatively safe (there shouldn't be conflicts). Let's do the same in io/, adding from .io import * to the init file of the io/ directory.[2] Once we've done this, nothing changes about the organization of our code, but the imports can instead be written as follows:

```
from io import some_io_func
from reduction.reduce_science import science_reduce
from plotting import plot_image
```

Now, we don't have to write from io.io or from plotting.plotting, because the init files have yanked those functions (in an access sense) into their namespace. Because we didn't add anything to the __init__.py of reduction/, we still need to specify which file inside reduction/ the reduce_science function is in.

At this juncture, we might ask ourselves whether we *should* add similar imports to the __init__.py of reduction/. If we did, imports for functions in those files would move to the reduction namespace, like this:

```
from reduction import reduce_dark_func, reduce_flat_func, science_reduce
```

In this example, I've made up some names for functions that might exist inside each of those Python files sitting in reduction/.

Do we want this? This is a key example of a decision we make as a software developer that determines the API of a program. The above imports seem reasonable to me, for example—if each of those three Python files only contains one or two primary functions, it will be more convenient, and not dangerous, to pull them into the reduction namespace to allow users (and us) to import it without separately calling reduction.reduce_science.science_reduce, reduction.reduce_darks.reduce_dark_func, etc. We now have the flexibility of the easier import, but the actual code is still nicely organized into separate Python files, so when we go to edit our code, we can easily find exactly where the code we need to work on is stored.

---

[2] Note the dot (.) in front of the file names.

On the other hand, if the files inside a module are disparate and full of many functions, or, for example, there is a function inside reduce_darks.py and one in reduce_flats.py that share a name, but have different behavior, then it is best to leave it so that users are forced to import from the full path. That way in our code, we know the difference between calling reduce_darks. combine_images() and reduce_flats. combine_images(), which might use different statistical methods to combine said images (e.g., mean versus median versus biweight location).

What about the outermost __init__.py file, which is sitting in myreduce/? Assuming this is going to be a package named myreduce, we need to import all of our submodules into the namespace of myreduce/. However, this is a case where we *do* want to keep it so that imports of our submodules require actually calling them. This is not a hard rule—much in the same way multiple Python files can be used to organize code that will ultimately share a submodule's namespace, the outer __init__.py can further pull all of those imports into the namespace of myreduce. In that case, the someplot function described inside the plotting.py file would actually be accessible all the way from the outermost import, import myreduce.someplot (or from myreduce import someplot). For some simple packages, that might be desirable. But often, we create modules (folders) because beyond simple organization, that submodule does something fundamentally different than the other modules in the package. While "short-cutting" the imports can be convenient to a point, there is also value in the organization provided by keeping modules within a package separate at the import stage. Once again, this represents a decision you make as a developer, and should be based on the software at hand.

To carry out the latter, "super shortcut" style import, we would add lines to the __init__.py of myreduce that mirrored those we added to plotting/: i.e., from .plotting import *. Because inside plotting/'s init we also did this to the functions inside the *file* plotting.py, they'll now be directly dot-accessible from myreduce.[3]

On the other hand, if we want to leave each module separate and force users to first call the module, we'd add the following to the myreduce __init__.py:

```
from .reducedata import *
from . import io
from . import reduction
from . import analysis
from . import plotting
```

Now, reduction, analysis, and plotting will be top-level, dot-accessible modules for myreduce, e.g., myreduce.plotting. The behavior of the files and functions in those files within each module will be set by however the init files for those modules were determined. Based on how we've done things here, myreduce.plotting.someplot would exist without having to call myreduce.plotting.plotting.someplot, but

---

[3] Apologies that the naming can get confusing as files and directories often share names.

myreduce.reduction.science_reduce would not (we'd need to call myreduce.reduction.reduce_science.science_reduce).

At first, dealing with setting up these imports in __init__.py files can be a headache, but ultimately, for code that has enough moving parts to "fill" a package (such as the example provided here) will ultimately benefit from the time taken for this setup.

Exercise 14.1:
In the myreduce example package, what would the __init__.py file inside reduction/ look like if we wanted to enable imports of all functions in the Python files in that directory directly from reduction, similar to the examples shown for io and plotting?

14.4 Custom Exceptions and Warnings

It is useful when writing code that is standalone, like a package, to provide customized exceptions and warnings when users are incorrectly using an aspect of our code (if we can catch it).

An example we've seen before is the NameResolveError from astropy—this is not one of Python's built-in exceptions, but is a subclass of the Exception class which provides more specific and useful information as to why our code execution failed.

At a basic level, defining our own exceptions to raise can be done by creating (empty) subclasses of the Exception superclass:

```
class MyCustomException(Exception):
    pass
```

```
raise MyCustomException('Helpful Exception Message')
```

```
MyCustomException
Traceback (most recent call last)
Cell In[4], line 1
----> 1 raise MyCustomException('Helpful Exception Message')

MyCustomException: Helpful Exception Message
```

Defining our own exceptions is generally most useful when writing code for others to use, but is a useful skill to have handy.

14.5 Installation and Development

Once you have your code organized properly, the next step is to install it on your system. As a reminder, installing means that you can import your package from any file or open interpreter (assuming the environment is the same as that in which it was installed), without you needing to be physically in the directory where the code exists.

You should be familiar with `pip`, the tool we use to install software hosted on PyPI. While `pip` is great for installing software from the web, it can also install our own software locally. There are two primary modes for this: regular, and development.

In a regular mode, the software will be installed to the same place all packages are stored in the current environment. This is usually a difficult to access location. For example, if you are using `anaconda`, packages are typically stored somewhere along the lines of `/Users/username/anaconda3/lib/python3.8/site-packages/`. For code installed from the web this is usually no issue, as you typically will not be *modifying* the actual code of the package.[4] But for our *own* packages, it is almost certain we will be modifying the code continuously as we test it and add features and fix bugs. Typically, we would have to reinstall our software using `pip` *every time* we made a change. Thankfully, `pip` has a development mode in which the install of the code points to the local directory where you are actively working on the code. When you make changes, it is immediately reflected in the package—so long as you reload the package when running code, or close and open Python (more on this in a moment).

The way to install code in development mode is to run, e.g., `pip install -e .`, where the -e flag sets the development mode. But we have one more file we need to add to our package before `pip` can install it successfully.

The file is called `setup.py`, and it is actually going to sit *outside* of `myreduce`. This file *can* become complicated, and has many options for you to describe the way software should be installed (including automatically installing needed dependencies —that is, if your code uses `Numpy` but the user does not have it, it will download and install `Numpy` before attempting to install your package). But in many cases, it can also be very simple:

```python
import setuptools

setuptools.setup(
    name="myreduce",
    version="0.0.1",
    author="Your Name",
    author_email="your.email@email.edu",
    description="package for observation image reduction",
    packages=["myreduce","myreduce/io","myreduce/reduction","myreduce/
  analysis","myreduce/plotting"]
)
```

Here we invoke the built-in `setuptools` package and set the **setup()** function. It takes some simple arguments, like the name of our package, version number, author,

---

[4] Most open source packages also have their code on Github or Gitlab for cases when you want to download and modify the code extensively.

etc. Important is the `packages` argument. This takes a list, and while you might think package would refer only to the entire `myreduce` package, it actually means (in practice) anything that has an `__init__.py`. Hence, we've added `myreduce` along with all the folder-level modules in it. As your packages get more complicated, there are ways to search for packages within a directory, so that you don't have to manually update this list when you add new modules.

I mentioned earlier that `setup.py` sits outside `myreduce`. But we still want the two to remain associated! Generally speaking, the full scope of a codebase will be structured something like this:

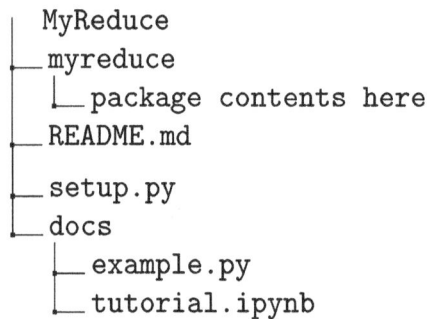

```
MyReduce
 ├─ myreduce
 │   └─ package contents here
 ├─ README.md
 ├─ setup.py
 └─ docs
     ├─ example.py
     └─ tutorial.ipynb
```

This structure will be what you see when you open many Github repositories of open source code. Note again that `setup.py` is sitting in the `MyReduce` folder, which need not share an exact name (but can) with the package itself. The naming of `MyReduce` is not important, while the naming of `myreduce` is, as the latter controls exactly what we type when we import the code.

With all that done, we're finally ready to install some software! With our shell inside the outermost directory (in this example, `MyReduce/`), we'll run

```
pip install -e .
```

Pip will then parse our setup file, and install (with the determined import structure) everything contained in `myreduce`.

14.6 Github and Version Control

Thus far, we have discussed software development in the context of a single user on one computer, coding and perhaps installing their own software for personal use. The next step in the direction toward capital-S software is to integrate `git`-style version control and online hosting. At this point, I will stop and note that there are multiple version control systems in existence, and multiple websites which host version-controlled files. Major ones include Github, Gitlab, and BitBucket. Each has their pros and cons, and offer slightly different feature sets. The *vast majority* of scientific open source software is hosted on Github, which is now owned by Microsoft. In this textbook, we'll exclusively use Github for examples—but all the other services have similar usages.

It is important to take a moment to distinguish between `git` as a version control tool, and Github as a hosting platform. The two are often referenced synonymously, but they are not quite the same.

Version Control specifically refers to the process of *tracking changes to a codebase over time*. That means when a code is updated, and new lines are added, or old lines removed or modified, those changes are gathered up and given a unique, identifying number (usually a hash). This constitutes a *version* of the code, and at any time, this version of the code can be restored and used, or compared to later versions.

You may be familiar with tools like Google Drive or Dropbox. These sync changes to your files every time they are *saved* across many devices and the cloud. Version control is a little different. While the "cloud" based saving is similar (i.e., code is hosted on Github, and if you lose your computer, you can retrieve the code from there), version control via `git` does not happen automatically every time you save a file.

Instead, *you* decide the coarseness with which versions are saved. You may be editing and testing code in your package, running it, fixing bugs, running it again, etc. You may then decide that you've fixed a bug, or added a "full" feature, or you're simply going home for the day and want a record of where you were. At this point, you can decide to tell `git` to create a new version that creates a permanent record of the codebase as it exists at that moment.

Functionally this looks like

1. Adding any files you wish included in this notation of changes (all is an option)
2. Committing those changes, usually with a short message describing what primarily has changed
3. Pushing those changes to Github.

Indeed, if you are diligent, the use of commit messages (which describe the nature of the specific version of the code is being saved) can both inform when a commit should be made, and make it so that later on, if things break or change, you can locate exactly the commit which edited the relevant code.

There are no hard and fast rules about how often to commit, but in general, it is recommended that version-controlled software be committed at least once per day in which significant modification is being performed, or whenever a developer feels they are about to embark on something which may break things.

Adds and Commits happen locally, while Pushes actually take the codebase and move it to the web, where on Github, the code shown in a repository (a page much like a directory containing all the code for a given package) will reflect the latest version. At any time, whether locally or on the Github site, you can open up the history of commits, and see how the code looked then, as well as things that have changed (known as "diffs"). It is also worth noting that git does not have to always be controlled from the command line — editors like VSCode have built in git support in a GUI interface, and some other tools are entirely designed to handle git operations.

14.7 Summary

Once you have a code you are ready to share with the world, if you have followed the above steps and have an organized code you can install locally, which is hosted on Github, you are nearly ready to release your package to a distribution service like PyPI. The few additional things to check include:

- Do all your functions and classes have documentation and type annotations? Have you built a documentation website?
- Do you have useful tutorials on the usage of the code?
- Do you have a `pyproject.toml` and `setup.cfg` file? These are (newer) entries to the Python package installation process, which help make your code robust when others install it.
- Do you have some automated tests to ensure future changes don't break the code, and guidelines for how others should contribute if they wish?

All of these steps are mostly bookkeeping, but make for a better and longer standing codebase. You can learn about these steps in full guides to Python package creation and installation, of which there are a multitude online.

Astronomical Python
An introduction to modern scientific programming
Imad Pasha

Chapter 15

Conclusions and Next Steps

15.1 Concluding Remarks

Congratulations! You've made it to the end of this whirlwind tour of the Python language and the libraries and tools used by astronomers and astrophysicists to carry out scientific research. If your head is spinning, that's okay—mine is spinning just from writing it. It is truly astonishing how many different tools, paradigms, and resources we have to leverage to make progress with astronomical research. This book covers around 0.1% of that—enough to get your feet under you, to get you moving in the right direction.

There are many directions you can go from here, whether you read this book in a classroom setting, used it as a reference, or self-taught using it. Research is a challenging and rewarding endeavour, and seeking out opportunities to take part is a great way to learn more coding, which in turn helps you garner and carry out more research. Ultimately, practice is the key driver of any learned skill, and with Python, it is especially true that more than any assignment or exercise, actual projects that you work through from start to finish—whether astronomy related or simply side projects for fun—are what will ultimately have you pondering algorithms, looking up new functions, and deepening your understanding of the language. I hope that the use of real astronomical data, and realistic walkthroughs of at least a first draft of what might pass for analysis, has given you a practical idea of what small portions of research might look like

If you are looking for more texts to dive into, there are several I can recommend. A logical next step from this book is the *Python Data Science Handbook* (VanderPlas 2016), which will provide complementary coverage of Numpy and Matplotlib, dig further into Pandas, and introduce machine learning libraries like Sklearn. Building on top of *that*, the *Statistics, Data Mining, and Machine Learning in Astronomy* text (Ivezić et al. 2020) is fantastic for having in-text examples with real code and realistic use cases, while covering the *actual* statistics one should know

15-1

when engaging with data science in an astronomical context. Rougier's Matplotlib text (Rougier 2021) is a great reference for all things plotting.

And, of course, there is the documentation and example pages for all of these (and many more) libraries. Some are better than others, but one way or another, at a certain point, most of your new coding acumen will come from parsing the source code for a function you need to understand, and picking up new tricks along the way.

Regardless of where your coding (or astronomy) journey takes you, I sincerely hope that this text has provided a useful, practical introduction to the subject. I would earnestly argue that even if you ultimately decide to pursue a path with no coding or astronomy in it, time spent learning to code is still of great value. Thinking about how to achieve a tangible goal, and how to implement an algorithm that can accomplish it within the bounds of a programming language is a fantastic skill, and one which dovetails with other modes of thinking about the world. And coding, in general, can be both a fun and rewarding way to tackle problems (or projects) in everyday life.

References

VanderPlas, J. T. 2016, Python Data Science Handbook: Essential Tools for Working with Data (Sebastopol, CA: O'Reilly Media, Inc.)

Ivezić, Ž., Connolly, A. J., VanderPlas, J. T., & Gray, A. 2020, Statistics, Data Mining, and Machine Learning in Astronomy. A Practical Python Guide for the Analysis of Survey Data, Updated Edition (Princeton, NJ: Princeton Univ. Press)

Rougier, N. P. 2021, Scientific Visualization: Python and Matplotlib https://www.labri.fr/perso/nrougier/scientific-visualization.html